T0133022

Multicore Computing

Algorithms, Architectures, and Applications

CHAPMAN & HALL/CRC
COMPUTER and INFORMATION SCIENCE SERIES

Series Editor: Sartaj Sahni

PUBLISHED TITLES

ADVERSARIAL REASONING: COMPUTATIONAL
APPROACHES TO READING THE OPPONENT'S MIND
Alexander Kott and William M. McEneaney

DELAUNAY MESH GENERATION
Siu-Wing Cheng, Tamal Krishna Dey, and
Jonathan Richard Shewchuk

DISTRIBUTED SENSOR NETWORKS, SECOND EDITION
S. Sitharama Iyengar and Richard R. Brooks

DISTRIBUTED SYSTEMS: AN ALGORITHMIC APPROACH
Sukumar Ghosh

ENERGY-AWARE MEMORY MANAGEMENT FOR EMBEDDED
MULTIMEDIA SYSTEMS: A COMPUTER-AIDED DESIGN APPROACH
Florin Balasa and Dhiraj K. Pradhan

ENERGY EFFICIENT HARDWARE-SOFTWARE
CO-SYNTHESIS USING RECONFIGURABLE HARDWARE
Jingzhao Ou and Viktor K. Prasanna

FUNDAMENTALS OF NATURAL COMPUTING: BASIC CONCEPTS,
ALGORITHMS, AND APPLICATIONS
Leandro Nunes de Castro

HANDBOOK OF ALGORITHMS FOR WIRELESS NETWORKING AND
MOBILE COMPUTING
Azzedine Boukerche

HANDBOOK OF APPROXIMATION ALGORITHMS
AND METAHEURISTICS
Teofilo F. Gonzalez

HANDBOOK OF BIOINSPIRED ALGORITHMS
AND APPLICATIONS
Stephan Olariu and Albert Y. Zomaya

HANDBOOK OF COMPUTATIONAL MOLECULAR BIOLOGY
Srinivas Aluru

HANDBOOK OF DATA STRUCTURES AND APPLICATIONS
Dinesh P. Mehta and Sartaj Sahni

HANDBOOK OF DYNAMIC SYSTEM MODELING
Paul A. Fishwick

HANDBOOK OF ENERGY-AWARE AND GREEN COMPUTING
Ishfaq Ahmad and Sanjay Ranka

HANDBOOK OF PARALLEL COMPUTING: MODELS, ALGORITHMS
AND APPLICATIONS
Sanguthevar Rajasekaran and John Reif

HANDBOOK OF REAL-TIME AND EMBEDDED SYSTEMS
Insup Lee, Joseph Y-T. Leung, and Sang H. Son

HANDBOOK OF SCHEDULING: ALGORITHMS, MODELS, AND
PERFORMANCE ANALYSIS
Joseph Y.-T. Leung

HIGH PERFORMANCE COMPUTING IN REMOTE SENSING
Antonio J. Plaza and Chein-I Chang

HUMAN ACTIVITY RECOGNITION: USING WEARABLE SENSORS
AND SMARTPHONES
Miguel A. Labrador and Oscar D. Lara Yejas

INTRODUCTION TO NETWORK SECURITY
Douglas Jacobson

LOCATION-BASED INFORMATION SYSTEMS:
DEVELOPING REAL-TIME TRACKING APPLICATIONS
Miguel A. Labrador, Alfredo J. Pérez, and Pedro M. Wightman

METHODS IN ALGORITHMIC ANALYSIS
Vladimir A. Dobrushkin

MULTICORE COMPUTING: ALGORITHMS, ARCHITECTURES,
AND APPLICATIONS
Sanguthevar Rajasekaran, Lance Fiondella, Mohamed Ahmed,
and Reda A. Ammar

PERFORMANCE ANALYSIS OF QUEUING AND COMPUTER
NETWORKS
G. R. Dattatreya

THE PRACTICAL HANDBOOK OF INTERNET COMPUTING
Munindar P. Singh

SCALABLE AND SECURE INTERNET SERVICES AND ARCHITECTURE
Cheng-Zhong Xu

SOFTWARE APPLICATION DEVELOPMENT: A VISUAL C++®, MFC,
AND STL TUTORIAL
Bud Fox, Zhang Wenzu, and Tan May Ling

SPECULATIVE EXECUTION IN HIGH PERFORMANCE COMPUTER
ARCHITECTURES
David Kaeli and Pen-Chung Yew

VEHICULAR NETWORKS: FROM THEORY TO PRACTICE
Stephan Olariu and Michele C. Weigle

Multicore Computing

Algorithms, Architectures, and Applications

Edited by

Sanguthevar Rajasekaran
Lance Fiondella
Mohamed Ahmed
Reda A. Ammar

CRC Press
Taylor & Francis Group
Boca Raton   London   New York

CRC Press is an imprint of the
Taylor & Francis Group, an **informa** business

A CHAPMAN & HALL BOOK

CRC Press
Taylor & Francis Group
6000 Broken Sound Parkway NW, Suite 300
Boca Raton, FL 33487-2742

© 2014 by Taylor & Francis Group, LLC
CRC Press is an imprint of Taylor & Francis Group, an Informa business

No claim to original U.S. Government works

Printed on acid-free paper
Version Date: 20130808

International Standard Book Number-13: 978-1-4398-5434-1 (Hardback)

This book contains information obtained from authentic and highly regarded sources. Reasonable efforts have been made to publish reliable data and information, but the author and publisher cannot assume responsibility for the validity of all materials or the consequences of their use. The authors and publishers have attempted to trace the copyright holders of all material reproduced in this publication and apologize to copyright holders if permission to publish in this form has not been obtained. If any copyright material has not been acknowledged please write and let us know so we may rectify in any future reprint.

Except as permitted under U.S. Copyright Law, no part of this book may be reprinted, reproduced, transmitted, or utilized in any form by any electronic, mechanical, or other means, now known or hereafter invented, including photocopying, microfilming, and recording, or in any information storage or retrieval system, without written permission from the publishers.

For permission to photocopy or use material electronically from this work, please access www.copyright. com (http://www.copyright.com/) or contact the Copyright Clearance Center, Inc. (CCC), 222 Rosewood Drive, Danvers, MA 01923, 978-750-8400. CCC is a not-for-profit organization that provides licenses and registration for a variety of users. For organizations that have been granted a photocopy license by the CCC, a separate system of payment has been arranged.

Trademark Notice: Product or corporate names may be trademarks or registered trademarks, and are used only for identification and explanation without intent to infringe.

Library of Congress Cataloging-in-Publication Data

Multicore computing : algorithms, architectures, and applications / edited by
    Sanguthevar Rajasekaran, Lance Fiondella, Mohamed Ahmed, Reda A. Ammar.
        pages cm -- (Chapman & Hall/CRC computer & information science series ; 29)
    Includes bibliographical references and index.
    ISBN 978-1-4398-5434-1 (hardback)
    1. Multiprocessors. I. Rajasekaran, Sanguthevar, editor.

QA76.5.M7932 2013
004.16--dc23   2013031929

Visit the Taylor & Francis Web site at
http://www.taylorandfrancis.com

and the CRC Press Web site at
http://www.crcpress.com

Dedication

To my teachers,

Esakki Rajan, P.S. Srinivasn, V. Krishnan, and John H. Reif

—Sanguthevar Rajasekaran

To my son,

Advika

—Lance Fiondella

To my wife,

Noha Nabawi

my parents, and my advisors

Professors Sanguthevar Rajasekaran and Reda Ammar

—Mohamed F. Ahmed

To my family,

Tahany Fergany, Rabab Ammar, Doaa Ammar and Mohamed Ammar

—Reda A. Ammar

கேடில் விழுச்செல்வம் கல்வி யொருவற்கு
மாடல்ல மற்றை யவை

திருவள்ளுவர்
(திருக்குறள்; பொருட்பால் - அதிகாரம் நாற்பது : கல்வி)

An indestructible and impeccable treasure to one is learning;
all the other things are not wealth.

Thiruvalluvar (circa 100 B.C.)
(Thirukkural; Section - Wealth; Chapter 40 - Education)

Contents

10 Sorting on the Cell Broadband Engine **285**

Shibdas Bandyopadhyay, Dolly Sharma, Reda A. Ammar, Sanguthevar Rajasekaran, and Sartaj Sahni

12 Backprojection Algorithms for Multicore and GPU Architectures   391

William Chapman, Sanjay Ranka, Sartaj Sahni, Mark Schmalz, Linda Moore, Uttam Majumder, and Bracy Elton

Preface

We live in an era of big data. Every area of science and engineering has to process voluminous data sets. Numerous problems in such critical areas as computational biology are intractable, and exact (or even approximation) algorithms for solving them take time that is exponential in some of the underlying parameters. As a result, parallel computing has become inevitable. Parallel computing has been made very affordable with the advent of multicore architectures such as Cell, Tesla, etc. On the other hand, programming these machines is much more difficult due to the oddities existing in these architectures. This volume addresses different facets of multicore computing and offers insights into them. The chapters in this handbook will help the readers in understanding the intricacies of these architectures and will prepare them to design efficient multicore algorithms. Topics covered range through architectures, algorithms, and applications.

Chapter 1 covers the memory hierarchy for multicore and many-core processors. The performance of computer systems depends on both the memory and the processor. In the beginning, the speed gap between the processor and memory was narrow. The honeymoon ended when the number of transistors per chip increased almost exponentially (the famous Moore's law). This transistor budget has translated to performance, at least until a decade ago! Memory system performance did not keep up with processor performance and has been improving at a much lower pace. When designs shifted from a single core to multicore, the memory system faced even more challenges. The challenges facing memory system designers, how to deal with them, and the future of this field are some of the issues discussed in this chapter.

In **Chapter 2**, the authors present Flexible Set Balancing (FSB), a caching strategy that exploits the large asymmetry in cache sets' usages on tiled CMPs. FSB attempts to retain cache lines evicted from highly pressured sets in underutilized sets using a very flexible many-from-many sharing so as to satisfy far-flung reuses. Simulation results show that FSB can minimize the L2 miss rate by an average of 36.6% for the tested benchmarks. This translates to an overall performance improvement of 13%. In addition, results show that FSB compares favorably with three closely related schemes and incurs minor storage, area, and energy overheads.

In **Chapter 3**, the authors describe the main features of the latest SPARC architecture specification, SPARCv9, and try to motivate the different design decisions behind them. They also look at each architectural feature in the context of a multicore processor implementation of the architecture. After describing the SPARC architecture, they present in detail one of its most successful implementations, the Sun UltraSPARC T1 (also known as Niagara) multicore processor.

Chapter 4 presents the Cilk and Cilk++ programming languages, which raise the level of abstraction of writing parallel programs. Organized around the concept of tasks, Cilk allows the programmer to reason about what set of tasks may execute in parallel. The Cilk runtime system is responsible for mapping tasks to processors. This chapter presents the Cilk language and elucidates the design of the Cilk runtime scheduler. Armed with an understanding of how the scheduler works, this chapter then continues to explain how to analyze the performance of Cilk programs. Next, hyperobjects, a powerful abstraction of common computational patterns is explained.

Chapter 5 introduces Parallel Linear Algebra Software for Multicore Architectures (PLASMA), a numerical software library for solving problems in dense linear algebra on systems of multicore processors and multisocket systems of multicore processors. PLASMA relies on a variety of multithreading mechanisms, including static and dynamic thread scheduling. PLASMA's superscalar scheduler, QUARK, offers powerful tools for parallel task composition, such as support for nested parallelism and provisions for task aggregation. The dynamic nature of PLASMA's operation exposes its user to an array of new capabilities, such as asynchronous mode of execution, where library function calls can be invoked in a non-blocking fashion.

Chapter 6 discusses Aho-Corasick, an exact multipattern string-matching algorithm that performs the search in a time linearly proportional to the length of the input text independently from pattern set size. However, in reality, software implementations suffer from significant performance variability with large pattern sets because of unpredictable memory latencies and caching effects. This chapter presents a study of the behavior of the Aho-Corasick string-matching algorithm on a set of modern multicore and multithreaded architectures. The authors discuss the implementation and the performance of the algorithm on modern x86 multicores, multithreaded Niagara 2 processors, and GPUs from the previous and current generation.

In **Chapter 7**, the authors first describe the architecture of NVIDIA Tesla GPU. They then describe some of the principles for designing efficient algorithms for GPUs. These principles are illustrated using recent parallel algorithms to sort numbers on a GPU. These sorting algorithms for numbers are then extended to sort large records. The authors also describe efficient strategies for moving records within GPU memory for various different layouts of a record in memory. Lastly, experimental results comparing the performance of these algorithms for sorting records are presented.

Chapter 8 discusses scheduling Directed Acyclic Graphs (DAGs) onto multi/many-core processors, which remains a fundamental challenge in parallel computing. The chapter utilizes the exact inference as an example to discuss scheduling techniques on multi/many-core processors. The authors introduce a modularized scheduling method on general-purpose multicore processors and develop lock-free data structures for reducing the overhead due to contention. Then, they extend the scheduling method to many-core processors using dynamic thread grouping, which dynamically adjusts the number of threads used for scheduling and task execution. It adapts to the input task graph and therefore improves the overall performance.

Chapter 9 evaluates design trade-offs among Intel and AMD multicore processors, IBM Cell Broadband Engine, and NVIDIA GPUs and their impact on dense numerical computations (kernels from computational statistics and the direct n-body problem). This chapter compares the core architectures and memory subsystems of various platforms; illustrates the software implementation process on each platform; measures and analyzes the performance, coding complexity, and energy efficiency of each implementation; and discusses the impact of different architectural design choices on each implementation.

In **Chapter 10**, the authors look at designing algorithms for Cell Broadband Engine, which is a heterogeneous multicore processor on a single chip. First, they describe the architecture of the Cell processor. They then describe the opportunities and challenges associated with programming the Cell. These opportunities and challenges are illustrated with different parallel algorithms for sorting numbers. Later, they extend these algorithms to sort large records. This latter discussion illustrates how to hide memory latency associated with moving large records. The authors end the chapter by comparing different algorithms for sorting records stored using different layouts in memory.

Chapter 11 begins by reviewing the architecture and programming model of the NVIDIA Tesla GPU. Then, the authors develop an efficient matrix multiplication algorithm for this GPU by going through a series of intermediate algorithms beginning with a straightforward GPU implementation of the single-core CPU algorithm. Extensive experimental results are provided. These results show the impact of the various optimization strategies (e.g., tiling, padding to eliminate shared memory bank conflicts, coalesced I/O from/to global memory) showing that our most efficient GPU algorithm for matrix multiplication is three orders of magnitude faster than the classical single-core algorithm.

Chapter 12 addresses Backprojection, an algorithm that generates images from Synthetic Aperture Radar (SAR) data. SAR data is collected by a radar device that moves around an area of interest, transmitting pulses and collecting the responses as a function of time. Backprojection produces each pixel of the output image by independently determining the contribution of every pulse, producing high-quality imagery while requiring significant data movement and computational costs. These costs can be mitigated through the use of Graphics Processing Units, as Backprojection is easily decomposed along its input and output dimensions.

Acknowledgements

We are very thankful to the authors for having contributed their chapters in a timely manner. We also thank the staff of Chapman & Hall/CRC. In addition, we gratefully acknowledge the partial support from the National Science Foundation (CCF 0829916) and the National Institutes of Health (NIH R01-LM010101).

Sanguthevar Rajasekaran
Lance Fiondella
Mohamed F. Ahmed
Reda A. Ammar

List of Contributing Editors

Sanguthevar Rajasekaran received his M.E. degree in Automation from the Indian Institute of Science (Bangalore) in 1983, and his Ph.D. degree in Computer Science from Harvard University in 1988. Currently, he is the UTC Chair Professor of Computer Science and Engineering at the University of Connecticut and the Director of Booth Engineering Center for Advanced Technologies (BECAT). Before joining UConn, he served as a faculty member in the CISE Department of the University of Florida and in the CIS Department of the University of Pennsylvania. During 2000–2002 he was the Chief Scientist for Arcot Systems. His research interests include Bioinformatics, Parallel Algorithms, Data Mining, Randomized Computing, Computer Simulations, and Combinatorial Optimization. He has published over 250 research articles in journals and conferences. He has coauthored two texts on algorithms and coedited five books on algorithms and related topics. He is a Fellow of the Institute of Electrical and Electronics Engineers (IEEE) and the American Association for the Advancement of Science (AAAS). He is also an elected member of the Connecticut Academy of Science and Engineering.

Lance Fiondella received a B.S. in Computer Science from Eastern Connecticut State University and his M.S. and Ph.D. degrees in Computer Science and Engineering from the University of Connecticut. He is presently an assistant professor in the Department of Electrical and Computer Engineering at the University of Massachusetts Dartmouth. His research interests include algorithms, reliability engineering, and risk analysis. He has published over 40 research papers in peer-reviewed journals and conferences.

Mohamed F. Ahmed received his B.Sc. and M.Sc. degrees from the American University in Cairo, Egypt, in May 2001 and January 2004, respectively. He received his Ph.D. degree in Computer Science and Engineering from the University of Connecticut in September 2009. Dr. Ahmed served as an Assistant Professor at the German University in Cairo from September 2009 to August 2010 and as an Assistant Professor at the American University in Cairo from September 2010 to January 2011. Since 2011, he has served as a Program Manager at Microsoft. His research interests include multi/many-core technologies, high performance computing, parallel programming, cloud computing, GPUs programming, etc. He has published many papers in these areas.

Reda A. Ammar is a professor and the head of the Computer Science and Engineering department, University of Connecticut. He received his B.Sc. in Electrical Engineering from Cairo University in 1973, B.Sc. in Mathematics from Ain Shams University in 1975, M.Sc. (1981) and Ph.D. (1983) in Computer Science from the University of Connecticut. His primary research interests are in the area of distributed and high-performance computing and real-time systems. His publication record exceeds 300 journal articles, book chapters, and conference papers. Dr. Ammar was the chairman of eight international conferences. He also served at many conferences as a member of the steering committees, an organizing officer, and as a member of program technical committees. He is the president of the International Society of Computers and Their Applications. He has been the Editor in Chief of the *International Journal on Computers and Their Applications* for six years. Dr. Ammar is the major advisor for more than 20 Ph.D. graduates. More details can be found at http://www.engr.uconn.edu/~reda/.

List of Contributing Authors

Bushra Ahsan is currently working with Intel as a Silicon Architecture Engineer. She received her Ph.D. in Electrical Engineering in 2010 from the City University of New York. Ahsan then joined the University of Cyprus as a research scientist post-doc for a year before joining Intel. Ahsan finished her bachelors in Electrical Engineering from the University of Engineering and Technology, Peshawar, Pakistan.

Reda A. Ammar is a professor and the head of the Computer Science and Engineering department, University of Connecticut. He received his B.Sc in Electrical Engineering from Cairo University in 1973, B.Sc. in Mathematics from Ain Shams University in 1975, M.Sc. (1981) and Ph.D. (1983) in Computer Science from the University of Connecticut. His primary research interests are in the area of distributed and high-performance computing and real-time systems. His publication record exceeds 300 journal articles, book chapters, and conference papers. Dr. Ammar was the chairman of eight international conferences. He also served at many conferences as a member of the steering committees, an organizing officer and as a member of program technical committees. He is the president of the International Society of Computers and Their Applications. He has been the Editor in Chief of the *International Journal on Computers and Their Applications* for six years. Dr. Ammar is the major advisor for more than 20 Ph.D. graduates. More details can be found at http://www.engr.uconn.edu/~reda/.

Nitin Arora is a Ph.D. student in Aerospace Engineering at Georgia Institute of Technology. He holds a Masters degree in Aerospace Engineering from Georgia Institute of Technology and a Bachelors degree in Civil Engineering from Punjab Engineering College, India. He has authored and coauthored various research papers in the field of high-performance computing with a focus on GPU computing and high-performance numerical algorithm development applied to orbital mechanics.

David A. Bader is a Full Professor in the School of Computational Science and Engineering, College of Computing, at Georgia Institute of Technology, and Executive Director for high-performance computing. He received his Ph.D. in 1996 from the University of Maryland, and his research is supported through highly competitive research awards primarily from NSF, NIH, DARPA, and DOE. Dr. Bader serves on the Steering Committees of the IPDPS and HiPC conferences. He has served on the Research Advisory Council for Internet2, as the General Chair of IPDPS 2010, Chair of SIAM PP12, and Program Chair of IPDPS 2014. He is an Associate Editor in Chief of the *Journal of Parallel and Distributed Computing (JPDC)* and serves as an Associate Editor for several high-impact publications including *ACM Journal of Experimental Algorithmics (JEA)*, *IEEE DSOnline*, *Parallel Computing*, and the *Journal of Computational Science*, and has been an associate editor for the *IEEE Transactions on Parallel and Distributed Systems (TPDS)*. He was elected as chair of the IEEE Computer Society Technical Committee on Parallel Processing (TCPP) and as chair of the SIAM Activity Group in Supercomputing (SIAG/SC). Dr. Bader's interests are at the intersection of high-performance computing and real-world applications, including computational biology and genomics and massive-scale data analytics. He has cochaired a series of meetings, the IEEE International Workshop on High-Performance Computational Biology (HiCOMB), coorganized the NSF Workshop on Petascale Computing in the Biological Sciences, written several book chapters, and coedited special issues of the *Journal of Parallel and Distributed Computing (JPDC)* and *IEEE TPDS* on high-performance computational biology. He is also a leading expert on multicore, many-core, and multithreaded computing for data-intensive applications such as those in massive-scale graph analytics. He has coauthored over 100 articles in peer-reviewed journals and conferences, and his main areas of research are in parallel algorithms, combinatorial optimization, massive-scale social networks, and computational biology and genomics. Dr. Bader is a Fellow of the IEEE and AAAS, a National Science Foundation CAREER Award recipient, and has received numerous industrial awards from IBM, NVIDIA, Intel, Cray, Oracle/Sun Microsystems, and Microsoft Research. He served as a member of the IBM PERCS team for the DARPA High Productivity Computing Systems program, and the NVIDIA ECHELON team for the DARPA Ubiquitous High Performance Computing program, and was a distinguished speaker in the IEEE Computer Society Distinguished Visitors Program. He has also served as Director of the Sony-Toshiba-IBM Center of Competence for the Cell Broadband Engine Processor. Bader is a cofounder of the Graph500 List for benchmarking "Big Data" computing platforms. Bader is recognized as a "RockStar" of high-performance computing by InsideHPC and as HPCwire's People to Watch in 2012.

Shibdas Bandyopadhyay did his Ph.D. developing parallel algorithms for many-core architectures at the University of Florida. Before starting his Ph.D., he did his M.S. in computer science from Indian Statistical Institute and his B.A. in Information Technology from Indian Institute of Information Technology, Kolkata. He previously worked at Nvidia as a part of their graphics driver and CUDA platforms team. Now, he works on the virtual graphics team at VMware enabling hardware-accelerated computer graphics inside a virtual machine.

Henricus Bouwmeester obtained a B.S. in Mathematics from Colorado Mesa University and then graduated from Colorado State University with a M.Sc. in Mathematics. After having spent some time working in the corporate world interacting with private parties and government entities, he reentered the academic arena to complete his doctoral degree. From 2009 to 2012, he was a Research Assistant at the University of Colorado Denver where he helped develop mathematical software libraries. Henricus recently defended his doctoral thesis titled *Tiled Algorithms for Matrix Computations on Multicore Architectures.*

William Chapman is a Ph.D. student at the University of Florida. His interest is in the implementation of data-intensive signal processing applications on GPU and hybrid CPU–GPU architectures. He received a B.S. degree from the University of Florida in May 2009.

Daniel Chavarría-Miranda is a senior scientist in the High-Performance Computing group at the Pacific Northwest National Laboratory (PNNL). He is the institutional Principal Investigator for the DOE ASCR X-Stack COMPOSE-HPC project as well as for the Center for Scalable Parallel Programming Models (PModels2). Chavarría-Miranda also served as co-PI for the DOD-funded Center for Adaptive Supercomputing Software-Multithreaded Architectures (CASS-MT), with a focus on scalable highly irregular applications and systems software, from 2008 to 2010. In addition, he has been the Principal Investigator of a PNNL-funded Laboratory Directed Research & Development project on intra-node optimization for scientific applications, as well as on the use of reconfigurable computing technology to accelerate scientific computations. Chavarría-Miranda received a combined Ph.D. and M.S. in Computer Science from Rice University in 2004 under Prof. John Mellor-Crummey. As part of his doctoral training, he participated in designing and developing key portions of Rice University's dHPF research compiler.

Sangyeun Cho received a B.S. degree in Computer Engineering from Seoul National University in 1994 and a Ph.D. degree in Computer Science from the University of Minnesota in 2002. In 1999, he joined the System LSI Division of Samsung Electronics Co., Giheung, Korea, and contributed to the development of Samsung's flagship embedded processor core family Calm-RISC(TM). He was a lead architect of CalmRISC-32, a 32-bit microprocessor core, and designed its memory hierarchy including caches, DMA, and stream buffers. Since 2004, he has been with the Computer Science Department at the University of Pittsburgh, where he is currently an Associate Professor. His research interests are in the area of computer architecture and system software with particular focus on performance, power, and reliability of memory- and storage-hierarchy design for next-generation multicore systems.

Jack Dongarra holds an appointment at the University of Tennessee, Oak Ridge National Laboratory, and the University of Manchester. He specializes in numerical algorithms in linear algebra, parallel computing, the use of advanced-computer architectures, programming methodology, and tools for parallel computers. He was awarded the IEEE Sid Fernbach Award in 2004 for his contributions in the application of high-performance computers using innovative approaches; in 2008, he was the recipient of the first IEEE Medal of Excellence in Scalable Computing; in 2010, he was the first recipient of the SIAM Special Interest Group on Supercomputing award for Career Achievement; and in 2011, he was the recipient of the IEEE IPDPS 2011 Charles Babbage Award. He is a Fellow of the AAAS, ACM, IEEE, and SIAM and a member of the National Academy of Engineering.

Bracy H. Elton received a B.Sc. cum laude in 1983 with a double major in Computer Science and Mathematics from Pacific Lutheran University (Tacoma, Washington) and an M.Sc. and Ph.D., in 1985 and 1990, both in Computer Science, from the University of California, Davis. Dr. Elton has authored 27 publications. Since receiving the doctorate, he has held computational and research scientist positions at Fujitsu America, Inc. (SuperComputer Group), and Cray Inc. (Scientific Libraries Group). He was a Participating Guest at Lawrence Livermore National Laboratory. More recently, he was a Technical Fellow at The Ohio State University (Ohio Supercomputer Center). He is presently a Senior Research Scientist at Dynamics Research Corporation (High Performance Technologies Group). He is a member of the IEEE Computer Society, the IEEE Signal Processing Society, the Society for Industrial and Applied Mathematics, the American Mathematical Society, Sigma Xi, and the American Association for the Advancement of Science.

Mathieu Faverge is an Associate Professor at IPB-ENSEIRB-Matmeca in Bordeaux, France and is part of the Inria HiePACS team. Before that, he received a Ph.D. degree in Computer Science from the University of Bordeaux, France, and has been a Postdoctoral Research Associate at the University of Tennessee Knoxville's Innovative Computing Laboratory for three years. His main research interests are numerical linear algebra algorithms for sparse and dense problems on massively parallel architectures and DAG algorithms relying on dynamic schedulers. He has experience with hierarchical shared memory and heterogeneous and distributed systems, and his contributions to the scientific community include efficient linear algebra algorithms for those systems.

Mohammad Hammoud is a Visiting Assistant Professor at Carnegie Mellon University in Qatar (CMU-Q). He received his Ph.D. degree in Computer Science from the University of Pittsburgh in 2010. Hammoud has a broad interest in computer systems with an emphasis on computer architecture and cloud computing. For his Ph.D. thesis, he focused on L2 cache design of multicore processors. After joining CMU-Q in 2011, he extended his work to cloud computing where he devised multiple MapReduce scheduling techniques and characterized task concurrency in Hadoop for improved performance. Recently, he started exploring ways to offer a cloud support for emerging Big Graph applications. In addition to research, Hammoud teaches various courses, including distributed systems, operating systems, computer architecture and cloud computing.

Seunghwa Kang is a research scientist at Pacific Northwest National Laboratory. He completed his Ph.D. at Georgia Institute of Technology in 2011 and joined Pacific Northwest National Laboratory as a Postdoctoral Researcher. He was also affiliated with the Institute for Systems Biology during his postdoctoral training. He has worked on optimizing various kernels and applications for multicore processors and accelerators and analyzing the match between a spectrum of applications and programming models and architectures. His recent research interest is in solving computational biology problems by designing novel models and algorithms and exploiting high-performance computing.

Jakub Kurzak received his M.Sc. degree in Electrical and Computer Engineering from Wroclaw University of Technology, Poland, and his Ph.D. degree in Computer Science from the University of Houston. He is a Research Director at the Innovative Computing Laboratory in the Department of Electrical Engineering and Computer Science at the University of Tennessee, Knoxville. His research interests include parallel algorithms, specifically in the area of numerical linear algebra, and also parallel programming models and performance optimization for parallel architectures, multicore processors, and GPU accelerators.

Julien Langou received an M.S. in Aerospace Engineering from Supaéro (Toulouse, France) in 1999 and a Ph.D. in Applied Mathematics from CERFACS in 2003. He is currently an Associate Professor in the Department of Mathematical and Statistical Sciences at the University of Colorado Denver.

Junjie Li received a B.Eng. of Posts and Telecommunications from Beijing University in 2009 and joined the University of Florida after that as a Ph.D. student in the Department of Computer and Information Science and Engineering. His research interests include high-performance computing, GPGPU computing and bioinformatics. His main focus is on developing GPU algorithms for bioinformatics.

Piotr Luszczek received his doctorate degree for independent research work on sparse direct methods for matrix factorizations that leverage existing optimized linear algebra kernel codes. In addition, he published papers on other matrix-factorization scenarios such as out-of-core solvers. Hardware and software benchmarking has also been the focus of his professional activities, primarily, the codes for numerical linear algebra. Variability of the bench-marked computer architectures and algorithmic approaches was a natural base for his investigation of the very broad topic of software self-adaptation. He investigated the language-design issues as they apply to scientific programmer productivity but also to resulting performance in the government-sponsored HPCS program. Luszczek has shared the acquired experience with collaborators and the general community at tutorials and talks across the US and has also contributed to book chapters on scientific computing. His work at MathWorks concentrated on parallel language design and its implementation with particular emphasis on high-performance programming. These activities resulted in three patent awards. His recent research focus is on performance modeling and evaluation in the context of tuning of parallelizing compilers as well as energy-conscious aspects of heterogeneous and embedded computing. Currently, he investigates how scientific codes are influenced by power and energy constraints and how to include performance-conscious optimizations into sustainable computational science. He influenced the high-performance computing and high-end computing fields by serving on program committees at conferences and workshops and organizing minisymposia for the dissemination of research results.

Uttam K. Majumder earned a B.S. in Computer Science from the City College of New York (Summa Cum Laude). He then earned an M.S. and an M.B.A. degree from the Air Force Institute of Technology and Wright State University, respectively. He is expected to earn a Doctoral Degree in Electrical Engineering from Purdue University by the end of 2013.

Majumder has been working with the Air Force Research Laboratory (AFRL) since August 2003. His research interests includes radar signal processing, synthetic aperture radar imaging, high-performance computing, inverse problems, and automatic target recognition.

Rami Melhem received a B.E. in Electrical Engineering from Cairo University in 1976, an M.A. degree in Mathematics, an M.S. degree in Computer Science from the University of Pittsburgh in 1981, and a Ph.D. degree in Computer Science from the University of Pittsburgh in 1983. He was an Assistant Professor at Purdue University prior to joining the faculty of the University of Pittsburgh in 1986, where he is currently a Professor in the Computer Science Department, which he chaired from 2000 to 2009.

His research interests include power management, real-time and fault-tolerant systems, optical networks, high-performance computing and parallel computer architectures. Dr. Melhem has served on program committees of numerous conferences and workshops. He was on the editorial board of the *IEEE Transactions on Computers* (1991–1996), the *IEEE Transactions* on *Parallel and Distributed systems* (1998–2002), the *Computer Architecture Letters* (2001–2010), and the *Journal of Parallel and Distributed Computing* (2003–2011). Dr. Melhem is a Fellow of IEEE and a member of the ACM.

Linda J. Moore is an electronics engineer at the Air Force Research Laboratory Sensors Directorate (AFRL/RY) at Wright-Patterson Air Force Base, Ohio, where she focuses on signal processing and exploitation for surveillance Synthetic Aperture Radar (SAR) systems. Her interests include the application of cutting-edge high-performance computing (HPC) technologies to enable real-time SAR processing. She obtained a B.S. in Computer Engineering (2004) and an M.S. in Electrical Engineering (2006) from Wright State University and Ohio State University, respectively. She is currently pursuing her Ph.D. in Electrical Engineering at the University of Dayton.

Viktor K. Prasanna (V. K. Prasanna Kumar) is a Charles Lee Powell Chair in Engineering and is a Professor of Electrical Engineering and Computer Science at the University of Southern California (USC) and serves as the director of the Center for Energy Informatics (CEI). He is the executive director of the USC-InfoSys Center for Advanced Software Technologies (CAST). He is an associate member of the Center for Applied Mathematical Sciences (CAMS). He leads the Integrated Optimizations (IO) efforts at the USC-Chevron Center of Excellence for Research and Academic Training on Interactive Smart Oilfield Technologies (CiSoft) at USC and the demand-response optimizations in the LA Smartgrid project. His research interests include high-performance computing, parallel and distributed systems, reconfigurable computing, cloud computing, and embedded systems. He received his B.S. in Electronics Engineering from Bangalore University, M.S. from the School of Automation, Indian Institute of Science, and Ph.D. in Computer Science from Pennsylvania State University.

Prasanna has published extensively and consulted for industries in the above areas. He is the Steering Committee Cochair of the International Parallel & Distributed Processing Symposium (IPDPS) (merged IEEE International Parallel Processing Symposium [IPPS] and Symposium on Parallel and Distributed Processing [SPDP]). He is the Steering Committee Chair of the International Conference on High Performance Computing (HiPC). In the past, he has served on the editorial boards of the *IEEE Transactions on Very Large Scale Integration (VLSI) Systems*, *IEEE Transactions on Parallel and Distributed Systems (TPDS)*, *Journal of Pervasive and Mobile Computing*, and the *Proceedings of the IEEE*. He serves on the editorial boards of the *Journal of Parallel and Distributed Computing* and the *ACM Transactions on Reconfigurable Technology and Systems*. During 2003–2006, he was the Editor in Chief of the *IEEE Transactions on Computers*. He was the founding chair of the IEEE Computer Society Technical Committee on Parallel Processing. He is a Fellow of the IEEE, the Association for Computing Machinery (ACM), and the American Association for Advancement of Science (AAAS). He is a recipient of the 2005 Okawa Foundation Grant.

He received an Outstanding Engineering Alumnus Award from Pennsylvania State University in 2009. He has received best paper awards at several international forums including ACM Computing Frontiers (CF), IEEE International Parallel and Distributed Processing Symposium (IPDPS), International Conference on Parallel and Distributed Systems (ICPADS), International Symposium on Computer Architecture and High Performance Computing (SBAC-PAD), International Conference on Parallel and Distributed Computing and Systems (PDCS), IEEE International Conference on High Performance Switches and Routers (HPSR), among others. He currently serves as the Editor in Chief of the *Journal of Parallel and Distributed Computing (JPDC)*.

Sanguthevar Rajasekaran received his M.E. in Automation from the Indian Institute of Science (Bangalore) in 1983 and his Ph.D. in Computer Science from Harvard University in 1988. Currently he is the UTC Chair Professor of Computer Science and Engineering at the University of Connecticut and the Director of Booth Engineering Center for Advanced Technologies (BECAT). Before joining UConn, he served as a faculty member in the CISE Department of the University of Florida and in the CIS Department of the University of Pennsylvania. During 2000–2002 he was the Chief Scientist for Arcot Systems. His research interests include Bioinformatics, Parallel Algorithms, Data Mining, Randomized Computing, Computer Simulations, and Combinatorial Optimization. He has published over 250 research articles in journals and conferences. He has coauthored two texts on algorithms and coedited five books on algorithms and related topics. He is a Fellow of the Institute of Electrical and Electronics Engineers (IEEE) and the American Association for the Advancement of Science (AAAS). He is also an elected member of the Connecticut Academy of Science and Engineering.

Sanjay Ranka is Professor of Computer and Information Science and Engineering at the University of Florida, and has over 20 years experience in Government-funded development of theory, algorithms, and software for high-performance and parallel computing applications including computer vision applications for military and medical imaging. He has developed a number of algorithms for performance and energy trade-offs of workflows for parallel and multicore architectures. Dr. Ranka's current research interests are in the areas of parallel algorithms, parallel computing models, and multiprocessor scheduling, data mining, and grid computing. His research has been funded by NSF, NIH, DOE, and DoD.

Sartaj Sahni is a Distinguished Professor of Computer and Information Sciences and Engineering at the University of Florida. He is also a member of the European Academy of Sciences, a Fellow of IEEE, ACM, AAAS, and Minnesota Supercomputer Institute and is a Distinguished Alumnus of the Indian

Institute of Technology, Kanpur. In 1997, he was awarded the IEEE Computer Society Taylor L. Booth Education Award "for contributions to Computer Science and Engineering education in the areas of data structures, algorithms, and parallel algorithms," and in 2003, he was awarded the IEEE Computer Society W. Wallace McDowell Award "for contributions to the theory of NP-hard and NP-complete problems." Dr. Sahni was awarded the 2003 ACM Karl Karlstrom Outstanding Educator Award for "outstanding contributions to computing education through inspired teaching, development of courses and curricula for distance education, contributions to professional societies, and authoring significant textbooks in several areas including discrete mathematics, data structures, algorithms, and parallel and distributed computing." Dr. Sahni received his B.Tech. (Electrical Engineering) from the Indian Institute of Technology, Kanpur, and his M.S. and Ph.D. in Computer Science from Cornell University. Dr. Sahni has published over 300 research papers and written fifteen texts. His research publications are on the design and analysis of efficient algorithms, parallel computing, interconnection networks, design automation, and medical algorithms.

Dr. Sahni is a managing editor of the International Journal of Foundations of Computer Science, and a member of the editorial boards of *Computer Systems: Science and Engineering, International Journal of High Performance Computing and Networking, International Journal of Distributed Sensor Networks and Parallel Processing Letters.* He is a past Coeditor in Chief of the *Journal of Parallel and Distributed Computing.* He has served as program committee chair, general chair, and been a keynote speaker at many conferences. Dr. Sahni has served on several NSF and NIH panels, and he has been involved as an external evaluator of several computer science and engineering departments nationally and internationally.

Mark Schmalz has over 25 years experience in the design and implementation of large-scale software systems for physical, optical, and oceanographic modeling and simulation, high-performance computing, image/signal processing, pattern recognition, and computational intelligence. His research has been variously and extensively funded by DARPA, ONR, AFOSR, Army, USMC, DOE, NIH, and NSF. Dr. Schmalz is author or coauthor of over 150 papers in conference proceedings and journals in the areas of high-performance computing, simulation and modeling, signal and image processing, and compression.

Simone Secchi received an M.Sc. in Electronic Engineering in 2007 and a Ph.D. in Electronic Engineering in 2011 from the University of Cagliari, Italy. He has been with Pacific Northwest National Laboratory (PNNL) since July 2010 as a postmaster research associate, and since June 2011 he holds a postdoctoral research associate position there. His main research interests include the areas of modeling and parallel software simulation of high-performance computing architectures, FPGA-based energy-aware emulation of multiprocessor systems and advanced Network-on-Chip architectures.

Dolly Sharma is an Assistant Professor at Amity Institute of Information Technology, Amity University, India. She obtained her Ph.D. in Computer Science and Engineering from the University of Connecticut. Her research interests include data mining, bioinformatics and computational biology.

Aashay Shringarpure graduated from Georgia Institute of Technology with his M.Sc. in 2010. He is now a software engineer at Google.

Antonino Tumeo received an M.S. degree in Informatic Engineering in 2005 and a Ph.D. degree in Computer Engineering in 2009 from Politecnico di Milano in Italy. Since February 2011, he has been a research scientist at Pacific Northwest National Laboratory (PNNL). He joined PNNL in 2009 as a postdoctoral research associate. Previously, he was a postdoctoral researcher at Politecnico di Milano. His research interests are modeling and simulation of high-performance architectures, hardware-software codesign, power/performance characterization of high-performance embedded systems, FPGA prototyping and GPGPU computing.

Hans Vandierendonck is a Lecturer (Assistant Professor) in High-Performance and Distributed Computing at Queen's University Belfast. He obtained an M.S. in Computer Science from the Faculty of Engineering at Ghent University, Belgium, in 2000 and a Ph.D. from the same university in 2004. Before moving to Belfast, he was a visiting Postdoctoral Researcher at INRIA, Rennes, France, and the Foundation of Research and Technology Hellas, Heraklion, Crete, Greece. His research focusses on parallel programming models and runtime systems to support these programming models. Current projects investigate task-oriented programming models incorporating dataflow aspects and runtime system support for exploiting nonvolatile memory hierarchies.

Hans has served in several program committees including IEEE International on Parallel and Distributed Processing Symposium (IPDPS), the Symposium on Application Acceleration in High-Performance Computing (SAAHPC) and Design Automation and Test Europe (DATE). He is a regular reviewer for the main IEEE and ACM conferences in his field. Hans holds a Marie Curie Fellowship from the European Research Council. He is a Senior Member of the ACM and a Senior Member of the IEEE.

Oreste Villa is a senior research scientist at NVIDIA Research in the Architecture Research Group (ARG). Previously, from 2008 to 2013, he was a research scientist at the Pacific Northwest National Laboratory (PNNL) in the high-performance computing group. He received a Ph.D. in Computer Science in 2008 from Politecnico di Milano in Italy and an M.S. in Electronic Engineering in 2003 from the University of Cagliari in Italy. He also received an M.E. degree in 2004 in Embedded Systems Design from the University of

Lugano in Switzerland. His research interests include computer architecture, simulation and modeling of architectures, high performance computing and large scale irregular applications.

Richard (Rich) Vuduc is an assistant professor in the School of Computational Science and Engineering at the Georgia Institute of Technology. His research lab, the High-Performance Computing (HPC) Garage (http://hpcgarage.org), is broadly interested in performance programming and analysis. Most recently, they have been pursuing a new theory to relate the time, energy, and power consumed by algorithms and new practical engineering methods for automated performance tuning ("autotuning"). This work has received numerous accolades, including membership in the US Defense Advanced Research Projects Agency (DARPA) Computer Science Study Group (2009); the US National Science Foundation's CAREER Award (2010); a jointly awarded Gordon Bell Prize (2010), which is supercomputing's highest honor for setting new high-water marks in scalable computing; and several best paper and best presentation awards, including a best paper prize at the SIAM Conference on Data Mining (SDM, 2012).

Yinglong Xia is currently a Research Staff Member in the IBM T.J. Watson Research Center. He is working on high-performance computing (HPC) and big data analytics for large-scale graphical models, dynamic graphs, and anomalous behavior detection in social networks. Prior to this, he was Computing Innovation Postdoctoral Research at the IBM T.J. Watson Research Center, working on distributed graph algorithms. Yinglong received a Ph.D. in Computer Science from the University of Southern California (USC) in 2010, an M.S. from Tsinghua University in 2006, and a B.Eng. from the University of Electronic Science and Technology of China (UESTC) in 2003.

With a solid background in both Parallel Computing and Machine Learning/Data Mining, Yinglong has been developing various parallel algorithms and middlewares for large-scale graph computations and graphical models. Yinglong explored parallelism at multiple granularity levels. He developed architecture-aware scalable solutions for probabilistic inference in Graphical Models that scale on various HPC platforms/ architectures, including Intel Xeon AMD Opteron, Sun UltraSPARC T2, IBM Cell/B.E., GPGPU, and Clusters. He also has some research experience in Gaussian mixture models, EM algorithms, meta-analysis for literature data mining, etc.

Some of his research findings have been published in several book chapters, and over 30 papers in international conferences/journals. Two best paper awards were received. Yinglong serves as a general cochair, publicity cochair, and TPC chair/member for several conferences and workshops, including IPDPS, HiPC, IEEE CSE, IEEE Cluster, IEEE CloudCom, PDSEC and ParLearning. He has been CCC/CRA Computing Innovation Fellow (CIFellow) since 2010.

Asim YarKhan is currently a Senior Research Associate at the Innovative Computing Laboratories at the University of Tennessee, where he is doing research on the efficient execution of data-driven task-based programs on modern hardware. His interests include runtime systems, distributed computing, grid computing, knowledge discovery and data mining, artificial intelligence, and neural and connectionist computing. He has an M.S. in Computer Science from the Pennsylvania State University, an M.S in Applied Mathematics from the University of Akron, and a Ph.D. in Computer Science from the University of Tennessee.

Mohamed Zahran is currently a faculty member with the Computer Science Department at New York University. He received his Ph.D. in Electrical and Computer Engineering from the University of Maryland at College Park in 2003. After finishing his Ph.D., he worked as a research scientist at the George Washington University for a year before joining the faculty of the City University of New York. Then he moved to New York University. His research interests span several aspects of computer architecture, such as architecture of heterogeneous systems, hardware/software interaction, and biologically inspired architectures. Zahran is a Senior Member of IEEE and a Senior Member of ACM.

Chapter 1

Memory Hierarchy for Multicore and Many-Core Processors

The memory system is a pivotal element of any computer system. It has a direct effect on performance, power dissipation, and bandwidth requirement. However, the importance of the memory system has undergone several stages in the relatively short lifetime of digital computers.

At the dawn of computer systems, the ENIAC (Electronic Numerical Integrator And Computer) era, the memory system was nonexistent because programs were hardwired into the machine. To have the machine execute a different program, it had to be rewired. This scheme, of course, did not last long due to its inflexibility. The stored program concept, introduced by John von Neumann, took the computer world by storm and was the first milestone in the history of memory systems. Instead of rewiring the machine for a new program, programs are stored in memory inside the computer. The first stored program machine is the EDSAC (Electronic Delay Storage Automatic Calculator, 1949), and this marks the birth of memory systems.

With the introduction of memory, the performance of computer systems depends on both the memory and the processor. This was not a problem for the first several generations of computers because the speed of the processor and memory were somewhat matched. The honeymoon ended with the introduction of Moore's law by Gordon Moore, Intel cofounder. Moore said that the number of transistors per chip doubles every 18 months [39]. This law held true for more than three decades, after which the period changed from 18 to 24 months. In the first three decades, this huge transistor budget was translated into performance. Although these advances in process technology are great, things did not get very rosy, and this is the second milestone.

Memory system performance has not kept up with processor performance and has been improving at a much lower pace. This has been famously named by McKee and Wulf as *memory wall* [57, 56]. As a result, a huge body of research related to memory system performance has emerged. Setups like cache hierarchy [51], write-buffers [50], stream-buffers [37], and prefetching [29, 20, 10, 40, 9], are now considered commodity. When microprocessors shifted from a single core to multicore [21, 41], the memory system also underwent major changes. This designates the third and current milestone for memory systems. Now several cores, each with its own private L1 cache, are sharing one or more caches before going off-chip. There are some systems with no shared caches at all, but the majority of designs are sharing at least one level before going off-chip.

We will see that by having multiprocessor on-chip does not mean that we will blindly move a multiprocessor system and put it on-chip or that we are going to reinvent the wheel. Surely there are similarities, but the constraints are different. On-chip, we have higher bandwidth but less area and constrained power budget. Also, the type of applications running on a traditional multiprocessor system are different from the ones running on a multicore processor. On a multicore, you can run a scientific application, which is similar to what runs on a multiprocessor system, but you can also run nonscientific applications. Scientific applications are usually computation bound and involve a lot of loops; they are, hence, easier to parallelize. On the other hand, nonscientific applications include the software applications that we use on a day-to-day basis (e.g., editors, OS kernels, etc.). This type of application is much harder to parallelize due to its complicated control flow graph and can be memory bound. This means the memory footprint and memory access patterns we are dealing with in a multicore system are not the same as in typical multiprocessor systems.

With several applications, or threads, competing for cache real estate and causing interference, both constructive and destructive, a new set of problems is facing computer architects. This means we need a new way of designing memory hierarchy, and we need to rethink the status quo in many aspects of memory systems; this is the subject of this chapter. Figure 1.1 summarizes the topics that we will discuss. They are grouped into three main categories. The first, design issues, lays the foundation of our *measures of success* when designing a memory hierarchy for multicore. The second category shows the design space, which is huge. Finally, the third category gives the technological constraints that we are given when designing a memory hierarchy. Constrains help to reduce our design space a bit, while design issues guide us to explore potentially successful areas in the design space. Basically, in this chapter we try to answer the following question: **What are the challenges and research directions for memory systems in the multicore era.**

In a nutshell,

- This chapter is intended for both intermediate and advanced readers. This includes seniors, graduated students, and researchers.

- The main goals of this chapter are to give the reader a detailed look at the current status quo of memory hierarchy and the main challenges that evolved during to the movement to the multicore and many-core era.

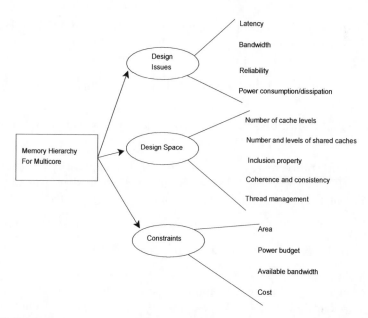

FIGURE 1.1: Memory hierarchy for multicore processors: Issues, design space, and constraints.

We first start by discussing design issues of the hierarchy, such as latency and power consumption. We then look at the physical design of the main memory, the last level of the hierarchy in most designs, which is dynamic RAM in most of devices, although this may change in the future. The next step is to look at the hierarchy organization issues, such as the number of caches versus the number of cores, shared versus private versus cooperative, etc. With multicore and many-core processors becoming de facto, we need effective and efficient techniques for sharing caches among cores, which is our next topic in the chapter. Then, we discuss different optimization techniques for the hierarchy. We cannot have a chapter about memory hierarchy for multicore processors without discussing coherence, which is discussed in the section after the coherence. Coherence is tightly coupled with the consistency model used. Therefore, we dedicate a section to discuss memory consistency models.

1.1 Design Issues

When designing memory hierarchy for multicore processors, there are many design issues that we must take into account. Before we discuss these design issues, let us summarize some facts about memory hierarchy for a multicore processor.

- The memory hierarchy involves memory components of different designs: SRAM (mainly for caches) and DRAMs (for system memory, and some large shared caches in some new systems).

- Some of the components of the hierarchy are on-chip and the others are off-chip.

- Off-chip bandwidth is a scarce resource due to contention on chip pins of pads, memory ports, and buses.

- Multicore processors may run multithreaded applications (dependent threads), multiprogramming environment (independent threads from different applications), or a mix of that.

- There is a very tight budget of power consumption/dissipation on-chip.

- The real estate on-chip is restricted.

With these facts in mind, let us see the different design issues related to multicore memory hierarchy. The importance of studying these design issues is to restrict our design space to those designs that are feasible to implement within specified technological constrains while reaching targeted performance. Therefore, any optimizations done to the memory hierarchy must be evaluated not only based on performance enhancements but also based on their effect on the following design issues.

1.1.1 Latency and Bandwidth

The ultimate goal of any computer system is to be fast. Although the definition of *fast* varies among different systems and applications, it is related to latency. That is, we want each core to finish executing the thread assigned to it in the shortest time possible. This speed is translated into shorter response time, higher throughput, or any other metric depending on the system. If we are talking about a single-core system, then one of the major factors affecting its execution time, beside of course instruction dependencies and high-latency instructions, is memory latency.[1] Memory latency depends on how fast our memory hierarchy is in providing the required data or instructions. The more L1 cache hits, the shorter the memory latency, and hence, the better execution time of the thread on the core. If we look at multicore processors, the situation becomes more complicated because in the case of multicore, we want the core to finish executing its thread as fast as possible *without* slowing down other threads.

For a core to see a short average memory latency, most memory accesses must be satisfied by L1 cache; if not, then by L2 cache. However, if L2 cache is shared, then the competition among cores for L2 real estate may cause some core requests to be satisfied by L2 while others will miss at L2 and have to go to L3 or memory. This means that short memory latency for one core has a negative effect on other cores. Sharing a cache is not the only way a thread can affect the performance of another thread. Coherence protocol used in the system is also another factor, especially invalidate protocols. In this chapter, we will discuss several methods on how to reduce interference among cores when sharing a cache. We will also discuss different coherence protocols.

The bottom line is that a simple average memory latency measure[2] used in single-core processors is no longer the best measure for the performance of the memory hierarchy. Here is a list of possible other ways of measuring latency of memory hierarchy for multicore processors.

- Sum of average memory latencies seen by each core

- Average memory latency seen by all cores (i.e., the sum calculated in the previous bullet divided by the number of cores)

- Sum of average memory latencies seen by each core divided by total number of memory accesses

Average memory latency does not depend only on cache hits and misses and access latency. It depends also on bandwidth used both on-chip and off-

---

[1]The extend to which memory latency affects total execution time depends on whether the thread is memory bound. The more memory accesses a thread needs to do, the higher the effect of memory latency on execution time.

[2]Which can be calculated experimentally by adding the latency of each memory access, then dividing it by total number of accesses. Or it can be calculated analytically using access latency of each cache and main memory, as well as the hit rate of each cache.

chip. A miss in the last-level cache means an off-chip access to main memory. This is constrained not only by the memory access latency but by the off-chip bandwidth available through buses, chip pins or pads, and memory ports. In some memory coherence protocols, when there is a miss on an L1 cache private to a core, the required cache block can come from another L1 cache private to another core. This is done though a network on-chip. So the bandwidth provided by this network affects the average memory latency.

The bottom line is that when designing or studying the average access latency of a memory hierarchy, we have to study also the available bandwidth budget.

1.1.2   Power Consumption

Designing a memory system with short latency is not enough. The wide usage of portable devices and their reliance on batteries for power has resulted in power being a major criteria in the design of any system for these devices. Therefore, we do not only need to know the latency of each cache and memory access, but also the *cost* of each cache and memory access in terms of power.

The memory hierarchy in general consumes a significant amount of power. If we look at the die pictures of many state-of-the-art processors, we will find that cache memory dominates the chip area, 60% or more in many designs. There are two main sources of power consumption: dynamic and static. The frequency of memory (SRAM or DRAM) access causes dynamic power consumption/dissipation. Static power loss is mainly dominated by leakage current. Leakage, as the name indicates, is due to imperfect transistors. Because transistors are getting smaller thanks to process technology, a transistor can not be totally turned off. So even in the low state, the transistor still leaks.

The sum of both dynamic and static power consumption can be presented by the following equation: $P = ACV^2f + VI_{leak}$. V is the supply voltage, f is the frequency of the system operation, A is the activity factor (because not every clock cycle there is an activity, so A has a maximum of 1), and C is the capacitance. The first term of the equation is the dynamic power and the second is the leakage. Because most of the cache is not accessed frequently, caches are a major source of leakage power loss (which is responsible for battery drain). The dynamic power still dominates the equation, although leakage can not longer be disregarded.

Reducing access to memory hierarchy reduces power loss. There are many techniques in literature whose aim is to reduce accesses to most of the memory parts. For instance banking is one scheme that reduces power loss. If memory, whether cache or memory, was only one bank, for instance, then every access to the memory will dissipate/consume power. However, with banking, only the bank that contains the required address is accessed.

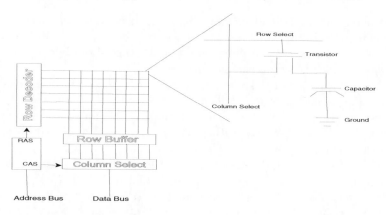

FIGURE 1.2: Dynamic random access memory (DRAM) array.

1.2 Physical Memory

The memory system of multicore processors consists of a per-core cache hierarchy, followed by one or more levels of shared caches, and then the physical memory (DRAM). The physical memory is the main bottleneck of the memory system due to its slow speed.

Dynamic RAM (DRAM), introduced in 1968 by IBM, is now the de facto of main memory. This is due to several factors. First, it is very dense because it uses only one transistor and one storage capacitor for each bit. Figure 1.2 shows the main design of a DRAM array.

Let us first look at a single DRAM cell, shown in the right part of the figure. The capacitor is the main storage entity for a bit. A charge in the capacitor means 1. Lack of charge means 0. Capacitors do a poor job of holding a charge. They constantly leak. This is why DRAM needs a refresh circuitry that continuously recharges the capacitors approximately 15 times per second.

The whole array works as follows. A row in the array is usually hundreds of bytes. The address submitted to DRAM array chooses a row and then a few bytes from that row. This is why a row buffer is needed, as shown in the figure. Any modifications done to the bytes of that row are done to the row buffer. Then the contents of the row buffer are written back to the row before another row is accessed. As we can see, accessing DRAM requires several steps. First, a row is copied to the row buffer. This is done through the row access signal (RAS). Then, the required bytes are read/written through the column select. This second step is triggered by a column access signal (CAS). The last step is to write back the row buffer. This step is called precharge.

A single DRAM array is not very scalable; therefore, several arrays are needed. If the memory is shared, among several cores, then the address space is shared, and we have a distributed shared memory. The address space is distributed among several memory banks. A memory bank consists of one or more DRAM arrays like the one shown in Figure 1.2. This distribution is needed in order to avoid having complicated, hence slow, decoders, and also to allow parallel access to the memory system.

At the beginning of this section, we mentioned that DRAM is the slowest part of the memory hierarchy. The first reason for this is that in most systems, the memory is placed on the board, off-chip. Off-chip access is usually very expensive and slow. Accessing the memory itself is slow due to its internal operation, which involves several steps, as we have seen earlier.

There are two recent technological advances that are affecting DRAMs in a positive way. The first one is embedded-DRAM (eDRAM) [38] used in IBM Power7 [31], and the second one is 3D stacking [58]. With eDRAM, we can put DRAM arrays on-chip. This will allow much faster access, and the eDRAM will act as another level of cache. 3D stacking will allow several layers of silicon to be stacked on top of each other. By using one of these layers for eDRAM, we can have a huge cache storage on-chip, which will decrease off-chip access. Decreasing off-chip access means higher performance and less power consumption.

In the next section, we will take a closer look not at the DRAM but at the cache memory. Cache memory is built in a way similar to the DRAM but with SRAM technology. SRAM stands for static RAM, which means it does not need to refresh, and hence, there is no need for refresh circuitry of row buffers. On the other hand, an SRAM cell requires 4 to 6 transistors, depending on its type, so it is much less dense than DRAM but has faster access time.

1.3   Cache Hierarchy Organization

Over the past decade, we have witnessed phenomenal improvements in process technology. This has led to an increase in the available on-chip transistor budget. This transistor budget resulted in a wave of innovations in microarchitecture, such as prefetching, deep pipelines, more aggressive execution, and of course more cores.

The increase in the number of on-chip cores, and the sophistication of each core, place significant demands on the memory system. We need high-bandwidth, low-latency, high-capacity, scalable memory systems. The current memory system hierarchy does not scale beyond a few cores due to technological and logical reasons.

On the technological side, increasing the number of ports per cache, or the number of cache banks, results in a tremendous increase in power, area, and wire delay. From a logical perspective, increasing the number of cores means increasing the number of applications/threads running simultaneously. If these threads share a cache, the interference between them will increase, leading to an increase in miss rate as well as bandwidth requirements. Therefore, many aspects of the memory system design need reconsideration and rethinking.

1.3.1   Caches versus Cores

One of the main issues that must be discussed in depth at the concept level of the life cycle of any new microprocessor is the number of caches versus the number of cores.[3] The architect needs to answer several questions. How many cores shall we include? Shall we use symmetric or asymmetric cores (a.k.a. homogeneous versus heterogeneous)? How many private caches? How many shared caches? What are the specifications of each cache (size, associativity, block size, etc.)? In order to answer these questions and achieve a balanced design, there are some factors that the designer must take into account.

1.3.1.1   Technological and usage factors

The first factor is the available real estate on-chip. Depending on the intended use or sector targeted by this processor, there will be some area constraints. For instance, a processor designed for embedded devices will be different in size than one used in notebooks, desktops, or servers, for example.

The second factor is the power budget. This is becoming a major technological constraint and must be addressed as early as possible in the design cycle. In the sub-micron era we are currently witnessing, we have to take care of static power dissipation—where caches are a major source—as well as dynamic power consumption.

The third factor is the targeted usage. What type of applications is this system intended to run? The importance of cache hierarchy specification depends on the application at hand. Streaming applications are different than scientific applications, which are different than non-scientific applications (like an operating system kernel for example). This is very tricky though. If the processor is intended to be general purpose, then a lot of simulations are needed to decide the best configuration. Statistical analysis and trace-driven simulations can help restrict the design-space exploration. However, extensive cycle-accurate simulations will be needed in order to assess the interaction between the different parts of the computer system and the cache hierarchy.

1.3.1.2   Application-related factors

There are several program characteristics that can help to reduce the cache hierarchy design-space exploration. The first is whether the program has enough parallelism. The second is whether the program is cache friendly. A cache-friendly program means it has good spatial and temporal locality. The

---

[3]The de facto definition of core is a central processing unit and its private L1 data cache and instruction cache.

last characteristic is the size of the program working set. These characteristics vary not only among different programs but also among phases of the same program [48].

A program can have parallelism at different granularities. If the parallelism is among instructions, that is instruction-level parallelism (ILP)[46], and to exploit it, we need *fat* pipelines with superscalar capabilities, out-of-order execution, speculative execution, and so on. This was the golden era of single-core processors. A core with such capabilities will be very demanding in terms of memory bandwidth. It needs to be fed with instructions and data at a fast rate. To satisfy such a power-hungry core, the computer architecture community proposed many solutions. Cache sizes and associativity have increased in the last two decades along with the number of caches and levels [23]. Then, lockup-free caches were introduced where a cache can keep track of several cache misses [34]. Therefore, a single cache miss will not bring the whole cache into halt waiting for the miss to be serviced. As cycle-time continued to decrease, designers soon found out the trade-off between associativity and latency in a cache design. Not only did cores need more data and instructions, they needed them *fast*. This sparkled several innovations such as the victim cache [30] that keeps blocks evicted from a cache at replacement for sometime in case they are referenced again instead of interrogating other caches in the hierarchy of the main memory off-chip. The victim cache is very small in size and hence is fast and fully associative. Because with such aggressive cores, several memory requests can exist simultaneously, the concept of memory level parallelism (MLP) was introduced [12]. MLP is very related to lockup-free memory because it generally is tightly coupled with the number of outstanding cache misses. The above techniques and many more are targeting single core with high ILP. Their main goal is to hide memory latency because memory system did not keep up with the decrease in processor cycle time. This widens the gap between processor speed and memory system speed.

If the parallelism in a program is of higher granularity, for instance, at the thread level, then the situation is different. We have several threads running simultaneously on different cores. This means that when there is enough parallelism, adding more cores translates into better performance. Let's assume for now that each core can run a single thread at a time. Usually every core has its own L1 instruction and data caches. The question is: how does the memory hierarchy look beyond L1 caches? How many levels shall we have? How many shared cache(s)? If we do not have shared caches at all, the private caches of each core will be competing to go off-chip to access external memory in the case of cache miss. Also, they will be competing for buses, memory ports, and controllers. Also, some threads may not need their whole caches, while other threads strive for more caches. This means we need to have a shared cache, but this shared cache can be distributed, as we will see later in this chapter. The more parallel threads we have, one of two scenarios will happen. The first scenario is that these threads have constructive interference and hence will make the best use of the shared cache. The second is the opposite scenario. Destructive interference among threads will cause more cache misses at the shared cache and will negatively affect overall performance. We will shortly discuss the different ways of dealing with shared caches in order to avoid this second scenario.

In addition to parallelism, cache friendliness is a major factor in determining the specifications of cache hierarchy. The short definition of cache friendliness is having temporal and spatial locality. Temporal locality means if X is used, it is likely that it will be used again in the near future. Spatial locality means if X is used, very likely X+1 will be used soon. X can be a byte, word, double word, or similar item. But this definition, although correct, is not enough and does not capture the whole definition. Cache-friendly behavior, in addition to localities, has to do with the distribution of cache misses through time. Cache misses are not all of equal cost [42]. If 10 instructions are waiting for a cache miss to be resolved, and 1 instruction is waiting for another cache miss, the first one is more costly. Moreover, cache misses that come in bursts tend to be less costly than individual cache misses, because in this case, the penalty in this mishandling will be leveraged among several misses.

The last characteristic of a program that can give a hint about the best memory hierarchy is the working set of that program. From a memory perspective, the working set of a program is the amount of information that must be present in memory at the same time in order for the program to execute correctly. The working sets are getting bigger because software applications are getting more complex. The bigger the working set, the more storage we need on-chip.

From the above discussion, we can deduce some rules of thumb about cache hierarchy design in a multicore chip.

- For every core added, there must be a corresponding storage added on-chip to reduce off-chip access.

- No cache hierarchy can do the best for all applications unless we are talking about limited applications in some embedded devices, in which case they may not even need a sophisticated cache hierarchy!

- Power budget, performance needed, and cost, pick any two.

- Sometimes it is wise to allow some cache misses in return for faster access time.

1.3.2   Private, Shared, and Cooperative Caching

There are several caches that coexist on-chip. Some of these caches are private. That is, they serve a single core. Some other caches are shared between two or more cores. Figure 1.3 shows the basic schemes with 4 cores and two-level cache hierarchy.

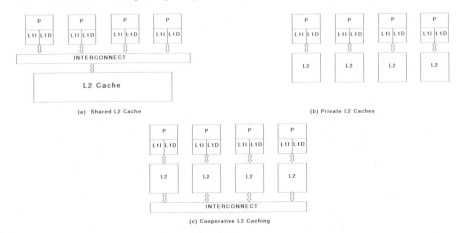

FIGURE 1.3: Private, shared, and cooperative caching.

Private caches have low access time, and they are usually close to the core they are serving. However, if we only have private caches, then we are not making the best use of the cache real estate. Some cores may not need their entire cache, while other cores are starving for more storage. For this situation, shared caches are used. A shared cache serves several cores. With that shared resource, storage is assigned dynamically among based. This leads to better usage of the storage real estate and translates to lower misses but at the expense of higher access time.

There is another scheme that tries to combine the strength of both private and shared caches. This scheme is called cooperative caching (CC) [7]. Cooperative caching starts from private L2 caches (the technique is applicable to other cache levels too). Each core has its own private L2 cache that can only be accessed by its own core. However, all these L2 caches give an aggregate behavior of a shared L2 cache whose associativity is the sum of the associativities of all the cooperative caches. For example, 4 caches each in an 8-way set associative will give the behavior of a 32-way associative cache. The main motivation behind CC is that off-chip access due to cache misses is order of magnitudes longer than accessing another on-chip cache.

The main goal of CC is to minimize average latency of memory requests. In order to accomplish its job, CC must allow local L2 cache misses to be served by remote on-chip caches. This requires the exchange of different information among caches using a coherence engine. Traditional coherence protocols allow different caches to exchange blocks only if these blocks are dirty, but in CC, clean blocks are allowed to be exchanged among caches. Of course, accessing the local cache is faster than accessing a remote cache, but both are much faster than accessing off-chip memory. This technique does not make sense in multiprocessor systems because remote caches are off-chip themselves.

Multicore processors make this clean cache-to-cache block transfer and access feasible.

In order to make this technique more efficient, the replacement policy in each of the L2 caches must take into account the different types of blocks. In CC, there are two types of blocks. A block that exists only in a single cache can be either clean or dirty. The other type is a duplicate block in different L2 caches, in which case, the copies are all clean. The modified replacement policy is replication-aware, so it discriminates against replicated blocks. It evicts single-copy blocks only if there are no replicated blocks or invalid blocks in the set.

In case a single-copy block is victimized, it can be spilled into another cache. There are two issues to consider here. The first is to which cache shall this victimized block go? The second issue is how to deal with the rippling effect if a spilled block causes another block to be spilled. Regarding the first issue, there are many schemes. The easiest one is to choose a random cache to accept the block. This can be the nearest cache to the source. The advantage of that scheme is that it is a low-cost solution. Other schemes include checking the cache with free slots. These content-aware schemes can be more efficient in using the caches' real estate but require global management among caches, which can be costly in terms of hardware and latency. Some blocks can become inactive or dead. It will be cost effective not to circulate these blocks. The most straightforward way of doing it is to assign a counter to each block. This counter is set to N when the block is spilled for the first time. Each time the block is recirculated, the counter is decremented. If the counter reaches 0, then the block is discarded. As for the second issue of the ripple effect, the system prohibits spilled blocks from triggering another spilled block.

Cooperative caching is one way of sharing large cache real estate among several cores. Whether or not to use this scheme is a question of trade-off between hardware cost and potential performance gain. Having a shared cache among cores is simpler in terms of hardware cost, but does it scale? This leads us to nonuniform cache access.

1.3.3   Nonuniform Cache Architecture (NUCA)

As the number of cores increases, the pressure on any shared cache also increases. Now this shared cache has to serve several cores within a reasonable amount of time. In the old days of single core and small caches, the cache access latency was more or less predictable. This is due to two main reasons. The first is that the cache was built with one or a few banks. The second is that the cycle time was large enough not to expose wire delays. As the cache increases in size, having several banks becomes a necessity. Without several banks, the contention on the cache will be huge, and the cache will be unable to serve several cores simultaneously unless the bank is multiported, which comes at a huge cost of area, complexity, and power consumption. Therefore, banking became a de facto of large cache design.

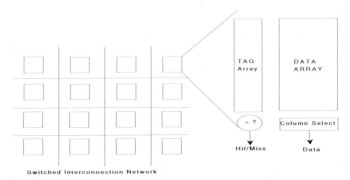

FIGURE 1.4: Nonuniform cache access.

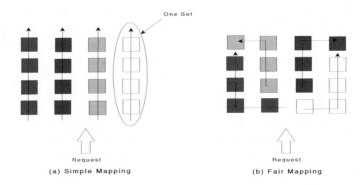

FIGURE 1.5: Set-sharing schemes.

Increasing the number of banks is a clever way to decrease the access time [32]. But as the number of banks increases and as the number of cores sharing the cache increases, the multibanked large shared cache cannot maintain a uniform access latency. This is due to the wires interconnecting the banks and the cores. When a core generates an address request, the access time of the cache depends on the location of the responding bank relative to the core. If the address maps to a bank in the vicinity of the requesting core, the access latency will be much shorter than if the bank is physically farther away. From this comes the name nonuniform cache architecture (NUCA).

Figure 1.4 shows the basic architecture of a NUCA which consists of several banks, presented as squares in the figure, connected with a two-dimensional switched network. These banks can be placed in the center of the chip so as to be equidistant from all cores, or each bank (or more) can be assigned to a core and all cores can access all banks, as in the case of tiled architectures [47]. A cache set spans several banks. This gives the ability to move a block among banks without affecting addressing. Figure 1.5 shows two ways of assigning sets to banks. The first method, shown in Figure 1.5(a), is called simple mapping. In this mapping, every column represents a set and every row a way, so the figure presents a 4-way set associative cache with 4 sets. This method exposes the main characteristic of NUCA. That is, the farther the bank from the request, the longer the delay. A different way of organizing sets is given in Figure 1.5(b). In this scheme, banks are placed in such a way that the average access time across all banks sets are equal. This is why it is called fair mapping. There is another scheme where the closer banks to the core are shared by several sets, while banks that are farther away are not shared. This is called shared mapping. These schemes were proposed in the single-core era. In the multicore era, things are more complicated because requests can come from many cores at different locations. This means NUCA needs to juggle several seemingly conflicting goals.

There are seemingly conflicting goals that NUCA needs to meet in order to ensure best performance. The first is to keep blocks in the nearest bank to the core that needs it. The second is to reduce on-chip traffic by decreasing block movement among banks. The third goal is to decrease its power consumption because NUCA usually requires most of the chip area. Finally, the fourth goal is to reduce the number of simultaneous accesses to a bank in order to avoid contention. Researchers have been trying to tackle one or more of the above challenges, usually reaching some goals at the expense of others.

One of the main techniques used within NUCA caches is *block migration*. As the name implies, blocks are moved closer to the cores that request them the most. Since a cache set spans several banks, if a block migrates from one bank to the other, it is still within its set. Basically it is a question of data placement: where to put each block so as to reduce the average access time. This is a very challenging problem because different blocks have different degrees of sharing (i.e., how many cores access this block?). The location of the cores that share the block relative to each other and to the bank containing the block is another issue. Moreover, the behavior of programs changes during their execution, and so does the sharing pattern. The simple scheme of moving a block to be at the bank[4] closest to the core that needs it is not very efficient because if two cores request that block, and these two cores are not close by, there will be a ping-pong effect in which the block will be moving back and forth among banks, resulting in a large amount of migration. Or if a block is shared by many cores, it may end up in a bank far away from all cores. One possible solution, called CMP-DNUCA [4], divides the banks into sets (i.e., sets of banks, which are different from a cache set). An address maps to one of the sets. A block can freely move among banks of the same set. These sets are *logical* separation. Banks are grouped *physically* into bank clusters. Every cluster contains one bank from every set. These clusters are placed on-chip in three different regions. The first region, called the *local region*, consists of the clusters near the cores. In the chip center resides the *center region*. Between

---

[4]Whatever we say about banks can also apply to a group of banks or slices.

these two regions resides the *intermediate region*. This setup allows each core to have a close cluster in the local region and another close (but farther than the first) in the intermediate region. The center region is usually equidistant from all cores. Ideally, hits in the local region give the least delay. Blocks that are not shared will usually be in the local region. As the number of sharers increases, a block moves toward the center region, or the other way around; thus, we have a gradual migration policy.

One final aspect of block migration has to do with bank design. Each bank can be n-way set associative. But if $(n + 1)$ blocks want to be in that bank in order to be close to the requesting core, then this bank will face conflicts. NuRapid NUCA, proposed in [11], uses centralized tags and a level of indirection. This way, they decouple data placement from set indexing and hence reduce conflicts.

Researchers have also tried *replication* as a way to reduce NUCA access time. The straightforward approach of duplicating a block based on the number of cores sharing that block has several drawbacks. First, it increases the hardware complexity because the cache needs to maintain coherence among these blocks. Second, if several blocks have a high degree of sharing, then a lot of the NUCA real estate is lost due to duplication. Another option is to duplicate only the clean blocks, so as to reduce coherence complexity, but then the cache controller has to keep track of these duplicated block accesses because a clean block can become dirty.

R-NUCA [22] cooperates with the operating system to classify memory accesses at the page granularity. By migrating some of the tasks to the software side, R-NUCA reduces hardware complexity and scales to a large number of cores. This scheme makes decisions about placement, migration, and duplication depending on memory accesses as follows. Modifiable blocks map to a single location in the cache. Read-only blocks, such as instructions, are shared among neighboring cores and replicated at distance ones. However, replication is not done at the level of a slice or bank because a block replicated to the neighboring slice will not yield significant savings in latency. Therefore, R-NUCA groups neighboring slices into logical clusters, and replication is done at the cluster level. Private blocks are placed in the bank or slice nearest to the requester. Private data are usually associated with stack space in memory.

1.4   Cache Hierarchy Sharing

Most multicore processors consist of several cores with one or more private caches followed by one or more shared caches. This is what we call a *logical* view of the hierarchy, as shown in Figure 1.6. The physical view depends on whether caches are single or multibanked, whether the shared cache is NUCA, and so on. In this section, we will study the aspect of sharing in a

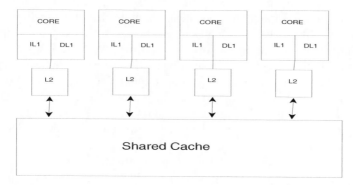

FIGURE 1.6: Logical view of a cache hierarchy in multicore processors.

cache hierarchy, which is somewhat orthogonal to the physical implementation of a cache.

1.4.1   At What Level to Share Caches?

The main goal of caches is to reduce expensive off-chip access. To achieve this goal, the cache must have a high hit rate and low access time. In the single-core era, and when the processor cycle time was not yet in the nano range, a single cache was able to do the job. When software became more complicated, caches increased in size and associativity. However, this made for longer cache access times. Longer cache access time and shorter processor cycle time were the invitation to include a second cache. A two-level cache hierarchy was quite satisfactory for most single-core processors and became a commodity. Most of the research done at the time of single-core cache hierarchy was centered around enhancing this memory hierarchy by introducing better replacement policies, exploiting memory level parallelism, increasing bandwidth among caches, and so on.

With the arrival of multicore chips, it became obvious that the cache hierarchy needed a rethinking. What does the cache hierarchy look like for a large number of cores? One possibility is for each core to have its own private cache hierarchy. On the one hand, a private cache gives shorter access times and avoids contention. On the other hand, private caches do not make the best use of the cache real estate because the same data may be duplicated in multiple caches. Also, private caches can result in some core starving for storage, while other cores are not making use of all their cache storage. Moreover, multiple caches will compete for the off-chip memory controller, bus, ports, and so on. The other option to avoid the above problems is to have a shared cache down the hierarchy and before off-chip access.

A shared cache at the last level, or a hierarchy of shared caches, is another possibility. Shared caches make better use of the cache real estate so that shared data are not replicated. The cache space is assigned based on the demand from multiple cores and hence is more flexible. But life is not that simple! Shared caches are usually bigger in size with higher associativity, so they suffer from higher nonuniform access time (check NUCA section above). Also, shared caches can suffer from destructive interference from the different threads or processes sharing them. This interference can lead to many conflict misses. As we can see, there are trade-offs in using shared versus private caches. This is why the most commonly used setup is the one shown in Figure 1.6, with private hierarchy for each core followed by a shared hierarchy. The question here is: at what level shall we put the shared cache(s)?

The answer to this question depends on several factors such as the number of cores, the expected application working sets, the type of applications expected to run on that processor, etc.

- For performance, L1 caches (both instruction and data) need to be private in order to reduce access time.

- If the number of cores is small (2–4), then the L2 cache can be a shared cache. This is because with a small number of cores, the likelihood of destructive interference is smaller than with a large number of cores. Also, the contention at L2 will be less severe.

- With a larger number of cores, a single shared cache may not do a satisfactory job at serving all the cores, in which case, different schemes are available. The first is to have several shared caches for each group of cores. Then, every few shared caches will share another cache farther down the hierarchy. We will end up with a tree-like hierarchy where the leaves are at the cores. The other setup is to have several levels of shared caches to accommodate the expected large number of conflict misses. Of course, for a large cache with high associativity, a NUCA design is required.

- If each core has a simultaneous multithreading capability, then several private caches are needed before the shared one. This is the case with processors like Intel's Nehalem, for example.

The above bullets are rules of thumb; depending on the target usage and market, the designer can decide on the best hierarchy.

1.4.2   Cache-Sharing Management

1.4.2.1   Fairness

A cache shared by several threads[5] needs careful management to avoid performance loss. If we implement shared caches the same way we build a private cache, we are missing a lot of opportunities. Shared caches must not

---

[5]A core can be running multiple threads if it incorporates simultaneous multithreading capability.

be thread-oblivious. For example, let us assume that a cache is shared by two threads A and B. Most caches use temporal replacement policies, such as the least recently used (LRU) policy or a variation of it. If thread A accesses many cache blocks and more frequently than thread B, the LRU will favor thread A, even though thread B maybe of higher priority. Moreover, thread A may cause victimization of many thread B blocks even though thread B may need only a very small cache space to achieve higher performance. Therefore, we need to have a *fair* sharing of the cache space. The operating system thread scheduler assumes that the hardware is providing fair sharing of coscheduled threads. If the hardware does not do that, some threads may starve for cache space, and worse, a higher priority thread may make less progress than a lower priority one.

In [33], the authors propose five definitions of fairness. Any algorithm that strives to reach fair cache sharing must try to minimize $M^{ij} = |X_i - X_j|$, where i and j are two coscheduled threads. The authors give five definitions for X_i.

1. $X_i = \frac{Miss\_shr_i}{Miss\_ded_i}$, where $miss\_shr$ is the number of misses when the thread is using a shared cache, and $miss\_ded$ is when using a dedicated (unshared) cache.

2. $X_i = Miss\_shr_i$

3. $X_i = \frac{Missr\_shr_i}{Missr\_ded_i}$, where $missr$ means miss rate

4. $X_i = Missr\_shr_i$

5. $X_i = Missr\_shr_i - Missr\_ded_i$

Any cache sharing algorithm must try to minimize M_x^{ij} for every two coscheduled threads i and j in order to reach fairness. Which one of the five definitions to use in minimization depends on the designer and how complex the hardware can be, etc. So when there are more than two threads, the fair metric can be the sum or average of the metric value of every two threads. The good thing about the above metrics is that they can be easily computed, and they depend on and affect the cache at hand and are not affected by any external behavior. Now the question is: once we calculate one of the above metrics, how can we use that in achieving fairness? The easiest and most straightforward method is to modify the replacement policy by allowing the thread that has experienced unfairness to keep more blocks. The main idea now is to calculate X_i for every thread sharing the cache, then calculate the chosen M_x every defined interval, and finally adjust the replacement policy accordingly. From the five definitions mentioned above, those that depend on misses when using dedicated caches need off-line profiling. So, if the off-line possibility is not possible, another metric must be chosen.

Contrary to fair sharing, some other researchers tried *unfair* sharing with the main goal of enhancing overall system performance [24]. In this paper, the authors called communist any policy that tries to achieve fairness like the one we discussed above. They call utilitarian a policy that strives to maximize the total benefit of the aggregate group, such as throughput. They have a capitalist policy, which is a free-for-all policy. This last one is the traditional unregulated cache we know. They found little difference in cache allocation between utilitarian and communist. However, they found some workloads for which a communist goal severely degrades global performance; and for some other workloads, a utilitarian goal severely impacts fairness, so the best strategy is to guarantee that neither fairness nor global performance is unduly sacrificed. This means the best cache allocation (i.e., varying replacement policy) is still an open question.

1.4.2.2 Quality of Service (QoS)

Sometimes fair sharing is not what we want. This is the case when several threads of different priorities are running. The higher priority threads need a higher *Quality of Service* (QoS) than lower priority threads. In this case, we need to improve the performance of a thread[6] at the expense of potential detriment to others. The authors in [26] studied several QoS policies for cache and memory in CMP processors. In order to ensure QoS for one or more threads, we need to answer several questions. First, who will make the decision about which threads are of higher priority? Second, what are the monitoring metrics to keep track of the resources used by each thread? Third, once the measurements are obtained, how can we decide on the policy to be used by the hardware to ensure the desired performance? Finally, how do we enforce the QoS after collecting measurements using metrics?

For the first question, the OS, or a virtual machine manager (VMM), is a logical choice to assign priorities to different threads, sometimes with intervention from administrators. Here, we are talking about software-guided priorities [26].

The choice of a monitoring metric is a bit more complicated because it depends on the different scheme and situations, but in general, metrics can be divided into three categories.

- Metrics related to resource usage, such as cache capacity used by each thread

- Metrics related to resource performance, such as cache misses

- Metrics related to overall performance, such as throughput

From this list, we can see that the choice of the metric depends on the platform used, the number and types of applications running, and the complexity of the hardware scheme the computer system designer can afford. Measuring

---

[6]Throughout our discussion, we will use application, thread, and process interchangibly. Although they have different meanings, the discussion here applies to all of them. An application, also known as a process, consists of one or more threads.

one or more of the above metrics varies in complexity in terms of hardware resources. For instance, using counters to measure cache misses is less complex than measuring throughput or instructions-per-cycle.

Once the designer decides which metric to use, an algorithm is needed to use this metric that decides the next action to be done. The authors in [26] propose two schemes, one static and one dynamic. In the static scheme, there are no readjustments of hardware resources. The metric is usually a resource usage, such as the number of ways in a cache set. Once decided by the OS or VMM, the job of the hardware is to enforce this metric. A dynamic scheme requires continuous reallocation of resources based on metric measurements. That metric can be resource performance or overall performance. If a measured metric is below some threshold, more resources are allocated. For example, if the number of misses per instruction increases beyond a threshold, then more cache ways are allocated to that thread.

Finally, the last step is how to take action to enforce the required QoS. The simplest way to do it is to alter the replacement policy of the cache to favor a thread of higher priority. This means that each cache block must be tagged with the thread ID to which it belongs.

1.4.3 Configurable Caches

Cache memory is designed with specifications chosen to provide good performance on average. Although these specifications can provide good average performance, they are far from delivering the best performance for most applications for several reasons. First, the memory requirements as well as the memory access pattern differ among applications, or even through the lifetime of a single application. Second, cache memory is built with many conflicting requirements, namely, low miss rate, low access time, and low power consumption. In order to have a low miss rate, the cache needs to be large, and this will have a negative impact on the access time, which is unacceptable given the continuous decrease in the cycle time. Moreover, increasing the hit rate implies increasing the associativity to overcome conflict misses. Higher associativity means slower cache. Beside the aforementioned performance trade-off, cache design parameters such as associativity, block size, and replacement policy are not easy to determine beforehand to deliver the best performance for all types of applications. Programs have different behaviors. For example, multimedia programs need a large block size, whereas general-purpose, nonscientific programs need a smaller block size. Another example related to replacement policy, least recently used (LRU), is the most commonly used. However, the LRU may not be the best policy all time, especially in a cache with small associativity, where the LRU block may be needed shortly in the near future. This makes the optimal design of cache system (for a general-purpose machine) an intractable problem. The designer generally tries to achieve a decent performance on the average. Although this was acceptable in the past, it is no longer wise to improve the average case only. This is because embedded systems, which are becoming increasingly popular, require real-time response, which requires improvement in the worst-case performance. Another very important reason for cache reconfiguration is power dissipation. Cache memories are big sources of power dissipation on-chip, especially static power dissipation. Because of all these aforementioned reasons, researchers have tried to build a *reconfigurable cache.*

A reconfigurable cache is a cache memory of which one or more aspects changes dynamically based on application behavior. There are many aspects of a cache that can be designed to be configurable: total size, associativity, block size, and replacement policy. In the previous section, we have seen how a replacement policy can react dynamically to reach a specified QoS. Albonesi proposed configuring the associativity [2]. He called the technique *selective cache way.* This technique provides the ability to disable some cache ways in times of modest cache activity. The data and tag arrays are partitioned into subarrays for each cache way. There is a software-visible register that enables/disables cache ways. This register is updated through special instructions. The main goal here is to save energy by disabling some cache ways, while sacrificing as little performance as possible. When some of the ways are disabled, the blocks residing in these ways can be flushed. Another more sophisticated approach is to move these blocks to some enabled ways, assuming there is room. In [55], the authors propose a cache memory where cache line is continuously adjusted by hardware based on observed application access. They change the size during line replacement. However, they need a counter for each word in the cache in order to determine the block fetch size. This may not be feasible for large caches. Another way of varying the fetch size is proposed in [35], where they are using two direct-mapped caches, main and shadow, with a small block size as well as a fully associative cache with a large block size. Based on the application behavior, one of the caches is chosen and the blocks migrate between them.

Beside the aspects to be reconfigured, the decision of when to reconfigure is another design issue. For example, in [2] Albonesi suggests having a special instruction inserted in the code to trigger the reconfiguration. These instructions can be inserted by a profiling tool, the compiler, or the programmer. Other designs use totally dynamic schemes to trigger reconfiguration. The challenge in doing this in hardware is to know when an application changed its behavior [15]. There are many techniques to measure changing behavior of an application. For example, every specific number of instructions, the hardware can keep track of the number of distinct cache blocks touched. If the difference between this period and the previous is higher than a specific threshold, a phase change is detected. The hardware can also store metadata about each phase (such as a signature [15]) together with the best configuration for that phase in case the phase is encountered in the future.

Reconfigurable caches has not gained big success, at least until now. The resulting hardware can become very complicated with higher access time, canceling the advantage we may have obtained. Another reason is that most of the proposals were targeting single-core processors. Using them in multicore is very challenging, especially with the dominance of wire delay as well as the interaction among the different threads.

Recently, most of the proposals about reconfigurable caches were targeting replacement policies at the last-level cache. Subramanian et al. [53] present another adaptive replacement policy: the cache switches between different replacement policies based on access behavior. They keep two tags in the cache. The extra tag, called the shadow tag, uses a different replacement policy than the other tag. Then, there are counters to keep track of which replacement policy is better performing. So basically, they choose the best replacement policy out of two. Another adaptive cache replacement policy, called *dynamic insertion policy* (DIP), has been proposed recently for single-core [43] and another version for multicore [27]. DIP is derived from LRU replacement policy and is based on *bimodal insertion policy* (BIP) [43]. BIP policy inserts the newly incoming block in the LRU position. This new block will move to the MRU (most recently used) position only if accessed a second time. The main idea is to victimize blocks that are referenced only once and have a tendency to pollute the cache. The DIP policy combines LRU and BIP to get the best of both policies. A small number of dedicated cache sets use a fixed replacement policy, half using LRU, the other half using BIP. These dedicated sets are used to update a counter indicating which, LRU or BIP, generates the fewest misses. The remaining cache sets (i.e., most of the sets) use the replacement policy that is best according to the counter value. The multicore version is almost the same, but there is a counter per thread.

1.5   Cache Hierarchy Optimization

Multilevel cache hierarchies have been known and used for single-core processors for over two decades now. There have been many optimizations to this single-core cache hierarchy. With the inclusion of multicore core processors, some of these optimizations need to be reconsidered, and other optimizations are needed to tackle different challenges. One of the obvious schemes that needs to be reconsidered is the inclusion property. Inclusive hierarchy has the advantage of simplifying the hardware needed for coherence, but with the multicore, duplicating the data in all the private L1 caches is a huge waste of storage. When and how to violate inclusion property is a major design issue and is the topic of this section.

1.5.1   Multilevel Inclusion

Inclusion in the cache hierarchy has attracted attention from researchers since the early days of cache memories. For almost two decades, cache hierarchies have largely been inclusive; that is, L1 is a subset of L2, which is a subset of L3, and so on. This organization worked well before the sub-micron era, especially when single-core chips were the main design choice. Uniprocessor cycle times were often large enough to hide the latency of accesses within the cache hierarchy, and execution was not that aggressive.

With the advent of multiple CPU cores on a chip [36, 14, 21], on-chip caches are increasing in number, size, and design sophistication. For instance, IBM's POWER 7 architecture [31] has eight cores each with its own private L1 and L2 caches. The private L2 is 256 KB. Then, there is an embedded DRAM acting as shared L3 cache of 32 MB of size. The older IBM POWER5 has a 1.875-MB L2 cache with a 36-MB off-chip L3 [49]. The six cores in AMD Opteron share a 6-MB L3 cache [13]. As the size and complexity of on-chip caches increase, the need to decrease miss rates gains additional significance, as does access time (even for L1 caches; single-cycle access times are no longer possible).

We have traditionally maintained inclusion in our cache hierarchies for several reasons: for instance, in multiprocessor systems, inclusion simplifies memory controller and processor design by limiting the effects of cache coherence messages to higher levels in the memory hierarchy. Unfortunately, cache designs that enforce inclusion are inherently wasteful of space and bandwidth: every cache line in a lower level is duplicated in the higher levels, and updates in lower levels trigger many more updates in other levels, wasting bandwidth. As the relative bandwidth on a multicore chip decreases with the number of on-chip CPUs and relatively smaller cache real estate per CPU, this problem has sparked a wave of proposals for noninclusive cache hierarchies.

We can violate inclusion in the cache hierarchy in two ways. The first is to have a noninclusive cache, and the second is to have a mutually exclusive cache. The former design simply does not enforce inclusion. Most of the proposals in this category apply a replacement algorithm that is local to individual caches. For instance, when a block is evicted from L2, its corresponding block is *not* evicted from L1. However, the motivation for such schemes is to develop innovative *local* replacement policies. Qureshi et al. [43] propose a replacement algorithm in which an incoming block is inserted in the LRU instead of the MRU position without enforcing inclusion. In other words, blocks brought into a cache have been observed to move from MRU to LRU without being referenced again. By bringing a new block into LRU and only making it MRU when referenced again improves efficiency. This approach improves efficiency but only at the levels of individual caches: each cache in the hierarchy acts individually, with no *global* view of the hierarchy.

The latter method for violating inclusion is to have mutually exclusive caches [59]. For instance, in a two-level hierarchy, the caches can hold the number of unique blocks that can fit into both L1 and L2. This approach obviously makes the best use of the on-chip cache real estate. In an exclusive hierarchy, the L2 acts as a victim cache [30] for L1. When both L1 and L2 miss, the new block comes into L1, and when evicted, it moves to L2. A block is promoted from L2 to L1 when an access hits in L2.

1.5.2 Global Placement

Designing an efficient cache hierarchy is anything but trivial and requires choosing among myriad parameters at each level. One pivotal design point is block placement—where to put a newly incoming block? Placement policies affect overall cache performance, not only in terms of hits and misses but also in terms of bandwidth utilization and response time: a poor policy can increase the number of misses, trigger high traffic to the next level of the hierarchy, and increase the miss penalty. Given these problems, much research and development effort has been devoted to finding effective cache placement and replacement policies. Almost all designs resulting from these studies deal with the placement and replacement policy *within a single cache*. Although such local policies can be efficient within a cache, they cannot take into account interactions among several caches in the (ever deeper) hierarchy. Given this, we advocate a *global view of the cache hierarchy* as we mentioned in the previous section.

Cache hierarchies usually assume that all blocks are of the same importance and hence deserve a place in all caches since *inclusive* policies are usually enforced. In our observation, this is not true. A block that is referenced only once does not need to be in cache, and the same holds for a block referenced very few times over a long period (especially for L1 cache). Overall performance depends not only on how much data the hierarchy holds, but also on *which* data it retains. The working sets of modern applications are much larger than all caches in most hierarchies (exceptions being very large, off-chip L3 caches, for instance), which makes deciding which blocks to keep where in the hierarchy of crucial importance.

The performance of a cache hierarchy and its effects on overall system performance inherently depend on cache-block behavior. For example, a block rarely accessed may evict a block very heavily accessed, causing in higher miss rates. Sometimes, if the evicted block is dirty, higher bandwidth requirements result.

The behavior of a cache block can be summarized by two main characteristics: the number of times it is accessed, and the number of times it has been evicted and refetched. The first is an indication of the importance of the block, and the second shows how block accesses are distributed in time. As an example, Figure 1.7 shows two benchmarks from SPEC2000: `twolf` from SPECINT and `art` from SPECFP [52]. These two benchmarks are known to be memory-bound applications [6]. The figure shows four histograms. Those on the left show the distribution of the total number of accesses to different blocks. For `twolf` the majority of blocks are accessed between 1,000 and 10,000 times, but for `art` the majority of blocks are accessed between 100 and 1,000 times. Some blocks are accessed very few times. For instance, more than 8,000 blocks are accessed fewer than 100 times.

The histograms on the right show number of block reuses, that is, the number of times a block is evicted and reloaded. More than 15,000 unique blocks in `twolf` and more than 25,000 unique blocks from `art` are loaded more than 1,000 times.

Based on the above observations, a block may be loaded very few times and then be accessed very lightly in each epoch (time between evictions). Other blocks can be loaded many times and be accessed very heavily in each. Many fall between these extremes. Success of any cache-placement policy depends on its ability to categorize block access behavior to determine correct block placement in the hierarchy based on this behavior. This placement policy must be global, that is, it must be able to place a block at any hierarchy level based on the block's behavior.

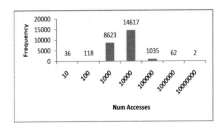 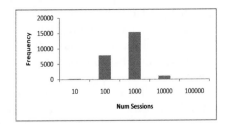

(a) twolf

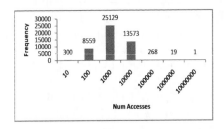 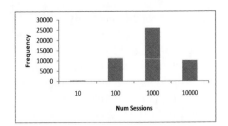

(b) art

FIGURE 1.7: `twolf` and `art` blocks behavior at the L1 data cache.

Observing block requests from the processor to the cache hierarchy allows us to classify each block into one of four categories:

- a block is accessed frequently, and time between consecutive accesses is small (high temporal locality);

- a block is accessed frequently in short duration, but then is not accessed for some time, and then is accessed again frequently within short periods (repetitive, bursty behavior);

- a block is accessed in a consistent manner, but the time between consecutive accesses is larger than for the first category; and

- a block is rarely accessed.

Figure 1.8 shows the four types. The best hierarchy should behave differently for each category. Blocks from the first category should be placed in both L1 and L2, since these are accessed frequently. Placing them in L1 allows them to be delivered to the processor as fast as possible. Evicting such blocks from L1 due to interference should allow them to reside in L2, since they are still needed. A block from the second category should be in L1 but not L2. This block will be heavily referenced for a while, so it should reside in L1, but once evicted it will not be needed again soon, and so it need not reside in L2. A block in the third category will be placed in L2 but not L1. L1 will thus be reserved for blocks that are heavily referenced, and for blocks not heavily referenced but that do not severely affect performance since they will only reside in L2 while being accessed. Finally, a block from the last category will bypass the cache hierarchy [28], and will be stored in no cache. Note that if consecutive accesses to blocks in the third category are very far apart, these blocks can be reduced to the fourth category.

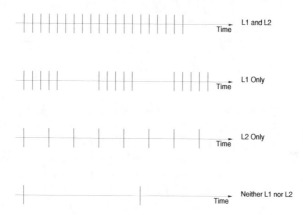

FIGURE 1.8: Different patterns for block accesses (vertical lines represent block accesses).

Let us see an example that shows the advantage of the global block placement. Assume the memory accesses shown in Figure 1.9(a). These instructions access four different cache blocks X, Y, Z, and W. For the sake of the example, assume a direct-mapped L1 and a two-way set associative L2 and that all blocks accessed in the example map to the same set in both caches. Figure 1.9(b) shows the timeline for accesses for each block. Every tick represents an access. If we map our four categories to these four blocks, then block X must be put in L1 because it has bursty access patterns and then periods of no accesses before it is touched again. Blocks Y and Z are placed in L2 because they are accessed consistently, but time between successive accesses is long. Finally, block W will not be cached because it is rarely accessed. Figure 1.9(c) shows hits and misses for both L1 and L2 for a traditional LRU scheme. We do not enforce inclusion: that is, a block victimized at L2 does not victimize

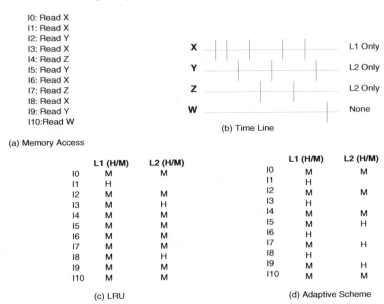

(a) Memory Access

(b) Time Line

(c) LRU

(d) Adaptive Scheme

FIGURE 1.9: Example of LRU versus Adaptive Scheme.

the corresponding block at L1. A quick look at hits and misses from a traditional LRU policy reveals two things: First, the hit rate at L1 is 1/11 and
at L2 is 2/10 (a hit at L1 needs no L2 access). Second, L2 has been accessed
ten times out of eleven references. Figure 1.9(d) illustrates ABP, which yields
better performance. Both caches start empty. When block X is loaded after
a compulsory miss, it is put into L1 but not L2. Blocks Y and Z are loaded
into L2 but not L1. Block W is not loaded into any cache. The hit rate at
L1 becomes 4/11 and at L2, 3/7, both higher than for the traditional LRU
hierarchy. Moreover, L2 is accessed only seven times, which reduces energy
consumption.

Global block placement is still an academic idea as of the writing of that
book. The main challenge is to design a cost-effective hardware scheme that
can categorize blocks at runtime.

1.6   Cache Coherence

Caches are now an indispensable component of almost all systems. However, to ensure the correctness of execution, we must deal with *the cache
coherence problem*. A memory is coherent if the value returned by a read operation is always the value that the programmer expected, which is, in most

cases, the value of the last write to the variable. The philosophy behind cache coherence protocol is to ensure the integrity of data so that an arbitrary number of caches can have a copy of the same block. These blocks will remain identical, provided that the processor associated with each of the caches does not attempt to modify its copy. Otherwise, the same block will have different data in different caches, which makes them inconsistent, a situation that must be avoided to ensure correctness of execution. Coherence protocols have been around for more than two decades in multiprocessor systems. This is why the main concept of cache coherence is the same for traditional multiprocessor systems and for multicore systems, but the very high cost of accessing off-chip memory and the low bandwidth available rendered many protocols impractical. The high bandwidth and low latency available in multicore processors give us the opportunity to use techniques that were not very efficient for conventional multiprocessors, such as write update.

The need for cache coherence arises when there is a shared address space among cores. If all cores are not sharing any memory or cache, that is, they do not have shared address space, then there is no need for coherence. Cores in that case use message passing to communicate instead of the shared memory. Therefore, in this section, we assume there is a shared address space among threads running on the different cores.

Coherence protocols can be categorized based on two criteria. The first is the type of interconnection among cores. The second is the action done after a core updates an item. If the interconnection is bus based or any medium that allows broadcasting, then the most common coherence protocol is the *snoopy protocols*. One such configuration is shown in Figure 1.10. Otherwise, we have *directory-based protocols*.

If a core updates a cache block,[7] there are two main actions. The first is called *write-invalidate* and the second is *write-update*. The write-invalidate family of coherence protocols are based on the action of similar blocks hosted by other caches when a core modifies its own block. The write-update family of protocols, on the other hand, updates blocks hosted by other core when a core modifies its own block.

Despite these categorizations, the main goal of cache coherence stays the same: to ensure the integrity of data.

1.6.1 Basics

Cache coherency protocols are usually described using a state-transition diagram. This diagram shows the state of a cache block in different situations and how it changes its state depending on different circumstances. The name of a protocol comes from the names of the different states. Each cache block

---

[7]We will assume that coherence is done at the cache block granularity, without loss of generality.

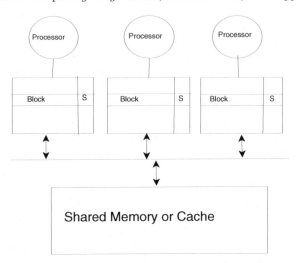

FIGURE 1.10: Bus-based multiprocessor system.

must be augmented with a few extra bits to indicate that block's status, as shown in Figure 1.10.

One of the first and most basic coherence protocols is MSI protocol. In this protocol, a block can be in one of three states: **M**odified, **S**hared, and **I**nvalid, hence the name of the protocol. Figure 1.11 shows the state transition diagram of the protocol. The dotted lines mean remote actions initiated by the different caches, and the solid lines mean local actions.

When a block is accessed for the first time by a cache, it is initially `invalid` and there is a cache miss. The type of access determines the next state. If the block is accessed for a read, the next state is `shared`. If, on the other hand, the access is for a write, the next state is `modified`. A block can be in the `modified` state in only one cache and must be `invalid` for the caches of the other cores, but it can be in the `shared` state in many caches at the same time.

If a block is in the `modified` state and another core needs it, since there can be only a single block in that state, the requesting core will experience a cache miss. The requested block is delivered by the cache that owns the modified block, or the shared storage if it has been updated with the new value by the modifying cache. If the remote core is needed for a write, it will be invalidated from all other cores and delivered to the requesting core in the `modified` state. If it is needed for a read, the block state in the initial cache and the requesting cache will be `shared`.

From the above description, you can see that the different cores must be able to know the requests of other cores. This can be implemented in one of two ways as we mentioned before. If there is a broadcast medium, such as a bus, different caches *snoop* on the bus. Otherwise, there is a directory that

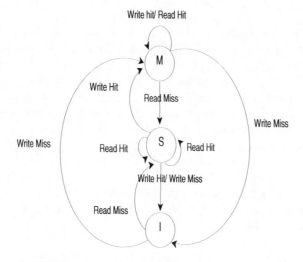

FIGURE 1.11: State transitions of MSI protocol.

keeps track of which caches have which blocks. This directory, which can be centralized of distributed, is responsible for sending the requests to the caches of interest.

1.6.2 Protocols for Traditional Multiprocessors

In the previous section, we discussed one of the most basic coherence protocols: the MSI protocol. There are many other coherence protocols proposed both in industry and academia. Figures 1.12–1.16 present the state transitions of five other protocols.

When designing a coherence protocol, the designer must take many factors into account.

- Reducing hardware cost: extra bits with each block, hardware needed to detect different cache requests, and so on

- Reducing traffic due to coherence, to change block states

- Reducing cache miss penalty

- Coherence protocol must be scalable to large number of cores

Each coherence protocol is trying to satisfy the above requirements. MESI protocol, Figure 1.12, was introduced in Intel Pentium processors. Pentium processors introduce the write-back caches. These caches need extra bits to tell whether a block is dirty. Because MESI protocol has a `dirty` (i.e., modified) state, it is suitable for write-back caches.

Figure 1.13 shows the *write-once* protocol, which is the first write-invalidate protocol proposed. All protocols we have seen until now are write-invalid, in the sense that for a core to modify its block, it has to have exclusive ownership of that block. This means it has to invalidate all other copies of that block in the other remote caches. In the write-once protocol, the `reserved` state means the block is the only copy of the memory, but it is still coherent, so write-back is needed if the block is victimized in a replacement. This protocol has an optimization to reduce off-chip traffic. It has write-update—that is, it updates other copies of the block if that block is modified—for the first write to the block but write-invalidate for subsequent writes.

For the Berkeley protocol, shown in Figure 1.14, the `valid` state is the same as the `shared` stated we have seen in the MSI protocol. Here, the block moves to `shared` only if it has been modified by a core then requested for a read by other core(s).

For the synapse protocol, shown in Figure 1.15, whenever there is a remote read/write miss, the block, if dirty, is written back to the memory (or to shared cache if the coherent caches are sharing a cache). This may increase traffic to the shared memory, but it tries to keep it as updated as possible. All write-invalidate protocols use the notion of block ownership, which means the block comes from an other cache if this cache has a dirty block; otherwise, the block comes from memory. The only exception is the Synapse protocol, in which the block always comes from memory; this increases off-chip traffic dramatically, which affects the performance. A detailed description of all these protocols is published extensively in the literature, for example [54, 3, 16].

Figure 1.16 is the only write-update protocol in this section. In write-update protocols, when a block is modified, the other copies are updated with the new modification instead of being invalidated. This can potentially reduce overall cache misses at the expense of extra traffic.

1.6.3   Protocols for Multicore Systems

In the previous section, we saw examples of coherence protocols that have been used in traditional multiprocessor systems. These protocols can still be used in multicore systems. However, with the extra bandwidth and low latency provided by multicore systems, a new generation of protocols have emerged.

Designing a coherence protocol for multicore processors must take into account technology trend. It is true that we have lower latency and higher bandwidth on-chip, but on the other hand, we have other constraints in area, power, and interconnection. Interconnection speeds do not scale very well with technology due to many technological factors including the trade-off between temperature and bandwidth (RC constant of wires).

The number of on-chip cores is increasing with technology. This means bus-based interconnection cannot be used for future multicore and many-core processors. Most of network-on-chip (NoC) used are not broadcast medium,

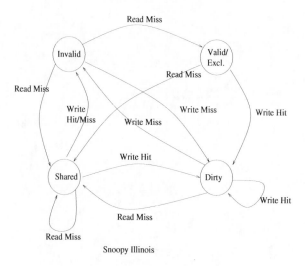

FIGURE 1.12: State transitions of Illinois protocol (also known as MESI dirty = modified).

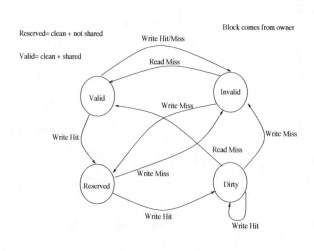

FIGURE 1.13: State transitions of Goodman protocol (write once).

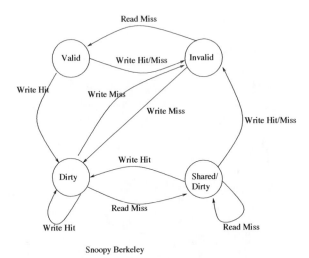

Snoopy Berkeley

FIGURE 1.14: State transitions of Berkeley protocol.

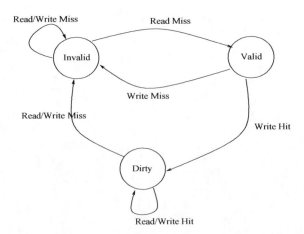

Snoopy Synapse

FIGURE 1.15: State transition of synapse protocol.

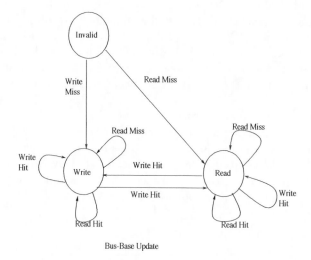

FIGURE 1.16: State transitions of write update.

and therefore snooping is not a very scalable solution here. The authors of [5] proposed a directory-based coherence protocol for multicore processors. Traditional directory-based protocols have been designed with traditional multiprocessor systems in mind. On traditional multiprocessor systems, the highest latency is among nodes. For multicore chips, latency among nodes is small relative to off-chip memory access. The proximity-aware directory-based coherence protocol [5] uses MESI protocol in an architecture similar to Figure 1.4. In that configuration, the directory itself is distributed among nodes (called tiles in many publications). Every node contains a core with its private L1 data and instruction caches, unified private L2 cache, and the part of the directory that keeps track of the location and states of each block. Also, each node contains a memory channel that gives access to a different range of memory addresses, so we have a distributed shared-memory scheme. This means that each block has a *home* node. This tiled architecture has been used extensively, with minor differences, both in academia (TRIPS [47]) and in industry (Tilera processors [1]). The home node always provides the block when another node requests it. The enhancement proposed in [5] is to make the closest sharer to the requester provide the data on a read or write request. This will reduce bandwidth requirement as well as off-chip access. For 16 cores, they reported 16% average performance improvement for the workload they used. One of the main reasons for performance loss in coherence protocols is false sharing. Two cores may share a block. If each core wants to modify different words on the same block, this operation needs to be serialized because coherence protocols allow only one instance of a block to be in a modified state. This increases the number of misses. One proposed solution to this problem is coherence decoupling [25]. In that method, there are two accesses. A speculative cache lookup protocol and a safe traditional coherence protocol. The speculative lookup loads values without consulting any coherence protocol, so the value is provided as soon as possible, typically from an invalid cache line. This allows the processor to compute with possibly incoherent data. In parallel, the coherence protocol obtains the correct value, which is compared to the speculative one. If they match, the processor commits the results. This method greatly reduces the effect of false sharing

Another optimization for directory-based protocol is to embody the directory within the router of the NoC [17]. This idea is efficient when we do not have tiled architecture but several cores connected by NoC, similar to Figure 1.3. This reduces bandwidth requirement because coherence messages will not be from end-to-end cores but only within the network.

1.7   Support for Memory Consistency Models

We cannot discuss coherence protocols without mentioning memory consistency models. Coherence protocols manage concurrent accesses to shared variables. Consistency models describe memory behavior to concurrent reads and writes. Simply speaking, memory consistency models define when a value written by a core must be seen by a following read instruction made by another core. There are many consistency models. In this section we will discuss two models.

One of the most widely used and known is called the *sequential consistency* model. In that model, every node of the system sees the (write) operations on the same memory part (page, virtual object, cell, etc.) in the same order.

Another *weaker* consistency model is called PRAM (pipelined random access memory). In the PRAM model, writes done by a single core[8] are received by all other cores in the order in which they were issued. However, Writes from different cores may be seen in a different order by different cores. For example, if core C0 makes writes A0 and A1, and core C1 writes B0 and B1, then all cores will see A0 and A1 in that order, and B0 and B1 in that order. However, some cores may see them as A0 A1 B0 B1 while other cores may see them as B0 B1 and A0 A1. PRAM is less strict, and hence weaker, than the sequential consistency model. The weaker the consistency model, the higher the expected performance.

---

[8]In the PRAM model, there is the notion of processes not nodes. Therefore, we will assume that every core is running a single process, for the sake of PRAM discussion.

1.8 Cache Hierarchy in Light of New Technologies

Until very recently, cache hierarchy was built using SRAM as the main technology, but SRAM starts to reach its limits and does not scale well due to its low density and high power dissipation (especially leakage). IBM Power7 includes eDRAM as its last-level cache. DRAM technology has higher density than SRAM and can be optimized for speed (to reduce the speed gap with other SRAM cache levels) or density. However, eDRAM is not without drawbacks. For instance, DRAM transistors are optimized for low leakage currents. This means low transistor performance. If we need to compromise, then fabrication cost may increase. Also, due to its dynamic nature, refresh time may result in nondeterministic response time from eDRAMs, although there has been some work in designing eDRAM with smart refresh [19]. Moreover, the increase in the number of cores as well as software complexity increase working set size much faster than the scalability of current memory technology. The main point here is that new technology is needed; otherwise the architecture community will hit a very solid memory wall!

The computer architecture community has been looking (and continues to do so) for a new memory technology. One of the strong candidates is resistive memory, such as phase-change memory (PCM) [45, 44]. PCM uses phase-change material (chalcogenide glass) that can exist in one of two states: high resistivity (representing 0) and low resistivity (representing 1). The switch from one state to the other is done by heating through electrical pulses. PCM has several advantages. First, it scales better than DRAM (which, in turn, scales better than SRAM). Second, It has higher density, as it can store more bits for different resistance values. Also, PCM is nonvolatile. However, there are several challenges. For instance, PCM has higher latency than DRAM and lower endurance (around 10 million writes per cell) and is bit constrained in write bandwidth. Nevertheless, technologies like PCM and other resistive memories such as STTRAM [8] and MRAM [18] are possible candidates for new memory technologies. So, with 3D stacking, we probably will witness hybrid memory hierarchy, involving several technologies, with new and different ways of management.

1.9 Concluding Remarks

In this chapter, we discussed the design and research topics related to memory hierarchy for multicore processors. We can summarize the main points of the chapter in the following bullets.

- We can make use of our experience with traditional multiprocessor systems when we design memory systems for multicore processors. But, due to different constraints, we have to tailor our designs to make use of higher bandwidth (relative to traditional multiprocessor systems) and more constrained power budget and area.

- Our measure of success in designing a memory hierarchy for multicore processors is short average memory access latency for *all* cores within the allowed budget of area, power, and bandwidth.

- Coherence protocols ultimate goal is to ensure integrity of data while not hurting overall memory hierarchy latency and while staying within budget on bandwidth, power, and area.

- We need to rethink many of our memory-hierarchy management algorithms in light of new memory technologies, 3D stacking, and the increase in the number of on-chip cores.

Bibliography

[1] Tilera homepage, http://www.tilera.com/, 2013.

[2] D. H. Albonesi. Selective cache ways: On-demand cache resource allocation. *Journal of Instruction-Level Parallelism*, 2002.

[3] James Archibald and Jean-Loup Baer. Cache coherence protocols: Evaluation using a multiprocessor simulation model. *ACM Transactions on Computer Systems*, May 1986.

[4] B. M. Beckmann and D. A. Wood. Managing wire delay in large chip-multiprocessor caches. In *Proceedings of the 37th Annual IEEE/ACM International Symposium on Microarchitecture*, 2004.

[5] J. A. Brown, R. Kumar, and D. Tullsen. Proximity-aware directory-based coherence for multi-core processor architectures. In *Proceedings of the Nineteenth Annual ACM Symposium on Parallel Algorithms and Architectures (SPAA)*, pages 126–134, 2007.

[6] F. J. Cazorla, E. Fernandez, A. Ramirez, and M. Valero. Dynamically controlled resource allocation in SMT processors. In *37th Annual IEEE/ACM International Symposium on Microarchitecture*, December 2004.

[7] J. Chang and G. S. Sohi. Cooperative caching for chip multiprocessors. In *Proceedings of the 33rd Annual International Symposium on Computer Architecture (ISCA)*, pages 264–276, 2006.

[8] Subho Chatterjee, Mitchelle Rasquinha, Sudhakar Yalamanchili, and Saibal Mukhopadhyay. A methodology for robust, energy efficient design of spin-torque-transfer RAM arrays at scaled technologies. In *Proceedings of the 2009 International Conference on Computer-Aided Design*, ICCAD '09, pages 474–477, 2009.

[9] T-F Chen and J-L Baer. A performance study of software and hardware data prefetching schemes. In *Proc. 21st Int'l Symposium on Computer Architecture*, 1994.

[10] T. Chilimbi and M. Hirzel. Dynamic hot data stream prefetching for general-purpose programs. In *Proc. ACM SIGPLAN Conference on Programming Language Design and Implementation*, pages 199–209, June 2002.

[11] Z. Chishti, M. D. Powell, and T. N. Vijaykumar. Distance associativity for high-performance energy-efficient non-uniform cache architectures. In *In Proceedings of the 36th Annual IEEE/ACM International Symposium on Microarchitecture*, 2003.

[12] Y. Chou, B. Fahs, and S. Abraham. Microarchitecture optimizations for exploiting memory-level parallelism. In *Proceedings. 31st Annual International Symposium on Computer Architecture, 2004*, pages 76–87, June 2004.

[13] P. Conway, N. Kalyanasundharam, G. Donley, K. Lepak, and B. Hughes. Cache hierarchy and memory subsystem of the AMD Opteron processor. *IEEE MICRO Magazine*, March/April 2010.

[14] J.D. Davis, J. Laudon, and K. Olukotun. Maximizing CMP throughput with mediocre cores. In *Proc. IEEE/ACM International Conference on Parallel Architectures and Compilation Techniques*, pages 51–62, October 2005.

[15] A. S. Dhodapkar and J. E. Smith. Managing multi-configuration hardware via dynamic working set analysis. In *Proc. 17th International Symposium on Computer Architecture*, 2002.

[16] Michel Dubois and Shreekant S. Thakkar. *Cache and Interconnect Architectures in Multiprocessors*. Kluwer Academic Publishers, 1990.

[17] N. Eisley, L-S. Peh, and L. Shang. In-network cache coherence. In *MICRO 39: Proceedings of the 39th Annual IEEE/ACM International Symposium on Microarchitecture*, 2006.

[18] W. J. Gallagher and S. S. P. Parkin. Development of the magnetic tunnel junction MRAM at IBM: From first junctions to a 16-MB MRAM demonstrator chip. *IBM J. Res. Dev.*, 50:5–23, January 2006.

[19] Mrinmoy Ghosh and Hsien-Hsin S. Lee. Smart refresh: An enhanced memory controller design for reducing energy in conventional and 3D die-stacked drams. In *Proceedings of the 40th Annual IEEE/ACM International Symposium on Microarchitecture*, MICRO 40, pages 134–145, 2007.

[20] Y. Guo, S. Chheda, I. Koren, C. M. Krishna, and C. A. Moritz. Energy Characterization of Hardware-Based Data Prefetching. In *ICCD04*, pages 518–523, 2004.

[21] Lance Hammond, Basem A. Nayfeh, and Kunle Olukotun. A Single-Chip Multiprocessor. *IEEE Computer*, 30(9):79–85, 1997.

[22] N. Hardavellas, M. Ferdman, B. Falsafi, and A. Ailamaki. Reactive nuca: Near-optimal block placement and replication in distributed caches. In *in Proceedings of the 36th Annual International Symposium on Computer Architecture (ISCA)*, June 2009.

[23] J. L. Hennessy and D. A. Patterson. *Computer Architecture: A Quantitative Approach*. Morgan Kaufmann, 2006.

[24] L. Hsu, S. Reinhardt, R. Iyer, and S. Makineni. Communist, utilitarian, and capitalist cache policies on CMPs: Caches as a shared resource. In *PACT '06: Proceedings of the 15th International Conference on Parallel Architectures and Compilation Techniques*, pages 13–22, 2006.

[25] J. Huhk, J. Chang, D. Burger, and G. Sohi. Coherence decoupling: Making use of incoherence. In *The 11th International Conference on Architectural Support for Programming Languages and Operating Systems (ASPLOS)*, 2004.

[26] R. Iyer, L. Zhao, F. Guo, R. Illikkal, S. Makineni, D. Newell, Y. Solihin, L. Hsu, and S. Reinhardt. QoS policies and architecture for cache/memory in CMP platforms. *SIGMETRICS Perform. Eval. Rev.*, 35(1):25–36, 2007.

[27] A. Jaleel, W. Hasenplaugh, M. Qureshi, J. Sebot, S. Steely, and J. Emer. Adaptive insertion policies for managing shared caches. In *PACT '08: Proceedings of the 17th International Conference on Parallel Architectures and Compilation Techniques*, pages 208–219, 2008.

[28] T. L. Johnson, D. A. Connors, M. C. Merten, and W-M. Hwu. Run-time cache bypassing. *IEEE Trans. Comput.*, 48(12):1338–1354, 1999.

[29] D. Joseph and D. Grunwald. Prefetching using Markov predictors. *IEEE Transactions on Computers*, 48(2):121–133, 1999.

[30] N. P. Jouppi. Improving direct-mapped cache performance by the addition of a small fully-associative cache and prefetch buffer. In *Proc. 17th*

International Symposium on Computer Architecture, pages 364–373, May 1990.

[31] R. Kalla, B. Sinharoy, W. J. Starke, and M. Floyd. Power7: IBM's next-generation server processor. *IEEE MICRO Magazine*, March/April 2010.

[32] C. Kim, D. Burger, and S. W. Keckler. Nonuniform cache architectures for wire-delay dominated on-chip caches. *IEEE Micro*, 23(6):99–107, 2003.

[33] S. Kim, D. Chandra, and Y. Solihin. Fair cache sharing and partitioning in a chip multiprocessor architecture. In *PACT '04: Proceedings of the 13th International Conference on Parallel Architectures and Compilation Techniques*, pages 111–122, 2004.

[34] D. Kroft. Lockup-free instruction fetch/prefetch cache organization. In *Proceedings of the 8th Annual International Symposium on Computer Architecture*, pages 81–85, 1981.

[35] J-H Lee and S-D Kim. Application-adaptive intelligent cache memory system. *ACM Transactions on Embedded Computing Systems*, 1(1), 2002.

[36] Y. Li, B. Lee, D. Brooks, Z. Hu, and K. Skadron. CMP design space exploration subject to physical constraints. In *Proc. 12th IEEE Symposium on High Performance Computer Architecture*, pages 15–26, February 2006.

[37] S.A. McKee. *Maximizing Memory Bandwidth for Streamed Computations*. PhD thesis, School of Engineering and Applied Science, University of Virginia, May 1995.

[38] Mesut Meterelliyoz, Jaydeep P. Kulkarni, and Kaushik Roy. Analysis of SRAM and eDRAM cache memories under spatial temperature variations. *Trans. Comp.-Aided Des. Integ. Cir. Sys.*, 29:2–13, January 2010.

[39] Gordon E. Moore. Cramming More Components onto Integrated Circuits. *Electronics*, April 1965.

[40] T. Mowry, M. S. Lam, and A. Gupta. Design and evaluation of a compiler algorithm for prefetching. In *Proc. 5th International Conference on Architectural Support for Programming Languages and Operating Systems*, 1992.

[41] K. Olukotun and et al. The case for a single-chip multiprocessor. In *Proc. 7th ACM Symposium on Architectural Support for Programming Languages and Operating Systems*, pages 2–11, October 1996.

[42] T. Puzak, A. Hartstein, P. Emma, and V. Srinivasan. Measuring the cost of a cache miss. In *Workshop on Modeling, Benchmarking and Simulation (MoBS)*, Jun. 2006.

[43] M. Qureshi, A. Jaleel, Y. Patt, S. Steely Jr., and J. Emer. Adaptive insertion policies for high performance caching. In *Proc. 34th International Symposium on Computer Architecture (ISCA)*, pages 381–391, Jun. 2007.

[44] Moinuddin K. Qureshi, Vijayalakshmi Srinivasan, and Jude A. Rivers. Scalable high performance main memory system using phase-change memory technology. In *Proceedings of the 36th Annual International Symposium on Computer Architecture*, ISCA '09, pages 24–33, 2009.

[45] S. Raoux, G. W. Burr, M. J. Breitwisch, C. T. Rettner, Y.-C. Chen, R. M. Shelby, M. Salinga, D. Krebs, S.-H. Chen, H.-L. Lung, and C. H. Lam. Phase-change random access memory: A scalable technology. *IBM J. Res. Dev.*, 52:465–479, July 2008.

[46] B. R. Rau and J. Fischer. Instruction-level parallel processing: History, overview and perspective. Technical report, HP Laboratories, 1992.

[47] K. Sankaralingam, R. Nagarajan, H. Liu, C. Kim, Huh J, N. Ranganathan, D. Burger, S.W. Keckler, R.G. McDonald, and C.R. Moore. TRIPS: A polymorphous architecture for exploiting ILP, TLP,and DLP. 1(1):62–93, March 2004.

[48] T. Sherwood and G. B. Calder. Time varying behavior of programs. Technical report, University of California at San Diego, 1999.

[49] B. Sinharoy, R. N. Kalla, J. M. Tendler andR. J. Eickemeyer, and J. B. Joyner. Power5 system microarchitecture. *IBM Journal or Research and Development*, 49(4/5), 2005.

[50] K. Skadron and D. W. Clark. Design issues and tradeoffs for write buffers. In *Proc. Int'l conf. on High Performance Computer Architecture (HPCA)*, pages 144–155, February 1997.

[51] A. Smith. Cache memories. *ACM Computing Surveys*, 14(3), 1982.

[52] Standard Performance Evaluation Corporation. SPEC CPU benchmark suite. http://www.specbench.org/osg/cpu2000/, 2000.

[53] R. Subramanian, Y. Smaragdakis, and G. Loh. Adaptive caches: Effective shaping of cache behavior to workloads. In *Proceedings of the 39th Annual IEEE/ACM International Symposium on Microarchitecture (MICRO 39)*, pages 385–396, Dec. 2006.

[54] Milo Tomasevic and Veljko Milutinovic. *The Cache Coherence Problem in Shared-Memory Multiprocessors: Hardware Solutions*. IEEE Computer Society Press, 1993.

[55] A. V. Veidenbaum, W. Tang, R. Gupta, A. Nicolau, and X. Ji. Adapting cache line size to application behavior. In *Proc. of the 1999 International Conference on Supercomputing*, 1999.

[56] M.V. Wilkes. The memory wall and the CMOS end-point. *ACM SIGArch Computer Architecture News*, 23(4):4–6, September 1995.

[57] Wm. A. Wulf and Sally A. McKee. Hitting the memory wall: Implications of the obvious. *Computer Architecture News*, 23(1):20–24, 1995.

[58] Y. Xie, G.H. Loh, B. Black, and K. Bernstein. Design space exploration for 3D architectures. *ACM Journal on Emerging Technologies in Computing Systems*, 2(2):65–103, 2006.

[59] Y. Zheng, B. T. Davis, and M. Jordan. Performance evaluation of exclusive cache hierarchies. In *ISPASS '04: Proceedings of the 2004 IEEE International Symposium on Performance Analysis of Systems and Software*, pages 89–96, Mar. 2004.

Chapter 2

FSB: A Flexible Set-Balancing Strategy for Last-Level Caches

Mohamed Zahran

New York University

Bushra Ahsan

Intel

Mohammad Hammoud

Carnegie Mellon University Qatar

Sangyeun Cho

University of Pittsburgh

Rami Melhem

University of Pittsburgh

This paper describes Flexible Set Balancing (FSB), a practical strategy for providing high-performance caching. Our work is motivated by a large asymmetry in the usage of cache sets. FSB extends the lifetime of cache lines via retaining some fraction of the working set at underutilized sets to satisfy far-flung reuses. FSB promotes a very flexible sharing among cache sets, referred to as many-from-many sharing, providing significant reduction in interference misses. Simulation results using a full-system simulator that models a 16-way tiled chip multiprocessor platform demonstrate that FSB achieves an average miss rate reduction of 36.6% on multithreading and multiprogramming benchmarks from SPEC2006, PARSEC, and SPLASH-2 suites. This translates into an average execution-time improvement of 13%. Furthermore, evaluations showed the outperformance of FSB over some recent proposals including DSBC [27] and V-WAY [25].

2.1   Introduction

Processor and memory speeds are increasing at about 60% and 10% per year, respectively [15]. Besides, after the emergence of chip multiprocessors (CMPs) as a mainstream architecture of choice, the off-chip bandwidth is expected to grow at a much slower rate than the number of processor cores on a CMP chip [7]. These factors together substantially increase the capacity pressure on the on-chip memory hierarchy, and, in particular, the last-level cache (LLC). Intelligent design and management of LLC continue, accordingly, to be very essential to bridge the increasing speed and bandwidth gaps between processor and memory.

In this work, we observe that more than two thirds of cache lines placed in an LLC logically shared by 16 CMP cores remain unused between placement and eviction. As such, these lines don't contribute to good utilization of the silicon estate devoted to the caches. One reason for this phenomenon is that cache lines might be rereferenced at distances greater than the cache associativity [23]. The problem is magnified on CMPs that share caches, as on-chip lifetimes of cache lines can become shorter due to the increasing interferences between coscheduled threads/processes. Cache performance can be improved by retaining some fraction of the working set long enough to provide cache hits on future reuses [23, 19].

Computer programs exhibit a nonuniform distribution of memory accesses across different cache sets [27, 25]. Figure 2.1 demonstrates this fact by

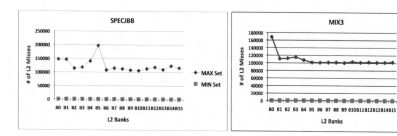

FIGURE 2.1: Number of misses experienced by two cache sets at different L2 banks for SPECJBB and MIX3 (MAX Set = the set that experiences the maximum misses and MIN Set = the set that experiences the minimum misses).

showing the number of misses experienced by cache sets at different physically distributed, logically shared L2 banks on a 16-way tiled CMP for two benchmarks, SpecJBB and MIX3.[1] Only the sets that exhibit the maximum and the minimum misses are shown. Clearly, some sets suffer from large local miss ratios while some others remain underutilized. Our work extends the lifetime of cache lines by exploiting this phenomenon via *flexibly* retaining cache lines evicted from highly pressured sets at underutilized ones.

Recently, Rolán et al. [27] proposed Dynamic Set Balancing Cache (DSBC) to mitigate the large asymmetry in cache sets' usages. DSBC suggests associating every two cache sets, making the capacity of an underutilized set available for a stressed one. When a cache line is evicted from a set whose working set seems not to fit in, the line can be stored at another set whose working set seems to fit in. They refer to the former and latter sets as *source* and *destination* sets, respectively. DSBC associates source and destination sets after a request made by a source set. The source and destination sets remain associated as long as the destination set hosts at least one line from the source set. Upon the eviction of the last line retained by the source set at the destination set, the association between the two sets is broken, and the destination set (assuming still underutilized) can be subsequently shared by another source set.

There are inherent drawbacks with DSBC. Once an association is established between two sets, the source S is not allowed to retain blocks at any other set but the destination set D. If the program phase changes after association and both S and D become stressed, they will compete on only D's capacity (i.e., retention is unidirectional), potentially causing significant thrashing at D. Besides, S is not permitted to request more capacity although many other underutilized sets might be available. Lastly, if D becomes underutilized, it is not allowed to share its space by more than one source set even though it

---

[1]Description of the adopted CMP platform, the experimental parameters, and the benchmark programs can be found in Section 2.4.1.

might be capable of doing so and even if many other sources are in need of its available space.

We refer to the capacity sharing provided by DSBC as *one-from-one* sharing. We propose Flexible Set Balancing (FSB) in which a highly pressured set is allowed to retain its lines with many underutilized sets. We refer to this sharing as *many-from-one* sharing because the capacity of many sets can be shared by a single set. Furthermore, we allow many pressured sets to retain their lines at a single underutilized set. This sharing is referred to as *one-from-many* sharing because the capacity of a single set can be shared by multiple sets. Consequently, FSB offers a very flexible (many-from-many) capacity sharing among cache sets. FSB adapts to phase changes in programs and doesn't solely associate any set with any other set (i.e., makes a set a sole owner of another set). As long as there is a space available at any set, any stressed set can immediately leverage that space.

The major contributions of our work are as follows:

- We propose FSB, a caching strategy that extends the lifetime of cache lines via exploiting the phenomenon of workload imbalance among cache sets. FSB suggests a very flexible sharing between cache sets, referred to as many-from-many sharing, that seeks to minimize interference misses and maximize system performance.
- We evaluate our work on a full-system simulator that models a 16-way tiled CMP and find that FSB reduces interference cache misses of a baseline shared last-level cache by an average of 36.6%. This translates into an average execution-time improvement of 13%.
- We employ the two recent and closely related works, DSBC [27] and V-WAY [25], on a CMP platform and compare them against FSB. On average, FSB provides 27.2% and 29.2% miss reductions against DSBC and V-WAY, respectively.

The rest of the paper is organized as follows: a motivational study and background are given in Section 2.2; we detail FSB in Section 2.3; we evaluate FSB and some related designs in Section 2.4; Section 2.5 summarizes prior work; and we conclude in Section 2.6.

2.2   Motivation and Background

2.2.1   Baseline Architecture

Our proposed scheme doesn't impose any limitation on employing any caching architecture. Without loss of generality, we assume a 16-way tiled CMP platform. Economic, manufacturing, and physical design considerations suggest tiled architectures (e.g., Tilera's Tile64 and Intel's Teraflops Research chip) that colocate distributed cores with distributed cache banks in tiles communicating via a network-on-chip (NoC) [14]. Each tile encompasses a core, private L1 caches (I/D), and an L2 cache bank. We assume block interleaved logically shared L2 cache banks. An in-cache directory coherence MESI-based protocol is employed [6, 13, 35]. Hence, each L2 cache line is associated with a bit vector indicating which cores have cached copies of that line in their L1 private caches. Lastly, we assume an LRU replacement policy.

2.2.2 A Caching Problem

To mitigate the high off-chip data access latency, the microprocessor industry has incorporated techniques such as deep cache hierarchies, large associative last-level caches (LLC), and sophisticated data prefetchers. But even with these techniques, a significant number of cache lines still miss in LLC [3]. Evaluations of 10 benchmarks from Spec2006, PARSEC, and Splash-2 (Section 2.4.1 describes the benchmark programs) manifested that more than two thirds of the cache lines are never reused before getting evicted. A similar observation appeared in [23]. These cache lines are referred to as *zero reuse lines.*

Many reasons cause the occurrence of zero reuse lines at LLC. First, memory references exhibit locality and are not evenly distributed across cache sets. This skew reduces the effectiveness of a cache and results in storing a considerable number of lines that are less likely to be rereferenced before replacement [22]. Second, the access stream visible to LLC is filtered through the higher level(s) caches on the memory hierarchy. Third, some cache lines reveal no temporal locality. Fourth, many cache lines exhibit far-flung reuses; that is, an evicted block may be used many times in the future, although not in the near future [8]. Fifth, the advent of CMPs exacerbates the problem due to interferences among coscheduled threads/processes on an underlying shared LLC. Recent research work on CMP cache management has recognized the importance of the shared CMP design [29, 13, 18, 35, 10]. Furthermore, many of today's multicore processors, the Intel Core^{TM}2 Duo processor family [26], Sun Niagara [21], and IBM Power5 [28], feature shared caches. We conducted a quantification study on different kinds of misses (i.e., compulsory, intra-processor, and inter-processor) on our adopted CMP model and found that 69.5% of misses are inter-processor (i.e., lines are replaced at earlier times by different processors).

2.2.3 Dynamic Set-Balancing Cache and Inherent Shortcomings

Cache sets' usages are typically asymmetric [27, 25]. An intuitive solution to the zero-reuse-lines phenomenon is to extend the lifetime of some cache lines long enough so that at least a portion of these lines can provide cache hits on future reuses. Dynamic Set Balancing Cache (DSBC) [27] extends the

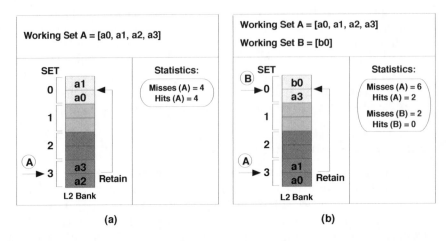

FIGURE 2.2: DSBC in operation. (a) A maps originally to set 3. The program executes A's references in the order of A, A. DSBC is able to save much of A's interference misses. (b) A and B map originally to sets 3 and 0, respectively. The program executes A's and B's references in the order of A, B, A, B. DSBC is incapable of adapting to the phase change in the program.

lifetime of some cache lines by exploiting the asymmetry in the usages of cache sets. Specifically, lines are shifted from sets with high local miss rates to sets with low local miss rates where they can be found later. Once a set reaches a saturation level (a set's miss rate hits a maximum value of $2K - 1$ where K is the associativity of the cache) it requests a free (not associated yet) underutilized set. If such a set is found, both sets, the highly pressured one (or the *source*) and the underutilized one (or the *destination*), are associated. As long as the two sets are associated, the source is allowed to retain its lines at the destination but not the reverse (i.e., unidirectional retention).

DSBC maintains a table with one entry per set called the Association Table (AT). AT stores in the ith entry AT(i).index, which corresponds to the index of the set associated with set i. Besides, AT stores a source/destination (s/d) bit (AT(i).s/d) that indicates whether the set is associated or not. Each AT entry can have three different values. First, if a cache set is not associated, its corresponding AT entry stores the set's index, and s/d = 0. Second, if a set is a source set, its corresponding AT entry stores the destination index, and s/d = 1. Lastly, if a set is a destination set, AT stores the source index, and s/d = 0. When a certain request misses at a source set, the destination set is looked up for either a secondary hit or a definitive miss.

DSBC has a number of shortcomings. First, once a destination set D is designated, it will continue receiving retained lines from a source set S until the association is broken. This overlooks the fact that D's pressure progressively increases while receiving more lines from S. Nevertheless, after association,

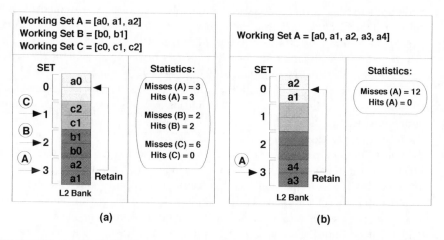

FIGURE 2.3: DSBC in operation. (a) The program executes A's, B's, and C's references in the order of A, B, C, A, B, C. DSBC doesn't allow one-from-many sharing. (b) The program executes A's references twice. DSBC doesn't allow many-from-one sharing.

a new program phase can start where S might remain pressured (and still associated with D), and D becomes highly pressured (due to receiving lines not only from S but further from a new large working set which now maps to it). As a result, S and D compete for only D's resources causing significant thrashing. We illustrate this problem in an example.

Consider a 2-way set associative cache shown in Figure 2.2. For simplicity we represent a cache by a linear array consisting of only 4 sets. Assume first (Figure 2.2(a)) that a working set A with reference pattern $[a0, a1, a2, a3]$ maps to set 3 and has been observed twice by a program. The sequence of references of A can't coreside in set 3. Accordingly, DSBC selects an under-utilized set, say set 0, in the cache and displaces the evicted blocks from set 3 to set 0. Figure 2.2(a) shows the final residences of lines in the cache after the completion of the program. A's resultant misses and hits are, consequently, 4 and 4, respectively (the cache is assumed to be initially empty). If the traditional caching strategy is to be followed, 4 more misses will be incurred.

In Figure 2.2(b), presumably at a different phase in the program, a new working set B with reference pattern $[b0]$ is considered and assumed to map to set 0. As in Figure 2.2(a), working set A still maps originally to set 3 and acts as a source set associated with set 0 as a destination set. We assume that the program executes A's and B's references in the order of A, B, A, B. The figure shows the final residences of lines after the completion of the program. A's resultant misses and hits are, consequently, 6 and 2, respectively. B, on the other hand, experienced two misses and got no hits. Note that DSBC didn't even attempt to break the association between sets 0 and 3 during the program's execution because there was always at least one retained block at set 0. *If DSBC would rather adapt to the phase change in the program, during the first execution of B's references, the evicted blocks from set 0 (i.e., a0) can be retained at another underutilized set (say set 1) so that in the second execution of A's and B's references, no misses will be incurred.*

We refer to the sharing policy employed by DSBC among cache sets as *one-from-one* sharing. That is, a destination set is shared by only a single source set. Figure 2.3(a) shows three working sets A, B, and C with reference patterns $[a0, a1, a2]$, $[b0, b1]$, and $[c0, c1, c2]$, respectively. We assume that A, B, and C map originally to sets 3, 2, and 1, respectively. The figure demonstrates two issuances of A's, B's, and C's reference patterns in the order of A, B, C, A, B, C. A's lines can't all coreside in set 3, and DSBC selects set 0 as a destination set for set 3. Also, C's lines can't all coexist in set 1; however, DSBC doesn't select any destination set for set 1 because no set that is both underutilized and not associated yet is found. As a result, C's references will experience *zero* hits during their two issuances (with this cache topology C is said to experience far-flung reuses). The cache depicts the final residences of all the cache lines after the completion of the program. The misses and hits counts of A are 3 and 3, respectively. On the other hand, B's references miss twice and hit twice. Lastly, C's references miss six times and get no hits. *If DSBC would allow set 0 to be shared by both sets 3 and 1, C's misses will be avoided when issued on the second time. We refer to this kind of flexible sharing as* **one-from-many** *sharing. That is, a single destination set can be shared by multiple source sets.*

Finally, as a consequence of the adopted one-from-one sharing strategy, DSBC doesn't allow a source set S to retain blocks in more than one destination set D. As such, if the working set that maps to S is large enough that both S and D are incapable of providing enough capacity as required, many conflict misses can be incurred. Figure 2.3(b) assumes a working set A with reference pattern $[a0, a1, a2, a3, a4]$ that maps to set 3. The program issues A's references twice. DSBC selects an underutilized set; say set 0, where evicted lines from set 3 can be retained. The cache in the figure depicts the final residences of A's lines after the completion of the program. The final miss and hit counts are 12 and 0, respectively (assuming that the cache was initially empty). In this case, DSBC didn't provide any benefit for A. *If DSBC will allow more than one destination set to be shared by set 3, A's misses will be avoided when issued on the second time. We refer to this kind of flexible sharing as* **many-from-one** *sharing. That is, many destination sets can be shared by a single source set.*

2.2.4   Our Solution

We propose Flexible Set Balancing (FSB), a caching strategy that adapts to phase changes in programs and allows many-from-many sharing among cache sets. The difference in this work compared to DSBC are two key in-

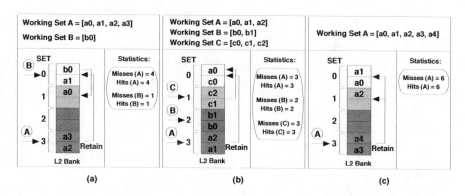

FIGURE 2.4: Our solution. (a) The program executes A's, and B's references in the order of A, B, A, B. We adapt to the phase change in the program. (b) The program executes A's, B's, and C's references in the order of A, B, C, A, B, C. We allow one-from-many sharing. (c) The program executes A's references twice. We allow many-from-one sharing.

sights: (1) retention should be efficiently and dynamically allowed at any point during the program's execution in any direction seeking for spare space to effectively minimize interference misses, (2) one-from-many and many-from-one (i.e., many-from-many) sharing should be allowed among cache sets for high flexibility. We demonstrate our solution with an example.

Figure 2.4(a) demonstrates the same example shown in Figure 2.2(b) but with FSB being incorporated instead of DSBC. Again, A and B are assumed to map to sets 3 and 0, respectively. The program executes A's and B's references in the order of A, B, A, B. In the first issuance of A's references, FSB selects set 0 as a destination set for set 3. Afterwards, when B is issued, line $a0$, which has already been retained at set 0, is evicted again. FSB doesn't discard $a0$ but retains it again at a new underutilized set, say set 1. In the second issuance of the working sets, A's and B's references hit on all their cache lines. As such, miss and hit outcomes become 4 and 4 for A, and 1 and 1 for B. Therefore, FSB saves 3 misses as compared to DSBC. The figure shows the final residences of all the cache lines after the program's completion. Clearly, this example illustrates FSB's capability to adapt to phase changes.

Figure 2.4(b) shows the same example illustrated in Figure 2.3(a). Again, we assume that A, B, and C map to sets 3, 2, and 1, respectively, and that the program observes A's, B's, and C's references in the order of A, B, C, A, B, C. FSB allows set 0 to be shared by many source sets. As such, in the first iteration of the working sets when lines of C can't all coreside in set 1, FSB retains $c0$ at set 0 (the current least pressured set). In the second iteration, the references of A, B, and C hit on all their cache lines. Miss and hit outcomes become, accordingly, 3 and 3 for A, 2 and 2 for B, and 3 and 3 for C. As compared to DSBC, FSB saves the three misses incurred by C in Figure 2.3(a).

The cache array in the figure displays the final residences of all the lines after the program's completion. Clearly, this example demonstrates FSB's efficiency in reducing conflict misses with one-from-many sharing.

Lastly, Figure 2.4(c) illustrates the same example demonstrated in Figure 2.3(b). Again, we assume that A maps to set 3 and that the program issues A's sequence of references twice. FSB allows many sets to be shared by a source set. As such, in the first issuance of A, FSB selects an underutilized set, say set 0, and retains $a1$ and $a0$ in it, then selects another underutilized set, say set 1, and retains $a2$ in it. In the second issuance, all references of A hit in the cache. FSB, consequently, saves the six misses incurred by DSBC in Figure 2.3(b). Clearly, this example magnifies the potential of FSB in reducing interference misses by employing many-from-one sharing among cache sets.

2.3 Flexible Set Balancing

FSB regulates cache allocation by flexibly retaining a fraction of a working set at underutilized cache sets to minimize interference misses and maximize system performance. FSB is extensible and practical in that it can be employed on single-core as well as multicore architectures. FSB is oriented towards last-level caches (in our case L2). FSB requires three main capabilities: (1) deciding upon source and destination sets, (2) retaining working sets of source sets at destination sets in a many-from-many sharing fashion, and (3) locating retained blocks on destination sets when requested. We next describe each capability in turn and close with an analysis of FSB's hardware storage, area, energy, and latency requirements.

2.3.1 Retention Limits

FSB is a pressure-aware strategy where lines evicted from highly pressured sets (source sets) are retained at low pressured sets (destination sets). The pressure at a cache set can be measured in terms of cache misses or hits. In this work, we adopt cache misses as a pressure function but provide a study on a variety of pressure functions in Section 2.4. The pressure information can be recorded at an array embedded within the L2 controller of a cache bank. Each cache set corresponds to an entry in the pressure array, and the indexes of the cache sets are used to index the array. Each time a miss occurs at a certain set, the array can be updated accordingly (by incrementing the corresponding array slot). In order to allow the array to accurately represent pressures at sets, after every time interval, we keep only part of the pressure values (e.g., 0.25 of values by shifting each value 2 bits to the right). This permits FSB to adapt to undergoing phase changes in programs. The collected pressures can be utilized to guide the retention process.

Clearly, the set that corresponds to the maximum value in the pressure array is the most highly pressured set. In contrast, the lowest pressured set is the one that corresponds to the minimum value in the array. In this work, we define two limits, the low pressure limit (LPL) and the high pressure limit (HPL), to allow a *range* of highly pressured sets to retain their blocks at a range of low pressured sets. A range can encompass one set or many. When the pressure of a set is below LPL, the set is deemed to be within the limit of the destination sets and can receive lines from *any* source set. In contrast, when the pressure of a set is above HPL, the set is considered to be within the limit of source sets and is permitted, accordingly, to retain its lines at *multiple* destinations sets. Clearly, this allows many-from-many sharing among cache sets. LPL and HPL are defined in Equations (2.1) and (2.2). The range of source and destination sets can be expanded or contracted by altering α. The max and min parameters are the maximum and minimum pressures on the pressure array.

$$LowPressureLimit(LPL) = min + (\alpha \times (max - min)) \qquad (2.1)$$
$$HighPressureLimit(HPL) = max - (\alpha \times (max - min)) \qquad (2.2)$$

2.3.2 Retention Policy

FSB maintains a small retention table (RT) per each L2 bank. Each cache set has a corresponding RT entry. As such, the number of entries in RT equals the number of cache sets in the L2 bank. RT can store in the ith entry *many* RT(i).index values, each pointing to a destination set with a different index. In Section 2.4, we empirically show that four RT(i).index pointers are enough to attain an efficient FSB. RT(i).index pointers can be used by FSB to locate retained blocks upon future reuses (more on this shortly).

When an LRU line L is evicted from a set i, our retention policy proceeds as follows:

1. We look up i's corresponding pressure value in the pressure array, generate minimum (MIN) and maximum (MAX) values, and calculate HPL and LPL.
2. If i's pressure is greater than HPL, i becomes a source set and L is deemed eligible for retention. Otherwise, L is discarded.
3. In parallel, RT(i) entry is looked up. If L is eligible for retention and RT(i) entry has no pointers to destination sets, we check if MIN is less than LPL. If satisfied, we retain L at the cache set corresponding to MIN and create an equivalent RT(i).index pointer. Otherwise, we discard L.
4. If RT(i) entry, on the other hand, has pointers (or at least one pointer), we use these pointers to index the pressure array, generate the minimum value out of the indexed values, and compare it against LPL. If satisfied,

    we retain L at the corresponding cache set and no RT(i).index pointer is created. Otherwise, we check if an invalid RT(i).index exists.

5. If an invalid RT(i).index is found and MIN satisfies LPL, we retain L at the corresponding set and create an equivalent RT(i).index pointer. Otherwise, we discard L.

Note that upon retention, we insert L as the most recently used (MRU) line in the selected destination set. The LRU line evicted at the destination set to make room for L is discarded simply because the destination set doesn't satisfy HPL. As such, FSB avoids ripple effects.

The LRU-evicted line L at the source set can be either *native* or *retained*. If L is native, FSB simply proceeds with our previously suggested retention policy. Otherwise, we check if L is *active*. We define L to be active if at least one core on the CMP platform has cached a copy of L (in its L1). This can be easily determined from L's associated directory bit vector. We assume that an active L is currently in use by the caching core(s) and, accordingly, attempt to retain it again. If L is retained and not active, we assume that it has been kept long enough in the cache without providing a cache hit, and, as such, avoid retaining it over again (although it is eligible for retention).

The pressure array is updated not only at a miss/hit but further when retaining a line at a destination set. When a destination set receives a retained line, its corresponding pressure value is incremented. This is critical so as to reflect the progressive increasing pressure on a destination set each time it receives a retained line. This makes FSB very flexible and attentive, as it allows selecting a different destination set once the pressure of the current destination set surpasses LPL.

Retaining cache lines at destination sets requires extending lines' tags. This is due to the fact that a cache line must have a one-to-one correspondence with a unique address. For instance, assume a line E is retained at a destination set S and that S has a line F which has an identical tag field as E. E and F addresses are, in fact, only distinct because they differ in their index fields. Now E and F coreside at S and thus become indistinguishable. Nevertheless, this suggests a simple solution: that is, augmenting each line's tag with the index field. Finally, upon discarding a retained line R from S, we match R's augmented index j with the augmented indexes of S's resident lines. A "no match" outcome means that R is the last retained line at S from the source set j. Consequently, we index RT(j) entry and invalidate the RT(j).index pointer that points to S.

To that end, we note that the retention process is activated in parallel with the resolution of a definitive miss. As such, the latency required to retain a cache block becomes completely hidden because resolving an L2 miss usually takes hundreds of cycles. However, in principle, FSB's retention policy requires hardware to compute MAX, MIN, HPL, and LPL, which may appear expensive. Smart implementation strategies exist. We need not, for instance, compute upon every eviction the exact MAX and MIN values. The L2 misses are usually infrequent, and the MAX and MIN values don't henceforth vary much upon a single L2 miss. As an approximation, we can compute MAX and MIN after a reasonable amount of misses (e.g., 1000 L2 misses) and perform partial comparisons incrementally. In another design, one could employ comparators combined with multiplexors in a tree structure as adopted in [27]. Lastly, α can simply be set to 0.25 (i.e., power of 2) (corroborated by the sensitivity study presented in Section 2.4.4). With this setting, a multiplier is not needed to compute HPL and LPL but a simple shifter. Section 2.3.4 describes FSB's storage, area, energy, and latency requirements.

2.3.3   Lookup Policy

Upon a request to a cache line L, the cache starts by always looking up the set i that L's index designates. RT(i) entry is also looked up concurrently. If a hit occurs at set i, the request is satisfied and the pressure array is updated (only if the pressure function involves hits). If, on the other hand, a miss occurs at set i, the cache sets identified by the RT(i).index pointers (if any) are *serially* looked up until either a secondary hit is acquired or a definitive miss is proclaimed. Sets' lookups are serialized in order to keep FSB simple, avoid port contention, and reduce power dissipation.[2] Section 2.4 demonstrates that such a serial policy doesn't hurt performance because the gain from hits on retained lines more than offsets the loss from sequential lookups. Upon a secondary hit, the request is satisfied and the pressure array is updated (only if the pressure function involves hits). If a definitive miss is asserted, the pressure array is updated at slot i (if the pressure function involves misses). On a definitive miss, the retention policy is triggered and, in parallel, the requested cache line is fetched from the main memory and inserted in set i.

FSB doesn't swap retained lines upon hits to return them to their original sets for several reasons. First, this simplifies management. Second, FSB is oriented towards last-level caches; once a hit is obtained on a retained line, the line is moved to the upper cache where successive accesses can find it. Third, swapping is undesirable because it requires four accesses to the tag-store, consumes energy, and increases port contention [25].

2.3.4   FSB Cost

FSB comes at little storage, area, latency, and energy overheads. In this work, we assume a 32 KB 2-way associative I/D L1 cache and a 512 KB 16-way associative L2 bank (512 cache sets) per each CMP tile. Section 2.4 shows that four pointers per each RT entry are enough for an effectively performing FSB. Each RT pointer requires 10 bits (1 valid bit and 9 bits to index the 512 L2 sets). Table 2.1 shows that ~2% storage overhead is required by FSB.

---

[2]Prior research has made use of serialization to increase flexibility and improve performance in large caches [9, 12]. Existing processors have also adopted serialization for looking up tag and data arrays seeking to reduce power dissipation [11, 32].

COMPONENT	BITS PER ENTRY	K ENTRIES	KB PER TILE
RT Entry	10	2	2.5
Augmented Bits Per an L2 Line	9	8	9.2
Total KBytes			11.7
% Increase of On-Chip Cache Capacity			2%

TABLE 2.1: FSB storage overhead.

TECHNOLOGY	BASELINE Energy	FSB Energy	BASELINE Area	FSB Area
45nm	1.23nJ	1.26nJ	5.36mm^2	5.47mm^2

TABLE 2.2: Baseline and FSB required energy and area in a 512 KB/16-way/64 B/LRU L2 bank.

To model area and energy, we use CACTI v5.3 [16]. We assume a 45 nm technology. Table 2.2 demonstrates the area and energy per access required for both a baseline L2 bank and an L2 bank with FSB being incorporated. The TR table, in addition, requires 0.14 mm^2 and 0.015 nJ area and energy per access, respectively. Note that the energy savings due to reducing off-chip accesses is not considered. Such savings are expected, in fact, to counterbalance our calculated energy overhead and further provide advantages as chip crossings are one of the greediest energy consumers [17]. Finally, and due to augmenting lines' tags by indexes, FSB incurs a negligible increase in latency (only 0.02 ns) per each L2 bank access.

2.4   Quantitative Evaluation

2.4.1   Methodology

We present our results based on a detailed full-system simulation using Virtutech's Simics 3.0.29 [31]. We use our own CMP cache modules fully developed in house. We implement the XY-routing algorithm and accurately model congestion for both coherence and data messages. A tiled CMP architecture comprised of 16 UltraSPARC-III Cu processors is simulated running with Solaris 10 OS. Each processor uses an in-order core model with an issue width of 2. The tiles are organized as a 4 × 4 grid connected by a 2D mesh NoC. Each tile encompasses a switch, 32KB I/D L1 caches, and a 512KB L2 cache bank. A distributed MESI-based directory protocol is employed. After every 20 million instructions, we keep only 0.25 of the pressure values (see Section 2.3.1). Table 2.3 shows our configuration's experimental parameters.

We use a mixture of multithreaded and multiprogramming workloads to study FSB and related designs. For multithreaded workloads, we use the com-

COMPONENT	PARAMETER
Cache Line Size	64 B
L1 I/D-Cache Size/Associativity	32 KB/2-way
L1 Hit Latency	1 cycle
L1 Replacement Policy	LRU
L2 Cache Size/Associativity	512 KB per L2 bank or 8 MB aggregate/16-way
L2 Bank Access Penalty	12 cycles
L2 Replacement Policy	LRU
Latency Per NoC Hop	3 cycles
Memory Latency	320 cycles

TABLE 2.3: System parameters.

NAME	INPUT
SPECJbb	Java HotSpot (TM) server VM v 1.5, 4 warehouses
Bodytrack	4 frames and 1K particles (16 threads)
Fluidanimate	5 frames and 300K particles (16 threads)
Swaptions	64 swaptions and 20K simulations (16 threads)
Barnes	64K particles (16 threads)
Lu	2048 × 2048 matrix (16 threads)
MIX1	Hmmer (reference) (16 copies)
MIX2	Sphinx (reference) (16 copies)
MIX3	Barnes, Ocean(1026 × 1026 grid), Radix (3M Int), Lu, Milc (ref), Mcf (ref), Bzip2 (ref), and Hmmer (2 threads/copies each)
MIX4	Barnes, FFT (4M complex numbers), Lu, and Radix (4 threads each)

TABLE 2.4: Benchmark programs.

mercial benchmark SpecJBB [30], five shared memory programs from the SPLASH-2 suite [33] (Ocean, Barnes, Lu, Radix, and FFT), and three applications from the PARSEC suite [4] (Bodytrack, Fluidanimate, and Swaptions). We composed multiprogramming workloads using the considered SPLASH-2 benchmarks and five other applications from SPEC2006 [30] (Hmmer, Sphinx, Milc, Mcf, and Bzip2). Table 2.4 shows the data sets and other important features of the simulated workloads. Lastly, the programs are fast-forwarded to get past their initialization phases. After various warm-up periods, each SPLASH-2 and PARSEC benchmark is run until the completion of its main loop, and each of SpecJBB, MIX1, MIX2, MIX3, and MIX4 is run for 8 billion user instructions.

2.4.2 Comparing FSB against Shared Baseline

Let us first compare FSB against the baseline shared (S) scheme. Figure 2.5(a) shows the L2 miss rates of S and four FSB configurations normalized to S. We denote FSB with retention tables (RT) storing 1, 2, 4, and 8 RT(i).index pointers per each entry i as FSB-1, FSB-2, FSB-4, and FSB-8, respectively. Furthermore, we assume a low-pressure limit (LPL) and a high-pressure limit (HPL) each with $\alpha = 0.2$. Section 2.4.4 offers a sensitivity study

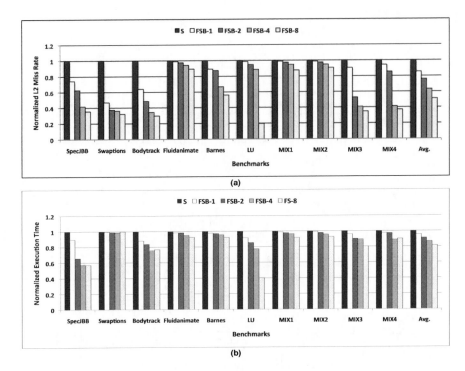

FIGURE 2.5: L2 miss rates and execution times of the baseline shared scheme (S), FSB-1, FSB-2, FSB-4, and FSB-8 (all normalized to S).

on different α values. We adopt cache misses as a pressure function but Section 2.4.3 provides a study on a variety of other functions. The figure demonstrates that as the number of pointers per RT entry increases, FSB achieves higher L2 miss-rate reductions. This behavior is apparent on all the examined benchmark programs. FSB centers around the flexible many-from-many sharing policy. More pointers indicate more exploitation to the many-from-many sharing strategy and, consequently, more alleviation to the imbalance across sets. On average, FSB-1, FSB-2, FSB-4, and FSB-8 accomplish average miss-rate reductions of 14.6%, 23.9%, 36.6%, and 48.7%, respectively.

FSB strategy adopts a serial lookup policy (see Section 2.3.3 for more details). Upon a miss on the original set i, RT(i).index pointers (if any) are utilized to serially index and look up corresponding L2 cache sets. Only the tag-stores are looked up until either a secondary hit is obtained or a definitive miss is asserted. Each tag-store access takes less than 0.68 ns, estimated by CACTI v5.3 [16] assuming a 45-nm technology. This incurs a higher latency per each L2 access that misses at the original set. As such, although more RT(i).index pointers result in more L2 miss-rate reductions, a latency cost must be paid. Figure 2.5 presents the execution times of S, FSB-1, FSB-2,

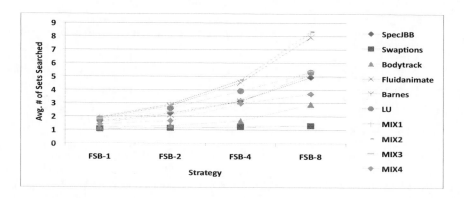

FIGURE 2.6: The average number of L2 cache sets searched under FSB-1, FSB-2, FSB-4, and FSB-8.

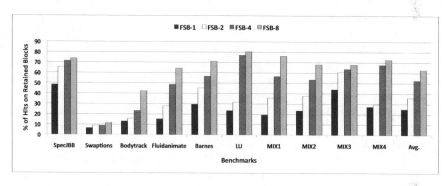

FIGURE 2.7: The percentage of hits on retained cache lines under FSB-1, FSB-2, FSB-4, and FSB-8.

FSB-4, and FSB-8 normalized to S. A main observation is that as we proceed through FSB configurations (FSB-1 to FSB-8), the performance of each application monotonically improves until FSB-8 is knocked. Under FSB-8, the case changes and programs are split into three categories: (1) no benefit is accomplished (e.g., SpecJBB), (2) a benefit is achieved (e.g., Fluidanimate, Barnes, Lu, MIX1, MIX2, and MIX3), and (3) a degradation is observed versus FSB-4 (e.g., Swaptions, Bodytrack, and MIX4).

Two factors define the eligibility of applications for accomplishing higher or lower performance when switching in between FSB's configurations: (1) the gain G from miss-rate reduction and (2) the loss L from increased access latency. Let $\triangle_i$ be defined as $G - L$ for FSB-i. When $\triangle_8$ exceeds $\triangle_4$, the performance of the application improves by switching from FSB-4 to FSB-8; otherwise, it degrades. Swaptions, Bodytrack, and MIX4 achieve miss-rate reductions of 6.1%, 7%, and 7%, respectively, after increasing RT pointers from

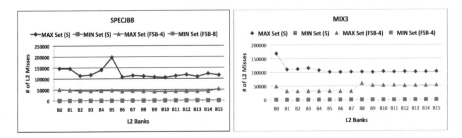

FIGURE 2.8: The number of L2 misses experienced by cache sets at different L2 banks for SpecJBB and MIX3 programs under the baseline shared scheme (S) and FSB-4. Only the sets that exhibit the maximum (MAX Set) and the minimum (MIN Set) misses are shown.

4 to 8. In fact, under FSB-8, these three applications reduce the L2 miss rates the least when compared to the other examined programs (see Figure 2.5(a)). Clearly, $\triangle_4$ of each of Swaptions, Bodytrack, and MIX4 overpasses $\triangle_8$; thus, they degrade under FSB-8 in comparison to FSB-4. FSB-1, FSB-2, FSB-4, and FSB-8 outperform S by averages of 4.3%, 8.8%, 13%, and 18.6%, respectively. Although FSB-8, on average, surpasses the remaining FSB's configurations, we consider FSB-4 more desirable for two main reasons. First, FSB-4 doesn't observe any degradation in performance for any application when compared against the preceding configurations. Second, FSB-4 offers a better tradeoff between hardware complexity, power dissipation, and performance.

To that end, Figure 2.6 depicts the average number of L2 cache sets searched for in all the applications under FSB-1, FSB-2, FSB-4, and FSB-8. Furthermore, Figure 2.7 displays the percentage of hits on retained cache lines for each program. In fact, the latter figure explores FSB's efficiency in satisfying far-flung reuses after retaining some fraction of the working set at underutilized sets. *With FSB-4, more than half of hits are satisfied by retained lines.* On average, the percentage of hits on retained lines provided by FSB-1, FSB-2, FSB-4, and FSB-8 are 25%, 35.8%, 52.6%, and 62.8%, respectively. Finally, Figure 2.8 explores FSB's effectiveness in mitigating nonuniformity across sets by showing the number of misses experienced by cache sets at different L2 banks for two benchmarks, a multithreading one (i.e., SpecJBB) and a multiprogramming one (i.e., MIX3). We present only the sets that exhibit the maximum and the minimum misses for the baseline shared S and FSB-4.

2.4.3   Sensitivity to Different Pressure Functions

In the previous section, we utilized cache misses as a pressure function. We tested other functions that can be used to measure pressures at cache sets. Figure 2.9 plots the results for only three functions F1, F2, and F3, which denote functions with *misses* only, *hits* only, and *spatial hits*, respectively. We

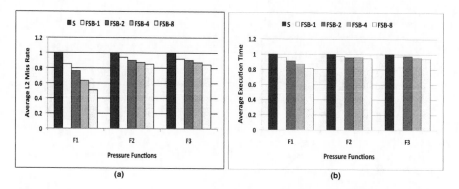

FIGURE 2.9: Average L2 miss rates and execution times of all the benchmark programs under the baseline shared scheme (S), FSB-1, FSB-2, FSB-4, and FSB-8 (all normalized to S) (F1, F2, and F3 are pressure functions that involve misses, hits, and spatial hits, respectively).

assume an LPL and an HPL each with $\alpha = 0.2$. The spatial hits function simply updates the pressure array with different values upon hits depending on lines' frames. That is, upon a hit on a line, L_{mru}, which exists at the MRU position, the function increments the bucket that corresponds to L_{mru}'s set by 1. However, upon a hit on a line, $L_{mru} - 1$, next to L_{mru}, the function increments the corresponding bucket by 2, and so on. The idea stems from the fact that a single highly contended line (say a lock) can result in a very high hit count at a particular set when, in fact, the pressure of lines competing for that set is very low. As depicted in Figure 2.9(b), on average, F2 produces performance improvements of 2.4%, 4.4%, 4.1%, and 5.4% for FSB-1, FSB-2, FSB-4, and FSB-8 over the baseline shared (S) scheme, respectively. F3, on the other hand, offers average performance improvements of 2.7%, 2.6%, 4.8%, and 6% for FSB-1, FSB-2, FSB-4, and FSB-8 over S, respectively. Lastly, F1 surpasses both F2 and F3 and provides average performance improvements of 4.3%, 8.8%, 13%, and 18.6% for FSB-1, FSB-2, FSB-4, and FSB-8 versus S, respectively. For the examined benchmarks, we conclude that cache misses are preferable among the tested functions to represent pressures at cache sets. More comprehensive functions can be considered in a future work.

2.4.4 Sensitivity to LPL and HPL

So far, we have been using $\alpha = 0.2$ for the low- and high-pressure limits, LPL and HPL. As Section 2.3.1 describes, by altering α, the range of source and destination sets can be expanded or contracted. We tested FSB-1, FSB-2, FSB-4, and FSB-8 with two more α values, particularly 0.1 and 0.3 for both LPL and HPL. Figure 2.10 shows the results. RL1, RL2, and RL3 denote the retention limits (i.e., LPL and HPL) with α values of 0.1, 0.2, and

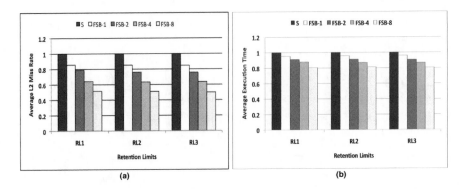

FIGURE 2.10: Average L2 miss rates and execution times of all the benchmark programs under the baseline shared scheme (S), FSB-1, FSB-2, FSB-4, and FSB-8 (all normalized to S) (RL1, RL2, and RL3 are the Retention Limits—HPL and LPL—with $\alpha = 0.1$, $\alpha = 0.2$, and $\alpha = 0.3$, respectively).

0.3, respectively. As demonstrated in Figure 2.10(a), on average, RL1 provides L2 miss-rate reductions of 14.4%, 21.3%, 35%, and 48.3% for FSB-1, FSB-2, FSB-4, and FSB-8 against the baseline shared (S) scheme, respectively. RL2, on the other hand, offers a little more enhancement and produces 14.6%, 23.9%, 39.4%, and 48.7% L2 miss-rate reductions for FSB-1, FSB-2, FSB-4, and FSB-8 versus S, respectively. Finally, RL3 achieves 15.2%, 24.3%, 36.2%, and 49.8% miss-rate reductions for FSB-1, FSB-2, FSB-4, and FSB-8 over S, respectively. Figure 2.10(b) depicts the performance outcome. For the simulated benchmarks, we conclude that FSB shows low sensitivity to the examined α values.

2.4.5   Impact of Increasing Cache Size and Associativity

We can improve cache performance not only by efficient cache management but also by increasing cache size and associativity. We note that increasing cache associativity is not equivalent to FSB. First, larger associativity results in fewer sets, which don't help much if the conflict on the sets varies widely. Second, increasing cache associativity equates to merging sets in an indiscriminate way [27], that is, which sets to merge is not an option. FSB, however, attempts to controllably and selectively increase the associativity of the sets that experience extensive conflicts without decreasing the number of sets (effectively decreasing associativity for underutilized sets).

In this section, we consider only FSB-4 (see Section 2.4.2 for a discussion on FSB's configurations). FSB-4 requires 11.5 KB storage overhead per tile (see Table 2.1). To justify FSB-4's incurred overhead, we optimistically augment each cache set of the baseline shared scheme S with two more ways. In total, this increases each L2 bank 64 KB in capacity. We refer to this configuration

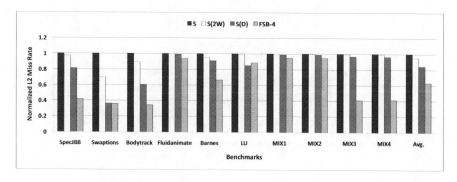

FIGURE 2.11: L2 miss rates of the baseline shared scheme (S), S with two more ways added (S(2W)), S with double sized cache (S(D)), and FSB-4 (all normalized to S).

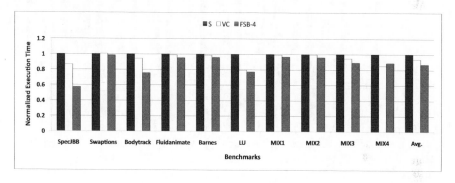

FIGURE 2.12: Execution times of the baseline shared scheme (S), victim cache (VC), and FSB-4 (all normalized to S).

as S(2W). Moreover, we examine S with a double-sized cache (i.e., 1 MB instead of 512 KB). We denote this latter configuration by S(D). Figure 2.11 shows the L2 miss rates of S, S(2W), S(D), and FSB-4 normalized to S. The figure demonstrates that doubling the size of the cache results in a greater miss reduction than increasing associativity by two ways. Nonetheless, FSB-4 surpasses S(D) for all the examined programs except Lu. On average, S(2W), S(D), and FSB-4 achieve L2 miss-rate reductions of 5.1%, 15.6%, and 36.6%, respectively. We conclude that FSB-4 is quite attractive as, with small design and storage overhead, it provides more than two times miss-rate reduction over S(D) which incurs 88.8% increase in the on-chip cache capacity.

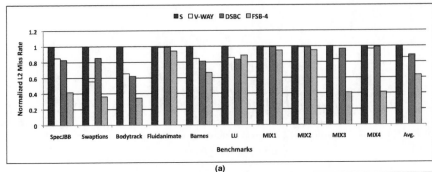

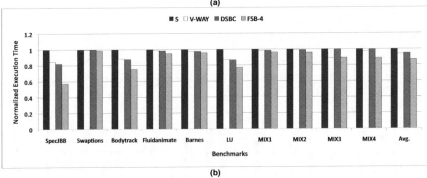

FIGURE 2.13: L2 miss rates and execution times of the baseline shared scheme (S), variable-way set associative cache (V-WAY), dynamic set-balancing cache (DSBC), and FSB-4 (all normalized to S).

2.4.6   FSB versus Victim Caching

In this section, we compare FSB against victim cache (VC) [20]. Again, we contrast only against FSB-4. VC effectively extends the associativity of hot sets in the cache to reduce conflict misses. For a fair comparison, we consider a fully associative 16-KB VC per tile to approximately match the storage overhead incurred by FSB-4. We furthermore optimistically assume only a six-cycle access time to VC after each miss on an L2 bank. Figure 2.12 depicts the execution times of S, VC, and FSB-4 normalized to S. VC outperforms S by an average of 6.3%. In contrast, FSB-4 improves upon S and VC by averages of 13% and 7.2%, respectively.

2.4.7   FSB versus DSBC and V-WAY

In addition to comparing with victim caching, we compare FSB against the closely related dynamic set-balancing cache (DSBC) [27] and variable-way set associative cache (V-WAY) [25] designs. Similar to FSB, both DSBC and V-WAY are directly extensible to CMPs. Section 2.2.3 details DSBC. V-WAY addresses the problem of workload imbalance among sets by varying the associativity of a cache by increasing the number of tag-store entries relative to the number of data lines. The tag and data stores are decoupled. The data store is structured as one large piece, and a *global* frequency-based replacement policy, referred to as *Reuse Replacement* is employed in order to achieve better replacements. In reverse, the tag-store keeps a conventional set granular (local) replacement strategy (e.g., LRU).

For the reuse replacement policy, V-WAY associates each data line in the cache with a *reuse counter*. A reuse count is defined as the number of L2 accesses to a cache line after its initial fill. Upon replacement, a line with a reuse counter equal to zero is replaced. To decide upon the number of bits required for reuse counters, we conducted a study to scrutinize the distribution of reuse counts for all evicted L2 cache lines from our benchmark programs. We observed that 99% of L2 cache lines are reused three or fewer times. Consequently, we chose to use two-bit saturating reuse counters.

Figure 2.13(a) depicts the L2 miss rates of S, V-WAY, DSBC, and FSB-4 normalized to S. On average, V-WAY and DSBC achieve miss-rate reductions of 14.7% and 11.3%, respectively. FSB-4 surpasses V-WAY and DSBC by averages of 27.2% and 29.2%, respectively. Figure 2.13(b) shows the execution-time results. V-WAY and DSBC outperform S by averages of 5.9% and 5%, respectively. FSB-4, however, improves upon V-WAY and DSBC by averages of 7.8% and 8.8%, respectively.

2.5   Related Work

Much work has been done to effectively minimize conflict misses in conventional cache designs. It is, in fact, quite impossible to do justice to this large body of work in this short article. As such, we briefly describe the proposals that are most relevant to FSB. The closest proposals to FSB are the dynamic set-balancing cache (DSBC) [27] and the variable-way set associative cache (V-WAY) [25] designs. Sections 2.2.3 and 2.4.7 describe DSBC and V-WAY in detail, and Section 2.4.7 compares them.

Many proposals suggest alternative indexing functions to achieve a more uniform distribution of memory accesses. Predictive Sequential Associative-Cache [5], Column Associative Cache [2], and Hash-Rehash [1] are proposed in the context of direct-mapped caches. They provide the capability of mapping a cache line at an alternative predetermined (using different hash functions) cache frame in order to provide performance similar to that of 2-way caches (schemes referred to as skewed caches). Rolán et al. [27] suggested a skewed set associative cache, denoted as static set-balancing cache (SSBC) and found it impractical. Consequently, they proposed DSBC as a superior scheme. Our work simply promotes FSB over DSBC.

Adaptive Group-Associative Cache (AGAC) [22] identifies underutilized cache frames and attempts to utilize them to approximate a global LRU policy while maintaining the fast access in direct-mapped caches. Indirect Index Cache (IIC) [12] suggests a fully associative, software-managed secondary cache system. IIC employs a generational replacement policy run by software. Simulation results in [25] manifested the outperformance of V-WAY versus AGAC. Besides, [25] shows that the miss reduction provided by V-WAY is comparable to that of IIC. In Section 2.4.7, we demonstrate the outperformance of FSB against V-WAY.

Utility-Based Cache Partitioning (UCP) [24] partitions at a way-granularity the last-level shared cache among concurrently running applications depending on how much each application is likely to benefit from the cache (i.e., utility) rather than the application's demand for the cache. Dynamic Insertion Policy (DIP) [23] makes a key observation that a large number of cache lines become dead on arrival. Thus, a Bimodal Insertion Policy (BIP) is proposed to insert incoming lines frequently in the LRU positions and infrequently (with a low probability) in the MRU positions. Lines inserted at the LRU positions are only promoted to the MRU positions upon hits while residing in the LRU positions. For LRU-friendly workloads (i.e., favoring MRU insertions), however, the changes to the insertion policy might become detrimental to cache performance. As such, a *Set Dueling* mechanism is proposed to select among BIP and LRU depending on which policy incurs fewer misses. Simulation results in [27] demonstrate the outperformance of DSBC versus DIP. As shown in Section 2.4.7, FSB surpasses DSBC .

DIP uses a single policy (LRU or BIP) for *all* the concurrently running applications. A subsequent proposal, namely, Thread-Aware Dynamic Insertion Policy (TADIP) [19], extends DIP to use a single policy for *each* application. Promotion/Insertion Pseudo-Partitioning (PIPP) [34] combines dynamic insertion and probabilistic promotion policies to provide the benefits of cache partitioning, adaptive insertion, and capacity stealing all with a single mechanism. Adaptive Set Pinning (ASP) [29] associates processors to cache sets and grants them sole permission to evict blocks from their sets on cache misses. Therefore, references that may potentially cause inter-processor misses are no longer allowed to interfere with each other even if they index to the same set.

Pseudo-Last-In-First-Out (Pseudo-LIFO) [8] proposes a family of replacement policies that manages each cache set as a fill stack. The replacement activities are restrained within a set to the upper part of the fill stack as much as possible. The lower part of the fill stack is left undistributed to extend the lifetime of the resident blocks. Among three members of the Pseudo-LIFO family, namely, dead block prediction LIFO (dbpLIFO), probabilistic escape LIFO (peLIFO), and probabilistic counter LIFO (pcounter-LIFO), peLIFO is central. peLIFO synergistically learns the probabilities of experiencing hits beyond each of the fill stack positions, and a set of highly preferred eviction positions is then deduced (based on this probability function) in the upper part of the fill stack.

Finally, Scavenger [3] partitions the total storage budget at the last-level cache (LLC) into a conventional cache and a novel victim file (VF). Block addresses missing at the LLC are prioritized based on the number of times they have been observed in the LLC miss stream. If a block is evicted from the conventional part of the cache and indicates a high priority (i.e., frequently missed in the recent past), it gets stored in the VF.

2.6 Conclusions and Future Work

Memory accesses are not evenly distributed across cache sets. Such a skew in sets' usages reduces the effectiveness of the conventional cache designs, and cache lines become less likely to be rereferenced before eviction. We propose Flexible Set Balancing (FSB), a strategy that exploits the demand imbalance across sets to retain cache lines evicted from highly pressured sets at underutilized sets so as to satisfy far-flung reuses. FSB adapts to phase changes in programs and promotes a very flexible sharing among cache sets. An underutilized set is allowed to share its space with any stressed set during any point in a program's execution, a policy that we refer to as *one-from-many* sharing. Besides, many sets are allowed to share their capacities with a highly utilized set, a policy that we refer to as *many-from-one* sharing. FSB incurs fewer storage, area, and energy overheads. FSB achieves an average miss-rate reduction of 36.5% versus a shared CMP cache organization for the tested benchmarks. This produces an average execution-time improvement of 13%. Furthermore, evaluations manifested the outperformance of FSB over some relevant designs including DSBC [27] and V-WAY [25].

FSB is extensible and practical in that it can be applied to single-core as well as multicore architectures. In this work, we evaluated FSB on a 16-way tiled CMP platform. FSB retains lines only at a bank granularity (intra-tile retention). When an L2 bank can't absorb anymore from the working set of a running program, the lines selected for replacements are simply discarded. In fact, in the meantime, other L2 banks might indicate the presence of some underutilized sets. As such, one may benefit from retaining lines across L2 banks (inter-tile retention) rather than only within a single L2 bank, so as to satisfy even more far-flung reuses. Exploring the promise of such a strategy is a main future goal.

Bibliography

[1] A. Agarwal, J. Hennessy, and M. Horowitz. "Cache performance of operating systems and multiprogramming," *In ACM Transactions on Computer Systems, 6,* Nov 1988.

[2] A. Agarwal and S. D. Pudar. "Column-associative caches: A technique for reducing the miss rate of direct-mapped caches," *ISCA*, May 1993.

[3] A. Basu, N. Kirman, M. Chaudhuri, and J. F. Martínez. "Scavenger: A new last level cache architecture with global block priority," *MICRO*, 2007.

[4] C. M. Bienia, S. Kumar, J. P. Singh, and K. Li. "The PARSEC benchmark suite: Characterization and architectural implications," *PACT*, Oct. 2008.

[5] B. Calder, D. Grunwald, and J. S. Emer. "Predictive sequential associative cache," *HPCA*, Feb. 1996.

[6] L. Censier and P. Feautrier. "A new solution to coherence problems in multicache systems," *IEEE Trans. Comput.* C-27 (12): 1112–1118, Dec. 1978.

[7] J. Chang. "Cooperative caching for chip multiprocessors," PhD diss., University of Wisconsin-Madison, 2007.

[8] M. Chaudhuri. "Pseudo-LIFO: The foundation of a new family of replacement policies for last-level caches," *MICRO*, Dec. 2009.

[9] Z. Chishti, M. D. Powell, and T. N. Vijaykumar. "Distance associativity for high-performance energy-efficient non-uniform cache architectures," *MICRO*, 2003.

[10] S. Cho and L. Jin "Managing distributed shared L2 caches through OS-level page allocation," *MICRO*, Dec 2006.

[11] Digital Equipment Corporation, Hudson, MA. "Digital Semiconductor 21164 AlphaMicroprocessor product brief," *Technical Document EC-QP97D-TE*, Mar. 1997.

[12] E. G. Hallnor and S. K. Reinhardt. "A fully associative software managed cache design," *ISCA*, 2000.

[13] M. Hammoud, S. Cho, and R. Melhem. "ACM: An efficient approach for managing shared caches in chip multiprocessors," *HiPEAC*, Jan. 2009.

[14] N. Hardavellas, M. Ferdman, B. Falsafi, and A. Ailamaki. "Reactive NUCA: Near-optimal block placement and replication in distributed caches," *ISCA*, June 2009.

[15] S. Harris. "Synergistic caching in single-chip multiprocessors," PhD diss., Stanford University, 2005.

[16] HP Labs. http://www.hpl.hp.com/research/cacti/, 2013.

[17] H. Huang, K. G. Shin, C. Lefurgy, and T. Keller. "Improving energy efficiency by making DRAM less randomly accessed," *ISLPED*, August 2005.

[18] J. Huh, C. Kim, H. Shafi, L. Zhang, D. Burger, and S. W. Keckler. "A NUCA substrate for flexible CMP cache sharing," *ICS*, June 2005.

[19] A. Jaleel, W. Hasenplaugh, M. K. Qureshi, J. Sebot, S. Steely Jr., and J. Emer. "Adaptive insertion policies for managing shared caches," *PACT*, 2008.

[20] N. P. Jouppi. "Improving direct-mapped cache performance by the addition of a small fully-associative cache and prefetch buffers," *ISCA*, 1990.

[21] P. Kongetira, K. Aingaran, and K. Olukotun. "Niagara: A 32-way multithreaded SPARC processor," *IEEE Micro*, March–April 2005.

[22] J. Peir, Y. Lee, and W. Hsu. "Capturing dynamic memory reference behavior with adaptive cache topology," *ASPLOS*, 1998.

[23] M. K. Qureshi, A. Jaleel, Y. N. Patt, and S. C. Steely Jr. "Adaptive insertion policies for high performance caching," *ISCA*, June 2007.

[24] M. K. Qureshi and Y. N. Patt. "Utility-based cache partitioning: A low-overhead, high-performance," *MICRO*, Dec. 2006.

[25] M. K. Qureshi, D. Thompson, and Y. N. Patt. "The V-WAY cache: Demand-based associativity via global replacement," *ISCA*, June 2005.

[26] Research at Intel. "Introducing the 45 nm next-generation Intel Core^{TM} microarchitecture," *White Paper*, 2012.

[27] D. Rolán, B. B. Fraguela, and R. Doallo "Adaptive line placement with the set balancing cache," *MICRO*, Dec. 2009.

[28] B. Sinharoy, R. N. Kalla, J. M. Tendler, R. J. Eickemeyer, and J. B. Joyner. "POWER5 System Microarchitecture," *IBM J. Res. & Dev.*, July 2005.

[29] S. Srikantaiah, M. Kandemir, and M. J. Irwin. "Adaptive set pinning: Managing shared caches in chip multiprocessors," *ASPLOS*, March 2008.

[30] Standard Performance Evaluation Corporation. http://www.specbench. org, 2013.

[31] Virtutech AB. "Simics Full System Simulator." http://www.simics.com/, 2013.

[32] D. Weiss, J. J. Wuu, and V. Chin. "The on-chip 3-MB subarray-based third-level cache on an Itanium microprocessor," *In IEEE journal of solid state circuits*, Nov. 2002.

[33] S. C. Woo, M. Ohara, E. Torrie, J. P. Singh, and A. Gupta. "The SPLASH-2 programs: Characterization and methodological considerations," *ISCA*, 1995.

[34] Y. Xie and G. H. Loh. "PIPP: Promotion/insertion pseudo-partitioning of multi-core shared caches," *ISCA*, June 2009.

[35] M. Zhang and K. Asanović. "Victim replication: Maximizing capacity while hiding wire delay in tiled chip multiprocessors," *ISCA*, June 2005.

Chapter 3

The SPARC Processor Architecture

Simone Secchi

Università di Cagliari

Antonino Tumeo

Pacific Northwest National Laboratory

Oreste Villa

NVIDIA

3.1   Introduction

The SPARC (Scalable Processor ARChitecture) instruction set architecture is one of the most employed RISC design for multicore processors.

The first 32-bit version of the architecture was introduced in 1986 with the name SPARC Version 7 (SPARCv7). In 1990, the next revision, SPARCv8, introduced integer multiply and divide instructions [1]. The current major revision, SPARCv9, moved the specification from a 32- to a 64-bit architecture [2]. Originally presented in 1993, the SPARCv9 architecture has since then been extended to support chip multithreading (CMT), hyperpriviliged processor instructions and mode of execution, SIMD extensions [3, 4, 5].

The very conception and the entire development of SPARCv9 has been focused around scalability. Many companies have licensed the SPARC architecture for their specific processor implementations, a range of computers that covers laptops as well as databases, web servers, and supercomputers. For example, the K Computer, currently (November 2011) the most powerful supercomputer in the world [6], uses a customized SPARC design to reach 10.51 Petaflop/s on the Linpack benchmark.

While preserving binary compatibility with the previous revisions, the SPARCv9 architecture specification includes the following main features:

- pure 64-bit addressing

- a RISC instruction set with register-to-register 32-bit instructions

- a configurable-length windowed register file

- multiprocess atomic synchronization instructions

- a relaxed memory order model

- support for nested traps

In this chapter, we describe the main features of the SPARCv9 architecture specification and the motivations behind the different design decisions. At the same time, we focus our attention on the relevance that each architectural feature has in the context of a multicore processor implementation of the SPARC architecture. The objective of this chapter is to provide insight into both the SPARC architecture as described by the specification and its implementation in the context of a multicore processor design. For this reason, after describing the SPARC architecture, we present in detail one of its most successful implementations, the Sun UltraSPARC T1 (also known as Niagara) multicore processor.

The Niagara processor is a low-power throughput-oriented multicore implementation of the SPARC architecture that relies heavily on Simultaneous Multi-Threading (SMT) to increase performance and to hide the memory-operation latencies. The Niagara CPU uses eight SPARC cores, each one natively capable of running four threads of execution, therefore supporting 32-way multithreading. Although further evolutions of the UltraSPARC T1 processor, such as the UltraSPARC T2 (codename Niagara 2) and UltraSPARC T3 (codename Niagara 3 or Rainbow Falls) increased the degree of multi-threading to 64 and 128 threads of execution, we focus on the UltraSPARC T1 design in the rest of this chapter.

The rest of this chapter is organized as follows. Section 3.2 describes the instruction set architecture as defined by the SPARC specification together with the description of how the register file is handled. Section 3.3 focuses on how the SPARCv9 specification regulates memory access and ordering of memory references. Section 3.4 describes the features of the ISA with respect to inter-process synchronization and mutual exclusion on data. We then analyze the Niagara multicore SPARC implementation, presenting in detail the single-core microarchitecture (Section 3.6), the interconnection layer of the multicore processor (Section 3.7), the memory hierarchy organization (Section 3.8), and the cache-coherence protocol (Section 3.8.1). Finally, Section 3.9 gives a brief overview of the main architecture evolutions of the Niagara processor line.

3.2 The SPARC Instruction-Set Architecture

The objective of this chapter is not to individually describe all the instructions of the SPARC ISA. For the readers who are interested in such a description, we refer them to [2].

The SPARCv9 architecture specification supports hardware integer and floating-point computations. The supported integer data types can be 32 or 64 bits wide, while the floating-point data types can be 32, 64, or 128 bits wide. The specification forces all the potential implementations to provide separate integer and floating-point units, each one with its own registers, therefore allowing concurrent execution of integer and floating-point instructions.

In order to enable isolation of user-level code execution from operating system execution (or resource management software more generally), the SPARC instruction set identifies two domains of instructions: *privileged* and *nonprivileged*. In turn, these two domains define the main two modes of operation of the processor. A processor running in privileged mode can execute any instruction, while a processor running in nonprivileged mode can only execute nonprivileged instructions. If a processor running in nonprivileged mode attempts to execute a privileged instruction, an exception will be triggered. On top of these two modes of execution, the UltraSPARC 2005 architecture specification [4] introduces a third execution mode, called *hyperpriviliged*.

The main categories of instructions supported by the SPARC specification are the following:

- memory access instructions

- integer arithmetic instructions

- floating-point arithmetic instructions

- control transfer instructions

- state register access instructions

- conditional move instructions

- register window management

The memory access instructions include load, store, and atomic access instructions. The load and store instructions calculate a 64-bit byte-aligned memory address either by using two registers or a register and a signed immediate. Byte, halfword (16-bit), word (32-bit), and doubleword (64-bit) access is supported for integer loads and stores, while word, doubleword, and quadword (128-bit) access is supported for floating-point loads and stores. The supported atomic access instructions implement the standard compare-and-swap (CASX), swap (SWAP) and test-and-set (LDSTUB) semantics, and will be described in more detail in Section 3.4.

The integer arithmetic category includes all those instructions that perform integer arithmetic, logical, and shifting operations. The instruction set includes 64- and 32-bit hardware multiplication and division. The floating-point category includes instructions that perform only floating-point calculations. They work on dedicated registers, both for the input operands and for the result.

The control transfer instructions include the PC-related branches, register-indirect jumps, and conditional traps. The SPARC specification intends most of the control transfer instructions to be delayed. In detail, the instruction immediately following the control transfer instruction (the one in the so-called *delay slot*) is fetched and executed before the control transfer instruction is executed. In case the delay slot is empty, the execution of its instruction can be avoided by setting a specific bit in the opcode.

The state register access instructions implement reading and writing of the processor state registers. Since different modes of execution are supported by the processor, different instructions are defined for privileged and nonprivileged state registers. The conditional move instructions move data between a source and a destination register upon an integer or floating-point condition to be verified on specific condition code registers. Finally, the register window management instructions operate on the register window structure. We describe in detail this structure in Section 3.2.1.

3.2.1   Registers and Register Windowing

The SPARCv9 specification includes two types of registers: the working registers (integer and floating-point) and the processor state registers. In this section, we focus on the working registers and how they are handled through the register windowing mechanism.

An implementation of the SPARC specification may have a variable number of general-purpose 64-bit working registers. In detail, every implementation contains at least eight global registers and an implementation-dependent number of 16-register sets. The 16-register sets are handled in such a way that a 24-register window can be defined by including the 16 registers and the first eight of the adjacent register set. The *register window* is thus the set of registers that, at any time, a process can address in addition to the eight global registers. It includes eight *in*, eight *local*, and eight *out* registers. These last eight physically belong to the adjacent set. The number of physical register windows is implementation-dependent and ranges from 3 to 32 windows.

A dedicated processor state register, the CWP register, stores at any time a pointer to the current register window. The total number of windows currently allocated in the processor is handled through two other state registers, the CANSAVE and CANRESTORE registers. If available, a new register window is allocated for the executing process through the execution of the SAVE instruction. Upon execution, the previous eight *out* registers will become the new eight *in* registers. A window can be deallocated by the executing process through the execution of the RESTORE instruction. Upon execution, the previous eight *in* registers will become the new eight *out* ones. The SAVE and RESTORE instructions also automatically update the CWP, CANSAVE, and CANRESTORE registers.

Figure 3.1 shows three register windows and their overlapping registers. It can be seen how, at any time, 32 general-purpose registers are visible to the executing context.

The registers that are shared between different windows can be easily used to pass parameters between different contexts of execution. The classical example of register-window switch is the procedure/function call. The caller will allocate a new register window for the callee using the SAVE instruction. The function parameters could be easily passed using the registers shared between the two adjacent windows (the *out* registers of the caller, the *in* registers of the callee). The register-windowing mechanism also allows the implementations to optimize subroutine calling and returning since, by providing specific hardware support to save and restore function frames, parameters and results do not always dictate a memory access, reducing register-spilling occurrences.

3.3 Memory Access

This section focuses on the features of memory access as described in the SPARCv9 architecture specification. In particular, we describe the main addressing SPARC conventions, the conditions that the implementing MMUs must satisfy, the requirements posed by the architecture specification for ordering the issue and execution of memory references, and specific SPARC instructions related to memory access.

As already discussed in Section 3.2, the load, store, and atomic load-store are the only instructions that access memory (along with the PREFETCH instruction, which is ignored for the purposes of this chapter). They build a 64-bit address by using two 64-bit registers or one 64-bit register and a 13-bit immediate field.

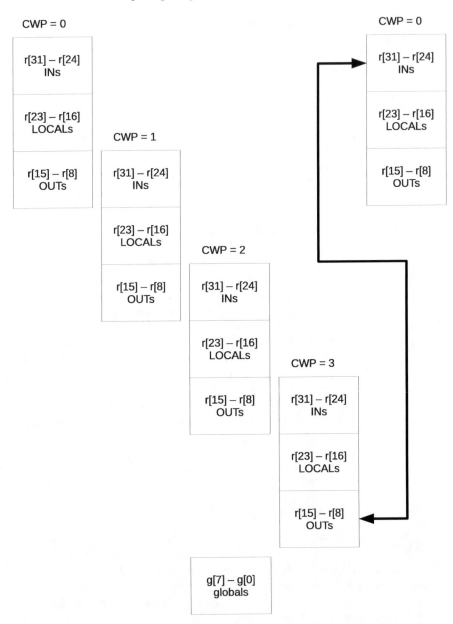

FIGURE 3.1: Register windows overlapping, with 4 register windows.

For each instruction or data memory access, an 8-bit field Address Space Identifier (ASI) is then attached to the 64-bit address by the integer unit. This field allows for explicit separation between different 2^{64}-words-wide address spaces. The ASI field is implicitly set by the integer unit for normal loads and stores or can also be arbitrarily set by the *load/store alternate* instructions. The implementations are intended to use the ASIs to provide access to system-critical registers or memory areas for privileged supervisor software.

The SPARC specification also defines the FLUSH instruction, which takes as input a 64-bit address operand, and ensures that this address is consistent across any cache. If the implementation is a multiprocessor system, it ensures that the data stored in that address are also coherent among all the processor caches.

3.3.1 MMU Requirements

Although it is possible to build a SPARC processor implementation without an MMU, the SPARCv9 specification defines a minimum set of requirements that the implemented MMUs must satisfy. The basic functions that the SPARCv9 implementation MMUs must provide are the following:

- translate 64-bit virtual addresses to physical addresses with 8-, 64-, 512-, or 4096-KB page sizes;

- provide operation for the processor Reset, Error, and Debug state (RED state) upon proper signaling from the processor;

- provide a method to disable the MMU. In this condition, the physical address will be an implementation-dependent, low-bits truncation of the virtual address;

- provide page-level protection, prefetch, and non-faulting load attributes;

- provide feedback to the processor in case a protection, prefetch, privilege or translation violation occurs;

- support multiple 64-bit address spaces through the ASIs, as specified by the processor;

- provide page-level statistics.

3.3.2 Memory Models

The SPARCv9 architecture specification also defines three different formal memory models. Such models specify the semantics of memory operations and the constraints on the order with which these are actually performed in memory. In fact, every SPARC implementation, regardless of the number of cores, is free to implement in any manner the order in which loads and stores are performed in memory as long as the behavior observed conforms to at least one of the memory models described in the specification.

The SPARCv9 architecture defines three different memory models:

- Total Store Order (TSO),

- Partial Store Order (PSO), and

- Relaxed Memory Order (RMO).

All the implementations must at least comply to the TSO memory model. The three memory models can be described with respect to the constraint that they add to the *processor self-consistency* property. This property defines a fundamental constraint on the possible reordering that can be made within an architecture implementation on issued instructions before they are actually executed. In detail, processor self-consistency allows any instruction reordering as long as the results of program execution are the same as they would be if the instructions were executed in program order.

The RMO model does not place further ordering constraints on memory references beyond those required for processor self-consistency. Therefore, RMO ensures that the final results of the program execution are the same as they would be if the instructions were executed in program order. The constraints posed by processor self-consistency translate into the classical functional correctness limitations of pipeline and out-of-order instruction execution, namely the avoidance of read-after-write, write-after-write, and write-after-read hazards both on register and memory references.

The PSO memory model defines two additional rules for memory reference execution:

- memory loads are blocking and ordered with respect to earlier loads;

- atomic load-stores are ordered with respect to loads.

Therefore, each load and atomic load-store instruction is guaranteed to be scheduled before other loads. However, the PSO model does not place any constraint on ordering with respect to memory stores or atomic load-stores. Thus, to achieve ordering with respect to stores or atomic load-stores, explicit memory synchronization instructions are needed in the PSO model.

The TSO memory model adds three rules to the constraints posed by processor self-consistency:

- loads are blocking and ordered with respect to earlier loads;

- stores are ordered with respect to stores;

- atomic load-stores are ordered with respect to both loads and stores.

The TSO is the strictest memory model of the three defined by the SPARCv9 specification. However, it is not as strict as the sequential consistency models (also known as strong-ordering models), in which the memory performs loads, stores, and atomic loads-stores in the same order in which they are issued by the processor. In fact, TSO does not place any constraint beyond those of processor self-consistency for the ordering of stores with respect to loads and vice versa.

From the previous description, it is worth noting how a program that executes correctly in the RMO memory model will execute correctly in the PSO and TSO models since it does not rely on any native ordering capability of the memory model beyond those ensured by processor self-consistency. This means that a program written for the RMO model must explicitly include the necessary memory fences and access ordering instructions to guarantee correctness, while a program written for PSO or TSO models can make use of some automatic ordering constraints that are guaranteed by the architecture implementation. Similarly, a program that executes correctly under the PSO model will execute correctly also under the TSO model.

3.3.3   The MEMBAR instruction

The MEMBAR instruction implements memory barrier in the SPARCv9 instruction set. It provides a way for the programmer to explicitly synchronize the issuing order or even the completion order of memory instructions. The first set of MEMBAR instructions, known as ordering MEMBARs, guarantees an ordering in the issuing of the memory instructions that appear before the MEMBAR with respect to those that follow the MEMBAR. The second set, also known as sequencing MEMBAR instructions, guarantees that the effect of the instructions is visible before instructions that follow the MEMBAR are issued.

For both ordering and sequencing MEMBARs, different parameters exist to indicate which type of instructions are affected by the barrier (loads, stores, all memory references, all instructions). For a precise definition of the instruction parameters, we refer readers to [2].

The MEMBAR instruction provides the programmer with necessary ordering control not offered by the implemented memory model when needed by the program semantics. For example, if the specific SPARC implementation uses a PSO memory model as we described in Section 3.3.2 but the program semantics require ordering with respect to stores, explicit MEMBAR instructions will have to be inserted.

3.4   Synchronization

As described in Section 3.3.3, the SPARCv9 architecture specification defines a set of primitives and models for memory-reference ordering that allow the programmer to implement thread- or process-level synchronization and data-access mutual exclusion. In addition to that, three instructions for synchronization are provided:

- Compare and Swap (`cas`)

- Load-Store Unsigned Byte (`ldstub`)

- Swap (`swap`)

All these instructions are atomic and have the semantics of both a load and a store with respect to the memory models defined in Section 3.3.2. When they are applied to memory-mapped I/O locations, the behavior is implementation-dependent.

The Load-Store-Unsigned-Byte instruction loads a byte from the accessed memory location to an output register and writes a constant value (FF_{16}) in the memory location. Considering its test-and-set behavior, it is clear how the `ldstub` instruction can be employed for the implementation of the mutual exclusion lock primitive. When two processes/threads access a synchronization variable using the `ldstub` instruction, the first one to acquire the lock will get a zero value (assuming that was the initialization state of the variable), the second one will instead get the FF_{16} value.

The Swap instruction swaps the value stored in the accessed memory location with a value (32-bit) present in an input register. The Compare-and-Swap instruction compares a value in an input register to a value stored in the accessed memory location and, if they are equal, swaps the value in the memory with the value stored in a second input register.

3.5   The NIAGARA Processor Architecture

The Niagara processor is one of the most widely known implementations of the SPARCv9 architecture. It was first announced in 2005 by SUN Microsystems [7] and complies to the UltraSPARC Architecture 2005 Specification [4]. In this section, we look in detail at its macroarchitecture organization, describing the main chip-level building blocks of the multicore processor and how they interconnect. Figure 3.2 shows a system-level block diagram of the Niagara processor macroarchitecture.

The Niagara includes eight 4-way multithreaded SPARC cores, supporting therefore a total of 32 simultaneous threads of execution in the CPU. The SPARC cores contain dedicated level-1 instruction and data caches. As we will describe in Section 3.6, each SPARC core provides hardware support to switch the execution among its 4 threads on a cycle-by-cycle basis. By doing so, the designers of the Niagara processor aimed at hiding the memory latency and the related pipeline stall. The Niagara 2 and Rainbow Falls designs then increased the number of integrated SPARC cores and threads per core in

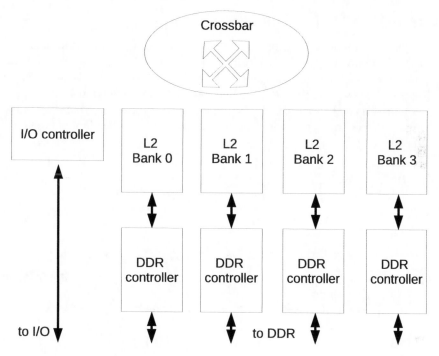

FIGURE 3.2: Niagara chip building blocks.

order to support a higher number of overall simultaneous threads, reaching a maximum of 128.

In the first Niagara processor, the eight cores shared a 3-MB level-2 pipelined cache, which was subdivided into four banks and set-associative with 12 ways. The evolutions of the processor doubled the number of banks and modified the set associativity in order to support an increasing number of threads. We describe in more detail the caching mechanism of the Niagara in Section 3.8. By integrating an on-chip shared cache, all the transactions due to L1-L2 coherence misses are confined within the chip and do not generate expensive high-latency traffic. In the Niagara design, all the cores of the chip also share a single floating-point unit (FPU), through a dedicated interconnect port.

The on-chip interconnect layer is based on a full crossbar that connects the eight cores, the L2 cache banks, and the on-chip peripheral controllers and provides a bandwidth 200 GB/s. The four DDR2 DRAM memory controllers, finally, provide an aggregate bandwidth of 20 GB/s.

3.6   Core Microarchitecture

This section deals with the description of the SPARC core as implemented in the Niagara design. As mentioned in Section 3.5, the Niagara integrates eight cores, each capable of running four simultaneous threads of execution. Figure 3.3 shows the SPARC pipeline block diagram. The hardware support for multithreaded is implemented through replication of some building blocks, namely the program counters, the instruction and data buffers, and the register files.

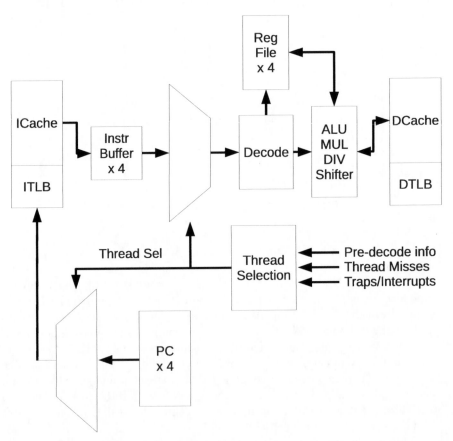

FIGURE 3.3: Niagara multithreaded pipeline building blocks.

The Niagara implementation of the SPARCv9 architecture features a six-stage single-issue pipeline. In the fetch stage, the instruction is fetched by accessing the Icache and the ITLB. The TLB has 64 entries and is fully associative. The instruction to be fetched is determined according to the program counter selected by the thread-selection logic, which is also the next stage of the pipeline.

In the thread-selection stage, the thread selector chooses which thread to execute and stores the previously fetched instruction in the instruction buffer in case the following stages of the pipeline are not available. The core was designed to implement thread-switching with a very fine granularity, potentially on a cycle-by-cycle basis, while maintaining fairness. Despite the fact that the hardware is able to support thread switching on a cycle-by-cycle basis, the scheduler does not necessarily recognize a single-cycle, round-robin switching. The execution is necessarily switched to a different thread only when the executing thread encounters a long-latency instruction. Long-latency instructions include loads/stores (even in the case of an L1 hit, which generates a three-cycle latency), branches, and multiply/divide operations. Once the thread scheduling is triggered, the scheduler chooses the next thread to execute, maximizing the fairness among the executable threads. The basic mechanism implements a least-recently-used arbitration policy. The set of threads' candidate for execution does not include unavailable threads.

Threads become unavailable because of pipeline stalls such as cache misses or traps. The thread-selection logic also makes use of some pre-decode information directly inserted in the instructions to decide which thread to schedule. For example, some long-latency instructions such as integer multiply or divide contain a *pre-decode* bit to identify a long latency. Memory access instructions are seen as long-latency operations since, in the best-case L1-hit scenario, they require three clock cycles before the data is available to following instructions. In some cases, threads that are unavailable because of long-latency instructions can still be scheduled to issue speculative instructions, but are assigned lower priority with respect to available threads. One of these cases involves cache misses; since the scheduler normally assumes that loads are cache hits, it issues dependent instructions in a speculative way, but with lower priority.

The decode stage performs the actual instruction decode and register file access. The working registers are selected according to the currently executing thread identifier and the current register-window pointer, as described in Section 3.2.1. The execution unit includes an arithmetic-logic unit (ALU), a shifter, a multiplier, and a divider. Also, forwarding is implemented with results that must be passed to dependent instructions of the same thread in the pipeline before the write-back into registers is performed. All the ALU and shift instructions have single-cycle latency, while multiply and divide are multiple-cycle operations. Therefore, issuing one of these two long-latency instructions causes the thread to switch.

The memory-access stage is handled through the load/store unit. It contains the Dcache, DTLB, and some buffers to save the store instructions. The unit features four store buffers, one per thread, with eight available entries each. The load/store unit also includes logic to check the store instructions for possible read-after-write (RAW) hazards with following loads. In the case where a RAW hazard is detected, the store buffer bypasses the data to the load instruction for write-back into the register file.

3.7   Core Interconnection

As already mentioned in Section 3.5, the Niagara interconnect architecture is based on a full crossbar, which connects the multithreaded SPARC cores, the L2 banks, and the shared I/O interfaces. The operating frequency is 1.2 GHz, and the peak overall physical bandwidth results in more than 200 GB/s, with a data bandwidth of 134.4 GB/s. The interconnect architecture is pipelined in three stages: implementing request, arbitration, and transmission. Two mirrored interfaces are implemented, the processor-to-cache (PCX) and the cache-to-processor (CPX).

Every possible source-destination pair has a dedicated full-duplex, input-queuing hardware support with two entries. The system is therefore able to support a total of 192 contemporary transactions (eight cores, four L2 cache banks, an FPU unit, I/O interface, all full-duplex). The arbitration policy is based on an aging mechanism in order to provide fairness and reduce the latency variability.

3.8   Memory Subsystem

This section presents the memory and cache hierarchy of the Niagara processor. Every SPARC core in the Niagara chip includes a dedicated 16-KB L1 instruction cache and 8-KB L1 data cache. The L1 instruction cache is shared among all the four threads per core and implements a 4-way set associativity with a random replacement policy. The L1 data cache also implements a 4-way set associativity, but the lines are half the width of the instruction cache lines (16 B instead of 32 B).

By including such small L1 caches, the Niagara architects have made a clear design choice. For all applications that use large data sets (e.g., commercial server applications), instead of relying on large caches to minimize the miss rate, the designers decided to rely on the single-core multithreading support to effectively hide the L1 cache miss latency.

The 3-MB shared L2 cache is 12-way set associative and is interleaved on four different physical banks with a 64-B granularity. Two interfaces per bank are included toward the external DRAM controller, providing an aggregate bandwidth of 20 GB/s. A 128-bit interface is used for cache line fill, and a 64-bit interface is used for memory writing. The L2 cache subsystem includes SRAM modules for data and tag storing, a CAM-based directory to store the L1 tags for coherence, and a register file to store the Valid, Used, Allocate and Dirty bits. The L2 is physically organized in an 8-stage pipeline. The performed operations are

- access to the tag SRAM and comparison to the requested address to generate the way-selection signal (stage 1),

- access to the data SRAM with way, address, data, and control signals (stages 2–3),

- reading and transmitting the data from the SRAM (stages 4–6),

- error correction (stage 7),

- sending the data to the crossbar (stage 8).

3.8.1   Cache-Coherence Protocol

The cache coherency protocol implements a directory-based mechanism. The L1 caches are write-through and not-allocate; therefore, they do not fetch the data on write misses. When a load miss occurs in the L1 cache of any core, the L2 CAM-based directory is accessed to get the missing cache line and properly updated to maintain a list of the cores that share the accessed data. One of the key aspects of the Niagara cache subsystem is that coherency and cache invalidations are completely performed in the shared L2 cache. For this reason, no snooping algorithms need to be implemented, and a single global address space may be easily exposed to the operating system. The single L2, in turn, implements write-back to minimize access to the off-chip memory.

3.8.1.1   Example 1

We now show, with a first example, the working principle of the directory-based coherency protocol. Let us assume that a first core performs a load instructions that generates a miss in the L1 cache. In this case, a transaction is generated in the pipeline memory unit and enqueued in the crossbar. The crossbar logic performs address decoding and routes the load to the appropriate L2 bank. The L2 loads the data from memory (we assume here an L2 miss happens), caches the line, sends the data back to the core, and creates an entry in the directory adding the requesting core as an owner. The data are now also in the L1 cache of the requesting core.

If a different core now issues another load instruction for the same cache line, its transaction will traverse the crossbar and arrive at the proper L2 cache bank. The transaction will now generate a hit in the L2 cache (therefore, no off-chip memory access is performed), and the data are returned to the requesting core. The L2 cache logic will now add the requesting core as a new owner of the same hit cache line. That cache line will now have two owners.

Let us assume that one of the two cores now generates a store for the same cache line. Because the L1 is a write-through cache, the transaction will be forwarded to the L2 cache bank. Once there, the directory logic will generate cache invalidations for all the owner cores and their L1 caches. This way, cache coherency control is centralized to the shared L2 and maintained without the need of a snooping algorithm. Finally, since the L2 cache implements a write-back mechanism, the data to be stored in memory will be actually transferred through one of the two DDR controllers attached to the proper bank later.

3.8.1.2   Example 2

In this second example, let us assume that, as before, a thread from a core issues a load instruction that generates an L1 cache miss and a subsequent L2 cache miss. Therefore, the data will be loaded from the off-chip memory into the L2 cache, then to the L1 cache of the requesting core. At the same time, a directory entry will be created for the cached data, adding the requesting core as an owner. If now a different thread from the same core generates a load for the same cache line, it will experience an L1 cache hit; therefore, the L2 cache will not be affected.

Let us assume that the second thread that loaded the data now issues a store for that same cache line. Its L1 will forward the data directly to the L2 cache, implementing the write-through policy. Once at the L2 cache, the transaction will hit the directory entry already created for the two previous loads. The store will then invalidate the L1 cache copy of the core. Therefore, even if the two threads contending for the same data belong to the same core, the L1 still behaves as a write-through cache, and the global visibility of the store is attained only through access to the L2 cache.

3.9   Niagara Evolutions

The Niagara processor architecture has had different successors. This section briefly highlights the main features and improvements of the most famous processor evolutions.

In 2007, Sun Microsystems released the UltraSPARC T2 processor, also known as Niagara 2 [8]. The Niagara 2 processor resembled its predecessor multicore architecture. The most significant improvement over Niagara is that the total number of threads that can be simultaneously supported increased from 32 to 64. In fact, while the number of cores per processor was kept at eight, each core was able to handle eight thread contexts, as opposed to the four of Niagara. Moreover, the single-thread performance was improved by a factor of 1.4 for the integer data pipeline and a factor of 5 for the floating-point data pipeline. The main contributions to these speed-ups are as follows:

- The operating frequency was raised from 1.2 to 1.6 GHz. Two stages were added to the pipeline.

- Instruction fetch was improved by letting each thread have its own instruction buffer.

- Each core included two execution units instead of one. Each one of the two execution units is shared by a group of four statically assigned threads.

- Each core has its own floating-point execution unit, while Niagara included a single unit shared among all the cores of the chip.

- Size, number of banks, and associativity of the L2 cache were changed to match the increased number of threads. Niagara 2 features a 4-MB, 16-way associative L2 cache organized in eight banks.

In 2010, the natural successor of the UltraSPARC T2 processor was announced. Known under the codename Rainbow Falls, the SPARC T3 processor improved the Niagara 2 performance by doubling the number of total threads from 64 to 128 [9]. This high thread count is essentially achieved by increasing the per-chip core count from 8 to 16, while maintaining the single-core organization and number of supported threads to 8. Because it was still designed for high throughput performance, other key design improvements were made to the number of 10 Gigabit Ethernet ports included in the chip (from one to two) and the number and functionality of cores for cryptography algorithms acceleration. Moreover, the L2 cache size was increased to 6 MB, and the memory interface was based on DDR3 instead of the Niagara2 DDR2.

Bibliography

[1] Inc. SPARC International. The SPARC Architecture Manual—Version 8. Technical report, 1990.

[2] Inc. SPARC International. The SPARC Architecture Manual—Version 9. Technical report, 1993.

[3] SUN MicroSystems and Fujitsu Limited. SPARC Joint Programming Specification (JPS1): Commonality. Technical report, 2002.

[4] Inc. SUN MicroSystems. UltraSPARC Architecture 2005—Draft D0.8.7. Technical report, 2006.

[5] Inc. SUN MicroSystems. UltraSPARC Architecture 2007—One Architecture ... Multiple Innovative Implementations—Draft D0.9.3b. Technical report, 2009.

[6] Top500 list of the world's most powerful supercomputers. http://www.top500.org, 2013.

[7] P. Kongetira, K. Aingaran, and K. Olukotun. Niagara: A 32-way multithreaded sparc processor. *Micro, IEEE*, 25(2):21–29, March–April 2005.

[8] Tim Johnson and Umesh Nawathe. An 8-core, 64-thread, 64-bit power efficient SPARC SOC (Niagara2). In *Proceedings of the 2007 International Symposium on Physical Design*, ISPD '07, pages 2–2, New York, 2007. ACM.

[9] J.L. Shin, K. Tam, D. Huang, B. Petrick, H. Pham, Changku Hwang, Hongping Li, A. Smith, T. Johnson, F. Schumacher, D. Greenhill, A.S. Leon, and A. Strong. A 40nm 16-core 128-thread CMT SPARC SoC processor. In *Solid-State Circuits Conference Digest of Technical Papers (ISSCC), 2010 IEEE International*, pages 98–99, Feb. 2010.

Chapter 4

The Cilk and Cilk++ Programming Languages

Hans Vandierendonck

Queens' University Belfast

4.1 Introduction

The Cilk language provides a simple parallel programming model that is a natural extension of the C language. The philosophy behind the Cilk language is that the programmer should not be concerned with runtime scheduling decisions in order to obtain good scheduling and load balancing. On the contrary, the task of the programmer is to construct a correct and functional program

```
cilk int fib(int n) {
  int a, b;

  if( n < 2 ) {
    return n;
  } else {
    a = spawn fib(n-1);
    b = spawn fib(n-2);
    sync;
    return a+b;
  }
}
```

LISTING 4.1: Cilk version of Fibonacci program.

where a large amount of parallelism has been identified. It is the responsability of the runtime scheduler to map parallel tasks to the processors of the system.

The power and ease-of-use of Cilk were demonstrated by implementing several real programs in the language. In particular, several world-class chess playing programs were developed, of which ⋆Socrates obtained a second place in the 1995 World Computer Chess Championship by running on a 1824-node Intel Paragon MPP.

The model of parallelism implemented in Cilk is sometimes referred to as *fork-join* parallelism. This term refers to the property that at some point during the execution, a main thread may fork off a second computational thread at a spawn point. Both threads may be executing in parallel until they join at a synchronization point.

The prototypical illustration of these concepts follows from parallelizing a program that computes the nth number in the sequence of Fibonacci numbers, shown in Listing 4.1.[1] The Fibonnaci sequence starts with the numbers 0 and 1 (case n < 2). After that, every number equals the sum of the previous two numbers (else case, utilizing recursion).

The Cilk language introduces minimal overhead in this code fragment. Only a few keywords are added: cilk, spawn, and sync. More details about these keywords are provided later in the chapter. For now, it suffices to know that spawn indicates that the called procedure may execute in parallel with the parent procedure and that sync forces the parent to wait untill all child procedures have finished execution.

The Cilk language has become synonymous with its scheduler, which orchestrates the parallel execution of procedures. The Cilk scheduler achieves at most a linear overhead in execution time and stack-space consumption over a serial execution. Linearity here means that the overhead grows linearly with the number of processors involved in the computation. This efficiency

---

[1] Although this is surely not the most efficient algorithm to compute the Fibonacci numbers, the example serves well to illustrate how Cilk works.

is obtained because of the work-first principle: the Cilk scheduler will sooner execute a spawned procedure *sequentially* than add it to a waiting list of procedures to execute.

Internally, Cilk divides procedures into *strands*. Each strand is a maximal sequence of instructions between spawn and sync statements. It is important to know that strands are the unit of scheduling in Cilk. Moreover, execution of a strand will never block: once the strand starts executing, it will not be unscheduled.

Cilk targets foremost shared memory systems where all processors have access to the same main memory. Cilk has, however, also been ported to distributed shared memory systems. In such systems, each processor (or small group of processors) has its own private main memory, but a runtime system presents the same view on memory for all processors. This chapter further considers shared memory systems only.

Cilk was developed under the supervision of Prof. Charles E. Leiserson at the MIT Laboratory for Computer Science. Cilk was subsequently licensed to the Cilk Arts startup that implemented the Cilk language extensions and scheduler in C++. In 2009, Intel acquired Cilk Arts and now sells Cilk under the name Intel Cilk Plus.[2] My experience with the Cilk language and runtime is to implement it from scratch as a C++ library and propose data flow extensions [11].

In the remainder of this chapter, we will first explain the Cilk language 4.2. Then, we will discuss how this language is implemented, touching upon the design of the Cilk scheduler in Section 4.3. We discuss how to measure and analyze the performance of Cilk programs in Section 4.4. Then, we discuss hyperobjects, an advanced mechanism for reconciling the use of global variables with parallel programs in Section 4.5. Finally, we make some closing remarks and point to references for further reading in Section 4.6.

4.2 The Cilk Language

4.2.1 Spawning and Syncing

The *cilk* keyword indicates that a procedure is a Cilk procedure. Cilk procedures are similar to C procedures except that they may spawn other Cilk procedures and synchronize with them. Cilk procedures have an argument list and a return value just like C procedures.

The *spawn* keyword is used to call a Cilk procedure. Just like calling a C procedure, control flow of a spawn continues in the child. But while in C the parent blocks until the child is finished and returns, in Cilk the parent may continue execution in parallel while the child is executing. This way,

---

[2]Intel® and Cilk™ Plus are trademarks of Intel Corporation in the U.S. and/or other countries.

parallelism is created dynamically. In fact, the parent may continue to spawn other children, resulting in potentially large degrees of parallelism.

Cilk procedures may return values, in which case the return value must be assigned to a local variable of the parent. Cilk procedure calls may not be used arbitrarily in expressions as is the case in C syntax. For instance, the statement

```
r = spawn fib(n-1) + spawn fib(n-2);
```

is illegal in Cilk.

A spawned child procedure will have some useful side effect such as returning a value or modifying memory. The parent has to make sure that these side effects have been fully executed before continuing dependent computations. This is possible with the *sync* keyword, which instructs the parent procedure to wait until all of its children have completed. In the Fibonacci example, we see that a sync statement preceeds the return statement that uses the return values of the spawned children. Without the sync statement, the returned value would take on unpredictable values.

4.2.2   Receiving Return Values: Inlets

Cilk provides two more keywords that are often useful: *inlet* and *abort*. The basic way of returning a value from a child is by assigning the value to a local variable. Sometimes it is necessary, however, to use more advanced code to incorporate the value returned by the child into the state of the parent. Hereto, Cilk provides *inlet* procedures. An inlet is a special procedure defined within the scope of a Cilk procedure. Normally, Cilk does not allow procedure spawns inside expressions, but an exception is made in case a spawn occurs as an argument to an inlet.

Listing 4.2 shows the Fibonacci example using inlets. The inlet summer() takes a return value as argument and adds it to the variable x in the parent procedure. Note that all local variables in fib() are available to summer() as it is a local (nested) procedure.

The inlet is called after the spawned child returns. However, it can happen that the inlet is accessing local variables at the same time as the parent, or other inlets are accessing those variables. This will never cause data races because Cilk guarantees that the strands of a procedure instance, including its inlets, operate atomically with respect to each other. In contrast, Cilk makes no guarantees about execution order between the strands of different procedure instances except that the ordering of strands implied by the spawn and sync statements is maintained.

Note that the syntax for updating variables with return values of spawned procedures are internally translated into inlets. For instance, Cilk generates a specialized inlet for the following statement:

```
x += spawn fib(n-1);
```

These inlets are called implicit inlets.

```
cilk int fib(int n) {
  int x;
  inlet void summer(int result) {
    x += result;
    return;
  }

  if( n < 2 ) {
    return n;
  } else {
    summer(spawn fib(n-1));
    summer(spawn fib(n-2));
    sync;
    return x;
  }
}
```

LISTING 4.2: The Fibonacci program using inlets.

4.2.3   Aborting Threads

Some algorithms are structured such that parallel work may be spawned off such that the completion of one work item makes the remaining work redundant. Cancelling the redundant work is possible using the *abort* statement. A typical use of abort is in parallel search algorithms where each processor searches through a different part of the search space. As soon as one processor has found the desired item, then there is no point in searching the remainder of the search space. It is then best to cancel all extant work as quickly as possible in order to allow the parent procedure to return.

Upon executing the abort statement, all of the already-spawned children of the procedure are terminated. Termination may not happen immediately because processors are not constantly communicating. As such, a child may still terminate normally after an abort statement is executed.

One should, however, be very careful when using abort statements because Cilk makes no guarantees whatsoever about the aborted children: there are no special return values to indicate abortion, and no efforts are made to clean up the state that a child constructed, such as heap-allocated data, locks, etc. Furthermore, it is important to be aware that abort applies only to children that are executing when the abort statement is executed. Children that have not yet been spawned are not aborted, so they will still be spawned. In order to stop these children from spawning, the programmer can set a flag in the procedure to remember that abortion has occurred and test this flag before spawn statements. Note that race conditions on this flag are impossible because all of the strands in the same procedure instance including its inlets are executed atomically.

```
#define CUTOFF 1024
cilk int is_sorted( int * A, int N ) {
  if( N < CUTOFF ) {
    int i;
    for( i=1; i < N; ++i ) {
      if( A[i-1] > A[i] )
        return 0;
    }
    return 1;
  } else {
    int sorted = 1;
    inlet void abort_if_fails( int is_sorted ) {
      if( !is_sorted ) {
        sorted = 0;
        abort;
      }
    }
    if( A[N/2-1] > A[N/2] )
      return 0;
    abort_if_fails( spawn is_sorted( A, N/2 ) );
    if( sorted )
      abort_if_fails( spawn is_sorted( &A[N/2], N-N/2 ) );
    sync;
    return sorted;
  }
}
```

LISTING 4.3: Example of abort keyword—verifying if an array is sorted.

Listing 4.3 shows an example of the abort keyword in a procedure that verifies whether an array is sorted. This verification is implemented in a recursive fashion. The array is recursively broken in two parts, and we test that the first part is sorted before the second part and that both parts are sorted by themselves. If any part is found to be not sorted, then verification of the other part is canceled. Note that the variable **sorted** tracks whether all parts of the array are sorted. It is also reused as a flag to avoid spawning the second child if the array is not sorted. This one test can make an order-of-magnitude performance difference for very large arrays that are nearly sorted.

4.2.4   The C Elision

Cilk is an extension of the C language that respects the semantics of C. As such, it is possible to easily revert Cilk programs to sequential programs by means of the C elision, i.e., to elide (or remove) the Cilk-specific keywords. In this way, it is possible to scale down parallel programs to sequential programs without changes to the source code. This is useful when debugging programs especially when bugs appear nondeterministically.

4.2.5   Cilk++

The Cilk++ implementation of Cilk follows exactly the same philosophy as Cilk, although some changes to the language are applied. The most visible of these changes are the following.

- To minimize the probability of name clashes, the spawn and sync keywords were replaced by `cilk_spawn` and `cilk_sync`. Also, the keyword `cilk_for` is introduced to indicate a counted parallel for loop. The keyword does not extend the expressiveness of the language, but it allows the compiler to optimize the spawning structure of the code. The nature of this optimization is dicussed in Section 4.3.7.

- In Cilk++, it is no longer necessary to add the cilk keyword to every procedure that may be spawned. The Cilk++ compiler infers these keywords itself. Moreover, Cilk++, uses distinct linkage rules for C++ and Cilk++ procedures to further differentiate between parallel and sequential code. Moreover, the keyword `cilk_run` is introduced to launch parallel code from within sequential code, i.e., to switch the program from C++ mode to Cilk++ mode.

- Cilk++ does not provide inlets or the abort statement. On the other hand, Cilk++ does provide powerfull alternatives to inlets that facilitate merging return values from spawned procedures in the form of hyperobjects. Hyperobjects allow us to do much more than merging return values. They are described in Section 4.5.

4.3   Implementation

The implementation of the Cilk language has seen major changes over the years. Prior versions, such as Cilk-3, scheduled strands in breadth-first order [2]. Strands that are independent might easily have executed in an order that differed from the order they were listed in the sequential program. This could happen even in the absence of work stealing, i.e., in a sequential execution. The Cilk-5 scheduler introduced the work-first principle [6]. This principle and the design of the Cilk-5 scheduler are described in this chapter.

4.3.1   The Cilk Model of Computation

Cilk divides a multithreaded procedure into strands where each strand is a maximal sequence of instructions between spawn and sync statements. Dependencies between strands are represented in a directed acyclic graph (DAG). In Figure 4.1, rectangular boxes represent procedures, and circles represent

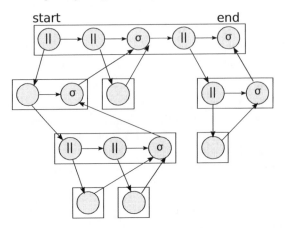

FIGURE 4.1: DAG view of a parallel program. The symbol "——" indicates creation of parallelism (spawn) and "σ" indicates synchronisation (sync).

strands. Vertices represent dependences between strands. Each strand has a successor strand in the same procedure and is represented by a horizontal vertex. Threads may spawn procedures, which is represented by a downwards vertex. Synchronization points are represented by upwards vertices.

The DAG limits the execution order of strands, as a strand can never start execution before all of the strands that it depends on have finished. The execution order is a partial order, which implies that there are multiple distinct ways of scheduling strands. The DAG is never explicitly stored. Rather it unfolds dynamically during execution. It is the task of the runtime scheduler to select which strands to execute next and thus what part of the DAG to extend next. The number of active procedures in the DAG determines the amount of state, and the number of dependencies between strands determines communication. As such, different scheduling decisions result in different time and space overheads. The Cilk scheduler, and by extension also the language, are designed such that both time and space overhead are constant compared to sequential execution.

4.3.2   Cactus Stacks

Sequential programs use a *linear stack* to organize procedure calls. The stack tracks control flow between procedures, namely, the order in which one procedure calls another one, and at the same time, the stack stores local variables for each procedure. Maintaining a linear stack is fairly simple: a *stack pointer* keeps track of the lowest occupied element on the stack (stacks typically grow "downwards," i.e., to lower addresses). The stack pointer is lowered on procedure calls to reserve space for local variables and return addresses. It is raised again when a procedure returns. Thus, the stack frames of parent and child procedures are placed adjacently in memory, and all (sequentially called) child procedures of a parent overwrite the same area on the stack.

Linear stacks are insufficient to execute Cilk programs because multiple child procedures may be executing concurrently. To solve this problem, Cilk uses a cactus stack. In a cactus stack, each procedure's stack frame is a contiguous memory area, but the stack frames of parent and child procedures are not necessarily placed in contiguous memory locations. Consequently, explicit pointers are stored in stack frames to link the stack frame of a child procedure to that of its parent. These links, however, allow each procedure to maintain the impression that it is operating on a linear stack. Consider the execution of a procedure A that spawns procedures B and C. In turn, procedure B spawns D and E, respectively (Figure 4.2). Each of the five spawned procedures is given the impression that it executes on a linear stack, but in reality the implemented stack is nonlinear.

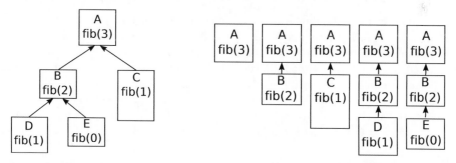

FIGURE 4.2: A cactus stack: Implementation with a tree (left) and each procedure's logical view of the stack (right).

Cactus stacks enforce the same restrictions as linear stacks. For instance, child procedures cannot pass pointers to their local variables to their parent, and sibling procedures cannot see each others' local variables. It is, however, possible to pass pointers to local variables to child procedures.

The implementation of cactus stacks is less efficient than the implementation of a linear stack. Because the cactus stack must be able to split at every procedure call, a common implementation idiom is to dynamically allocate a frame for each procedure call and to insert a link in a child frame to its parent frame in order to obtain the cactus structure.

4.3.3 Scheduling by Work Stealing

The Cilk scheduler is a work-stealing scheduler, which means that each processor maintains its own pool of strands to execute. When the pool runs empty, the processor will steal strands from another processors' pool. As such, the scheduling of strands across processors is demand-driven: processors communicate only when a pool runs empty. Furthermore, the Cilk scheduler steals strands in the outermost called procedures in order to increase the likelihood that a steal brings a large amount of work to a processor. Finally, the Cilk scheduler employs the work-first principle, meaning that it preferentially executes spawned strands over queueing up ready strands for later execution. These principles help to reduce the overhead of scheduling and communication.

The work-first principle enables a significant optimization that underlies much of the design of Cilk: in the absence of work stealing, execution proceeds serially, so spawns will appear as normal function calls, and sync statements will appear as no-ops. Therefore, Cilk is designed such that the common path, which is serial execution, is heavily optimized and that steals, which should be rare, are more expensive.

4.3.4   Runtime Data Structures

The central structures are the per-processor strand pools. Each strand pool is organized as a deque (double-ended queue) of call stacks. Figure 4.3 illustrates such a spawn deque. Each call stack in turn consists of stack frames. Each call stack is assigned a level in the deque. Call stacks are spread across levels according to the difference between procedure spawns and procedure calls: spawned frames are always inserted at a new level while call frames are inserted at the same level as the parent frame. Each stack frame also contains a pointer to its parent frame.

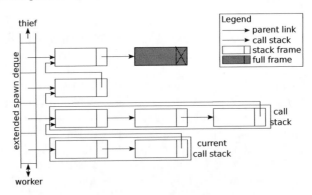

FIGURE 4.3: The extended spawn deque stores the stack frames of the executing program and organizes them by call stacks.

When considering a single spawn deque, the operation of the deque is not all that different from a linear stack. Because of the work-first principle, execution proceeds in a serial manner: stack frames are pushed at calls and spawns, and they are popped at returns. Calls and spawns have an effect on the creation and destruction of new call stacks, but when taking everything together, all frames in an extended deque form a linear stack, albeit not implemented in contiguous memory.

Each worker also has a current call stack, which contains the currently executing procedure's stack frame. The combination of spawn deque and current call stack is also refered to as the *extended spawn deque* for convenience.

All stack frames together are organized in a rooted tree by their parent links. This rooted tree implements the cactus stack. Figure 4.4 shows how the cactus stack is represented by the runtime system. Each spawn deque contains a linear leg of the cactus stack. Not all stack frames are necessarily contained in an extended spawn deque; some stack frames may live outside the extended spawn deques. These stack frames are suspended. They are blocking on a call operation, or they are waiting in a sync operation. Execution of these frames may continue when at least one of their children has finished execution.

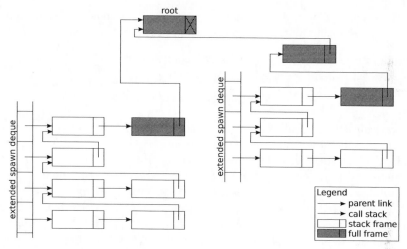

FIGURE 4.4: Runtime data structures of the schedulers. Two extended spawn deques are depicted, assuming two threads.

It is typical for the Cilk approach that the cactus stack is very thin, featuring a number of branches near the top that flow over in linear segments. Furthermore, the stack frames where branching occurs are either the oldest stack frame in a spawn deque, or they lie outside any spawn deque.

When a worker's extended deque is empty, then it will attempt work stealing. Work stealing is possible only from the head of the spawn deque. The spawn deques are accessed differently by the worker and the thiefs: the worker can insert and remove call stacks from its end of the deque, but other workers can only remove call stacks from the other end of the deque when work stealing. Work stealing always occurs in units of call stacks. The reasoning behind this is that all frames in a call stack except the oldest one are created by a call. As calls do not allow for parallel execution, there is no point in stealing a caller because it is blocking anyway. The youngest frame in the call stack has performed a spawn, and thus it makes sense to steal it and continue execution in this frame. The older frames in the call stack are stolen with it because this simplifies the design: each extended spawn deque contains a linear leg of the cactus stack starting at some spawn up to a leaf node of the cactus stack.

The stack frames are used by the runtime system to store return addresses after calls, local variables for procedures, function arguments, etc. In a multithreading system, the stack frames require additional fields to keep track of multiple executing children:

1. **parent link**: A link to the parent stack frame. These links link all stack frames in a cactus stack.

2. **lock**: A lock to guard against races when multiple workers attempt to access the same stack frame.

3. **continuation**: A continuation identifies where to resume execution after a suspension point, where to leave a procedures return value, etc.

4. **join counter**: A join counter to count the number of outstanding children. This is essential for implementing sync.

5. **children**: A doubly-linked list of child frames.

Because work stealing is, in principle, rare and execution proceeds serially as much as possible, most frames require only the parent link and continuation. The Cilk runtime system therefore makes a distinction between two types of frames: stack frames containing the parent link and the continuation and full frames containing all attributes necessary for multithreaded synchronization.

In principle, only the frames with parents and/or children that are involved in parallel execution are full frames. All other frames are stack frames. Full frames are depicted in Figure 4.4 by shaded rectangles. It can be observed that all frames in an extended deque except the oldest frame are stack frames because they have at most one child. The oldest frame in an extended deque may be accessed by multiple threads. As such, it requires a lock and must be a full frame. Frames that do not belong to any extended deque are also full frames because they may have multiple outstanding children.

Full frames are the top-level frames in the cactus stack: all ancestors of a full frame are full frames, and all descendants of a stack frame are stack frames. Furthermore, Cilk exploits the parallelism encoded by spawn statements only in full frames. Spawns executed in stack frames are executed as if they are sequential calls. These frames are, however, encoded with some overhead in the spawn deque in order to allow them to be stolen. Only then will their inherent parallelism be exploited.

One can also observe that Cilk exploits the parallelism present in the outermost levels of the program. This generally yields coarser-grain parallelism than in the inner levels.

4.3.5 Scheduling Algorithm

On program startup, each worker starts off with an empty deque. On its deque, one worker pushes a call stack with a stack frame for the main procedure of the program. Every worker now checks if its deque is empty. If so, then the worker will try to steal stack frames from another worker. If not, the worker executes the stack frame at the highest level in its deque. It performs the following operations at spawn, sync, function call, and return:

- **call**: When a procedure instance A calls a procedure instance B, the continuation in A is set such that the execution of A resumes at the instruction following the call to B. This is principally the same as pushing a return address on a linear stack. The worker then allocates a stack frame for B and pushes it on the current call stack as a child of A. Execution proceeds with B.

- **spawn**: When a procedure instance A spawns a procedure instance B, the continuation in A is set such that execution of A resumes at the instruction following the spawn statement. The worker then allocates a stack frame for B and pushes the current call stack on the tail of its deque. It starts a new current call stack and pushes B onto this stack. Execution then proceeds with B.

- **return from call**: It is necessary to distinguish between the case of returning from a stack frame and returning from a full frame. If control flow returns from a stack frame A, then the worker pops A from the current call stack and resumes execution of A at its stored continuation.

  If control flow returns from a full frame A, then the worker pops A from its current call stack. Both the current call stack and the spawn deque are now empty (because only the oldest frame on a spawn deque may be full). The worker is now left without work, and it must try to steal work. The Cilk scheduling policy is to perform an *unconditional steal* of the parent frame. An unconditional steal, defined below, is a steal that is always successful.

- **return from spawn**: Again we distinguish between stack frames and full frames. When returning from a stack frame A, the worker pops A from the current call stack. The current call stack is now empty, so the worker tries to pop a call stack from the tail of its deque. If the pop succeeds, then the popped call stack becomes the current call stacks, and execution continues from the continuation of A's parent. If the pop fails, then the spawn deque was empty, and the worker performs *random work stealing*. Random work stealing, defined below, is an attempt to steal work from another worker's spawn deque.

  If control flow returns from a full frame A, then the worker pops A from the current call stack, and both the current call stack and the spawn deque are now empty. The worker performs a *provably good steal* of the parent frame.

- **sync**: Sync operations are no-ops on stack frames because all spawned children have finished execution before reaching the sync statement by construction of the scheduler.

  For a sync on a full frame A, pop A from the current call stack. The current call stack and spawn deque are now empty. The worker now performs a *provably good steal* of A.

The Cilk scheduler uses three different work-stealing actions. The baseline work-stealing mechanism is *random work-stealing*. The other two mechanisms are specializations to specific scheduler states.

- **random work stealing**: A worker that is idle tries to advance the execution of the program by stealing work from another worker's spawn deque. The worker that attempts to steal is the *thief*. The thief selects a *victim* worker at random, making sure that the victim has a nonempty spawn deque. The thief then steals the oldest call stack from the victim's deque and will attempt to execute this call stack itself. Hereto, it turns every stack frame on that call stack into a full frame, which involves incrementing the join counter and inserting each frame in its parent's list of children.

  The youngest frame on the stolen call stack is now inserted in the thief's current call stack. The thief resumes execution at the continuation of this frame. The remaining frames are now suspended and live outside any spawn deque. Because each of these frames issued a call statement to create the child on the stolen call stack, these frames will be reconsidered for execution when execution of the youngest frame finishes, in which case, the worker will return from a call statement with a full frame and will perform an unconditional steal of the parent frame.

- **provably good steal**: A provably good steal is a work-stealing action that guarantees progress on a leaf frame. By always executing leaf frames first, it is possible to achieve worst-case linear overheads in execution time and stack space. A provably good steal of a frame A may occur only when A is a full frame and when the worker's extended deque is empty. If the join counter of A is zero and no other worker is working on A, then execution of frame A is resumed. Otherwise, it is not yet possible to make progress on A, or A is already executed. The worker then searches different work by performing random work stealing.

- **unconditional steal**: An unconditional steal of a frame A happens when returning from a call with a full frame. In this case, frame A is blocked until the call returns, so we know that no other worker could be executing A. The worker will thus steal frame A without further conditions, insert A in its current call stack and resume execution at the continuation stored in A.

4.3.6   Program Code Specialization

Cilk further optimizes the execution of a program by executing different versions of the code in stack frames and full frames. The rationale is that stack frames are executed (mostly) sequentially, and thus little overhead should be paid to overheads for multithreaded execution. Full frames on the other hand are directly involved in multithreaded execution, and their bookkeeping overhead is larger.

The Cilk compiler generates two versions of every Cilk procedure: a fast nano-scheduled version and a slower micro-scheduled version. The nano-scheduled version makes all possible assumptions valid for sequential execution: spawn statements are translated to conventional calls, and sync statements are treated as no-ops. The nano-scheduled version must however perform some bookkeeping to allow work stealing. In particular, it must allocate the cactus stack and maintain the spawn deque. Furthermore, it must make sure that all variables that are live across spawn and call statements are stored in the cactus stack, such that they can be properly restored when stealing a frame.

The micro-scheduled version of a procedure is used when resuming full frames. As such, the procedure will be called from a generic interface, changing the signature to accept only a pointer to its stack frame. Furthermore, the micro-scheduled version is modified to allow jumping to an indicated suspension point and to restore local variables after that. Obviously, code generation of the nano-scheduled and micro-scheduled versions of the procedures is strongly intertwined.

4.3.7   Efficient Multi-way Fork

A single spawn statement has the potential to increase the amount of parallelism in a program by at most one active strand. If the amount of parallelism is high at some point of the execution, e.g., when entering a DOALL loop, then many tasks should be spawned as quickly as possible.

Listing 4.4 shows a straightforward approach to implementing a DOALL loop using spawn/sync statements. A control loop iterates over all iterations and spawns a procedure consisting of the actual loop body. The control loop is followed by a sync statement to wait for all iterations to finish. This construction is correct but does not give the best performance because the spawn statements are executed sequentially. To understand why this is not good, we must analyze the work-stealing activities that are undertaken by the scheduler to exploit parallelism. We assume in this analysis that there is no nested parallelism.

Initially, one worker is executing the sequential code and the other workers are idle. The busy worker reaches the control loop and encounters the spawn statement in the first iteration. It now pushes a new call stack on its deque that evaluates the procedure body on iteration number 0. It starts executing

```
cilk void body(int i) {
  // Code for loop body
}

cilk void doall() {
  int i;

  for(i=0; i < N; ++i) {
    spawn body(i);
  }
  sync;
}
```

LISTING 4.4: Sequential spawns.

this procedure, and in the mean time, another worker can succeed in stealing the frame of the `doall` procedure. The second worker continues executing this procedure and reaches the spawn statement for the second iteration. It now likewise pushes a new call stack on its deque and starts executing the procedure `body` on iteration number 1. This allows another idle worker to steal the frame of the `doall` procedure, and the process continues in the same fashion.

The problem with this spawning approach is that only one worker at a time can steal the frame executing the control loop, and thus only one spawn is executed at any moment. It takes N steals before N strands can be executing.

A better approach is presented in Listing 4.5. Here, the domain of the DOALL loop is recursively split into two halves. As such, two strands are active after the first spawn statement, one iterating over the lower half of the domain and the other iterating over the upper half of the domain. Both these strands further divide their domain by recursing the procedure `split`. Thereby a new call stack is pushed on their spawn deque, allowing two other workers to steal a frame and become busy. In the next step, four workers will have stealable call stacks in their spawn deque, allowing four more workers to become nonidle. In short, it still takes N steals before N strands participate in the execution, but these steals can occur in parallel, giving a perceived delay of $O(log_2(N))$ steal operations.

Writing code this way is tedious, so the Cilk++ compiler provides the `cilk_for` keyword to describe DOALL loops and generates code like this on behalf of the programmer.

```
cilk void body(int i) {
  // Code for loop body
}

cilk void split(int lo, int hi) {
  if( lo+1 < hi ) {
    spawn split( lo, (hi-lo)/2 );
    split( (hi-lo)/2, hi );
    sync;
  } else {
    body( lo );
  }
}

cilk void doall() {
  spawn split( 0, N/2 );
  split( N/2, N );
  sync;
}
```

LISTING 4.5: Parallel spawns.

4.4 Analyzing Parallelism in Cilk Programs

The DAG model of computation is a handy means to estimate the amount of parallelism in a Cilk program, to predict potential speedups and for performance tuning. The model involves three parameters that describe the program: T_1, T_∞, and P. The parameter T_1 corresponds to the amount of work performed by a Cilk program and is measured by the serial execution time of the program on a single processor. The parameter T_∞ is the critical path length of the DAG. It is measured by timing all strands but does not include scheduling overheads or communication costs. The number of processors is denoted by P.

When executing the Cilk program on P processors, we measure the execution time T_P including scheduling and communication costs. We can make the following observations on T_P.

1. A parallel execution on P processors can never be faster than the critical path length in the DAG: $T_P \geq T_\infty$.

2. A parallel execution can never be more than P times faster than the sequential execution: $T_P \geq T_1/P$.

Consequently, these equations set an upper bound on the parallel execution time T_P. It is therefore possible to estimate how well a Cilk program performs. Figure 4.5 shows these bounds for the speedup $S = T_1/T_P$, where the bounds read as $S \leq T_1/T_\infty$ and $S \leq P$. Thus, speedup is bounded by a curve consisting of two straight line segments. The ratio T_1/T_∞ is the average parallelism in the Cilk program, so the bound $S \leq T_1/T_\infty$ says that the average speedup can never exceed the average parallelism. The other straight line segment indicates that speedup can never exceed the number of processors.

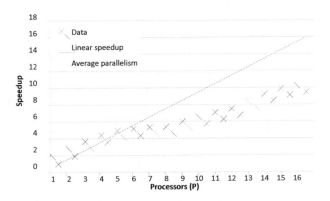

FIGURE 4.5: Speedup limitations for a Cilk program.

When a Cilk program suffers from less than desirable performance, this analysis allows the developer to analyze whether the program requires more processors to see a better speedup or whether it is the parallelism in the program that is an inherent limitation.

The speedup model has some limitations, as it assumes that the serial and parallel versions of the code perform exactly the same amount of work. This is not true for speculatively parallel programs that abort strands or for programs that allow sibling strands to update shared variables. In both cases, there is the potential of speeding up a strand by using values that are "in the future" of the computation for the corresponding strand in the C elision. Finally, secondary effects can influence the individual execution times of the strands, most notably cache effects.

The effectiveness of the Cilk scheduler can be read from the near-perfect scalability of programs. In particular, it has been shown that the runtime of a program on P processors can be accurately modeled as

$$T_P \approx T_1/P + c_\infty T_\infty$$

where the constant $c_\infty \approx 1.5$ [2]. The component T_1/P of the execution time model is the desired *perfect linear speedup* that any scheduler should approximate. The Cilk scheduler accrues some overhead that is shown to depend only on the critical path length. When $T_\infty \ll T_1/P$, then near-perfect linear speedup is obtained. When, on the other hand, $T_\infty > T_1/P$, the program contains too little parallelism to execute on P processors because the average parallelism T_1/T_∞ is less than P. Put differently, it is not a limitation of the scheduler that near-perfect linear speedup is not obtained in this case.

4.5 Hyperobjects

Discovering all inherent parallelism in a program is of utmost importance to obtain scalable speedup. The inherent parallelism in a program can often be hidden behind programming constructs that obfuscate the true semantics of the program. One such case is the use of global variables.[3] Often it appears as if a loop accessing global variables is not highly parallel or, if parallelization were attempted, then accesses to the global variables should at least be synchronized. In many cases, however, the global variables impose artificial limitations on parallelism, and it should be possible to parallelize the program without introducing synchronization.

Hyperobjects are a linguistic mechanism that allow us to lift such artificial limitations on parallelism. Hereto, hyperobjects allow different branches of a multithreaded program to maintain different but coordinated views of a global variable. Three types of hyperobjects are documented in the literature [5] corresponding to three particular uses of global variables, namely, *reducers*, *holders*, and *splitters*.

Reducer hyperobjects allow us to implement a reduction operation in parallel. A reduction operation is a computational pattern that reduces a set of values into a single datum. Reduction operations occur frequently in practical programs and often can also be performed in parallel. A common example of a reduction operation is to sum the elements of an array. Summing is defined inherently as a sequential process as each value is added to a *running sum*. When we take into account the associativity of addition, we see that parallelism does exist because multiple running sums can be maintained for different segments of the array. The total sum is obtained by summing the running sums, which is a smaller problem.

Holder hyperobjects correspond to the programming practice where variables are declared globally in order to save the burden of passing these variables to every procedure that uses them. Using a global variable locally within a loop iteration is enough to kill the parallelism of the loop; even when none of the iterations depends logically on the other iterations, all iterations are accessing the same global variables and hence they cannot execute in parallel.

Splitter hyperobjects are a bit more esoteric. They are useful in computations where a spawn statement is preceded by the modification of a nonlocal

---

[3]In general, hyperobjects apply to nonlocal variables, i.e., variables that are defined outside the current procedure's scope. Global variables are defined in the global program scope and are the best known and probably most heavily used form of nonlocal variables.

variable and followed by the restoration of that variable. These modifications could be increments and decrements of an integer, push and pop on a stack, etc. The underlying idea is that the parent can continue its execution on a different view than the child and that the parent and child views are initially identical copies. Because splitters are not so commonly needed in code as reducers, and because Cilk++ does not implement splitters [5], we will not discuss them further.

All three types of hyperobjects share a common theme: when spawning a child procedure, it is possible to let parent and child operate on distinct views of the hyperobject. Furthermore, the reduction, holder, and splitter patterns allow a formulation of rules to initialize new views when procedures are spawned and also to merge views of parent and child procedures on a sync statement.

A view is an intermediate value in the computation of a value, e.g., a partial sum in a sum reduction. When creating a new view, it means that the program starts to use additional intermediate values. When merging views, it means that intermediate values are reduced into a single intermediate value. A view on a hyperobject is in reality nothing more than a pointer to a distinct object of the same underlying type.

Figure 4.6 illustrates the creation and merging of views. The parent procedure P starts with a view $x_{P,1}$ on the object x. When spawning the child procedure C, the parent gets a view $x_{P,2}$, and the child gets a view x_C. Obviously, for efficiency reasons, one of those two views will equal the view $x_{P,1}$.

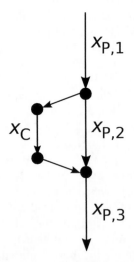

FIGURE 4.6: Depiction of strands in the parent frame spawning a child procedure with 1 strand. Each edge corresponds to a strand, while knots correspond to spawn and sync points. Views are created on spawns, and they are destroyed on syncs.

We will see later that it is best to transfer the view $x_{P,1}$ to the child procedure. Next, when parent and child procedures synchronize, the views x_C and $x_{P,2}$ are merged in the view $x_{P,3}$.

In the next three sections, we describe the hyperobject patterns in more detail and discuss the initialization and merging rules for each of the three hyperobjects.

4.5.1 Reducers

To understand how reducer hyperobjects work, we first need to consider the exact semantics of a *reduction* operation. Cilk++ defines reductions on algebraic monoids. An algebraic monoid is a triple $(T, \oplus, e)$ and consists of a set of values T and a binary operator $\oplus$ over T with the identity element e ($e \in T$). We consider that the operator $\oplus$ may be used in a reduction if it is associative, i.e., for any elements $a, b, c \in T$, we may change the evaluation order: $(a \oplus b) \oplus c = a \oplus (b \oplus c)$. Note that this property holds for integers in the mathematical sense as well as for the int data type in the C language. The relation is inexact in the case of floating-point numbers because of potential differences in rounding when changing the evaluation order.

Associativity stretches further: adding elements to the end of a list is also an associative operation! To see why, consider that adding an element d to the end of a list l is really concatenating the lists l and the single-element list (d). Now, concatenating lists is clearly associative, e.g., $((1,2) + +(3)) + +(4,5) = (1,2,3) + +(4,5) = (1,2,3,4,5)$ and $(1,2) + +((3) + +(4,5)) = (1,2) + +(3,4,5) = (1,2,3,4,5)$. In this example, we can choose to append the 3 first to the left sublist and then append the left sublist to the right sublist or vice versa. The end result is the same.

Now that we know what an algebraic monoid is, we can also cast it into C++ code. Hereto, we need to define the components of the triple $(T, \oplus, e)$. Here, T is a type and is defined by a typedef; $\oplus$ is a binary operator and is defined by a function of type $void\ reduce(T * l, T * r)$ where it is assumed that the left variable l is updated as $l := l \oplus r$; and the identity e is defined as a function that initializes a variable of type T to the identity value and has a signature $void\ identity(T * p)$. The type T and the two functions are defined as a member type and static member functions of a C++ class that defines all three components of the monoid.

So now we have defined a monoid class. All we need to do is to instantiate it. Assuming sum_monoid is a class defining the three monoid components, the line

```
cilk::reducer<sum_monoid> x;
```

defines x as a sum reducer over ints. By inheriting the template class cilk::monoid_base<>, the reducer is connected to the Cilk++ runtime system that creates and merges views as necessary. Furthermore, the member function `view()` returns a reference to the current view on the object. As such, operations such as these are allowed and do what you expect:

```
int y = x.view();
x.view() = 2*y;
x.view()++;
--x.view();
```

The Cilk++ library actually goes one step further and defines a template class **sum_reducer** that provides a wrapper around the **sum_monoid** class such that one can simply write

```
int y = x;
x = 2*y;
x++;
--x;
```

In other words, once you redefine the variable x from `int x` to `cilk::sum_reducer<int>` x, you do not even have to change the code that operates on variable x because the **sum_monoid** class implements all associative operations on type `int`. The only caveat is that you cannot take and store the address of x. That would be dangerous as the address may get out of sync with the actual view as it may be passed to a different thread in a way that the Cilk++ runtime system is not aware of. Luckily, the address of x is now of type **sum_monoid<int>*** and cannot be cast automatically to type `int *`, so the C++ semantics protect against this type of error.

4.5.2   Implementation of Views

Now we know how we can create and merge views. What remains is to discuss when the runtime system decides to create views, what view is passed to each strand, and how to merge those views again. In this discussion, it is important to remember the fundamental design of Cilk, where most of the execution time spent in nano-scheduled code and micro-scheduled code is used only when multiple threads interact in the execution of a procedure.

Procedure spawns mandate the creation of a new view. For reduction operations, the best organization passes the view that the parent holds before the spawn to the child and creates a new empty view for the continuation of the parent. The rationale is that in a serial interpretation of the code, the child "comes before" the continuation of the parent. Remember, Cilk programs spend most of their time in serial execution of the nano-scheduled versions of Cilk procedures! Figure 4.7 illustrates why these actions are correct using a list-append reducer. In the example, the list is l before the spawn statement. The child appends element c while the continuation of the parent appends element p. Under the discussed scheme, the child gets the view l and updates this view to $l + +(c)$. Concurrently, the parent updates its empty view to (p). When parent and child synchronize, it is possible to construct the list

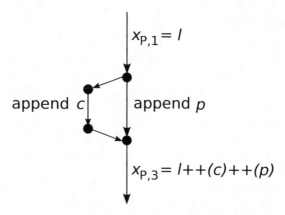

FIGURE 4.7: Depiction of strands in the parent frame spawning a child procedure with 1 strand. The parent starts with a view for a list with contents l. The spawned child appends c while the continuation of the parent appends p. Views must be managed such that the end results correspond to the value in a serial calculation, namely, $l + +(c) + +(p)$.

$l + +(c) + +(p)$, which corresponds to the list computed by the C++ elision of the program.

We have already seen in the examples of the monoids that merging views is implemented by merging one view into the other. The question is what view to retain and what view to execute. The rationale is again given by the nano-scheduling optimization: the runtime system merges the parent's view into the child's view, leaving the parent's view in an empty state, and passes the child's view back to the parent.

4.5.3 Holder Hyperobjects

Holder hyperobjects model thread-local storage, i.e., a variable that is locally used by one strand and whose value is then discarded. It turns out that holders can be easily implemented by utilizing the infrastructure we have for reducers. In particular, when a new view is created, we also pass the parent view to the child. The parent gets a new empty view. When merging views, we simply ignore the parent's view and pass back the child view to the parent.

Note that we might also create a new empty view for the child. This action would implement the same semantics for holders. However, the runtime system is implemented such that passing the parent view to the child is a no-op. Creating a new empty view entails at least allocating fresh memory that would incur some overhead.

4.5.4   Discussion

Hyperobjects are an important part of the Cilk++ programming language. They allow us to increase the parallelism in a program without significant rewriting of the program or the introduction of locks. Furthermore, they allow us to parallelize programs where certain actions need to occur in original program order. A good example is performing output in a parallel code region while the order of the output must adhere to the serial output order for the output to be meaningfull. Reducers allow us to enforce the correct sequencing of output operations by temporarily buffering output data from parallel strands to in-memory buffers. When a strand becomes the head of the computation (a property than can be easily detected), then its buffer is flushed. A good example is the parallelization of the bzip2 compression algorithm, where a reducer is defined for exactly this purpose [4]. For the interested reader, the Cilk-enhanced version of bzip2 is available through the given reference.

As a matter of interest, hyperobjects allow the programmer to observe if a strand has been stolen or not [9]. The clue is to create a holder hyperobject of a boolean where the identity element is *false*. Before entering a parallel region, the hyperobject is initialized to *true*. When a strand is stolen, a new view is created containing the identity value *false*, and this gives away the steal.

When developing your own hyperobjects, it is important to pay attention to the cost of merging views. These operations may happen quite frequently depending on the structure of your parallel program and the locations in it where hyperobjects are used. In any case, it is important to use constant-time operations to merge views. For instance, when constructing an append reducer over a linked list, the best practice is to append the lists by setting the link in the last element of the left argument list to point to the first element of the right argument list.

While hyperobjects solve many issues, some problems remain to be solved. For instance, it is problematic to apply hyperobjects to heap-allocated arrays with variable length because the length of the array is not known when a new view is created. The only way to solve this is by adding to the object a pointer to a global variable that holds the length of the array. When views are created or merged, it is possible to read the currently desired array length from the global variable. This mechanism works if the length of the array remains constant during the parallel region. If this length may change, then it may become necessary to place the length variable too in a holder hyperobject. It is easy to see this escalate into a myriad of pointers to hyperobjects.

Another weakness of Cilk is the execution of pipelines. A pipeline is a pattern of parallelism where the loop body is divided in stages. Executing the loop body corresponds to executing the first stage, followed by the second stage, and so on. Because this code is in a loop, parallelism can be exploited by overlapping the execution of stages from different iterations. For example, one might execute Stage 3 on iteration $i - 1$ concurrently with Stage 2 on iteration i concurrently with Stage 1 on iteration $i + 1$. Cilk is not able to capture this type of parallelism (it is possible to think up schemes for a two-stage pipeline). Granted, the amount of parallelism is limited to the number of stages in the pipeline, but the idiom occurs quite frequently in real programs.

4.6 Conclusion

The Cilk/Cilk++ model of parallel computation is quite general and fairly easy to use. Cilk Arts has published several cases where adding a few spawn statements to a program exposed significant amounts of scalable parallelism. Finding the right way to parallelize a program remains a nontrivial problem, but this effort is the same for any programming model. At least Cilk provides an easy way to specify the parallelism.

While the spawn/sync programming model is quite general, there are some styles of parallel programming that are not supported by Cilk, such as producer/consumer patterns, pipeline parallel-loops, and message passing. Some recent research considers the extension of Cilk with notions of task-graph execution, i.e., a program is broken up into a set of dependent tasks, and the tasks are inserted in a task graph where an edge from task S to task T means that S must execute after T. This task graph is then traversed by a Cilk program [1]. Such extensions generalize the model and allow us to implement some producer/consumer patterns and pipeline parallel-loops. A complete language extension of Cilk, including dataflow variables, is presented in [11].

Now that Intel has acquired the Cilk technology, it has been implemented on top of the same runtime system (read: scheduler) as other Intel products, such as the Threading Building Blocks and Array Building Blocks. As such, it should be possible to mix and match these programming models and express the constructs that are not allowed by Cilk using a different programming model.

4.6.1 Further Reading

Good introductions to Cilk/Cilk++ are the *Cilk Arts e-Book* on "How to survive the Multicore Software Revolution" [7] as well as the *MIT Cilk-5 Manual* [10].

The theoretical properties of Cilk have been described in a number of research papers. Blumofe et al. [3] formally show that Cilk's work stealing leads to constant overheads in space and time when executing Cilk programs.

The design of the Cilk-3 system is described in detail in [2]. Cilk-3 is a precursor to the Cilk-5 system described in this chapter. Its scheduler incurs a higher overhead than the Cilk-5 scheduler and does not implement the work-first principle, but other than that, it has all the properties of Cilk-5.

The Cilk-5 system is presented in detail in [6], and a detailed algorithmic description of the Cilk-5 scheduler is also presented in [5]. The latter paper also introduces Cilk++ hyperobjects and shows some usage cases.

Real programs that have been parallelized using Cilk are available through the Intel website, e.g., bzip2 [4] and MD6 encryption [8]. These examples illustrate how little effort is required to parallelize a program using Cilk.

Because of the regular structure of the Cilk language, it becomes fairly easy to devise a tool that analyses a Cilk program for data races, i.e., unsynchronized uses of variables in multiple strands. Such a tool is implemented in the cilkscreen race detector. See [7] for more information.

Bibliography

[1] K. Agrawal, C. E. Leiserson, and J. Sukha. Executing task graphs using work-stealing. In *Proceedings of the 24th IEEE International Parallel and Distributed Processing Symposium (IPDPS)*, pages 1–12, Atlanta, GA, Apr. 2010.

[2] R. D. Blumofe, C. F. Joerg, B. C. Kuszmaul, C. E. Leiserson, K. H. Randall, and Y. Zhou. Cilk: An efficient multithreaded runtime system. In *Proceedings of the Fifth ACM SIGPLAN Symposium on Principles and Practice of Parallel Programming*, PPOPP '95, pages 207–216, 1995.

[3] R. D. Blumofe and C. E. Leiserson. Scheduling multithreaded computations by work stealing. In *Proceedings of the 35th Annual Symposium on Foundations of Computer Science*, pages 356–368, Washington, DC, 1994. IEEE Computer Society.

[4] J. Carr. A parallel bzip2. http://software.intel.com/en-us/articles/a-parallel-bzip2/, Apr. 09. Retrieved Jan. 19th, 2011.

[5] M. Frigo, P. Halpern, C. E. Leiserson, and S. Lewin-Berlin. Reducers and other Cilk++ hyperobjects. In *SPAA '09: Proceedings of the Twenty-First Annual Symposium on Parallelism in Algorithms and Architectures*, pages 79–90, 2009.

[6] M. Frigo, C. E. Leiserson, and K. H. Randall. The implementation of the Cilk-5 multi-threaded language. In *PLDI '98: Proceedings of the 1998 ACM SIGPLAN Conference on Programming Language Design and Implementation*, pages 212–223, 1998.

[7] C. E. Leiserson and I. B. Mirman. How to survive the multicore revolution (or at least survive the hype). http://software.intel.com/en-us/articles/e-book-on-multicore-programming/, 2008.

[8] I. Mirman. Cilk++ sets world record for crypto hash function throughput. http://software.intel.com/en-us/articles/cilk-sets-world-record-for-crypto-hash-function-throughput/, Oct. 2009.

[9] A. Robinson. Detecting theft by hyperobject abuse. http://software.intel.com/en-us/blogs/2010/11/22/detecting-theft-by-hyperobject-abuse/, Nov. 10. Retrieved Jan. 19th, 2011.

[10] Cilk 5.4.6 Reference Manual. http://supertech.csail.mit.edu/cilk/manual-5.4.6.pdf, 1998.

[11] H. Vandierendonck, G. Tzenakis, D. S. Nikolopoulos. A Unified scheduler for recursive and task dataflow parallelism. In *PACT'11: IEEE International Conference on Parallel Architectures and Compilation Techniques*, pages 1–11, Sep. 2011.

Chapter 5

Multithreading in the PLASMA Library

Jakub Kurzak

University of Tennessee, Knoxville

Piotr Luszczek

University of Tennessee, Knoxville

Asim YarKhan

University of Tennessee, Knoxville

Mathieu Faverge

University of Tennessee, Knoxville

Julien Langou

University of Colorado, Denver

Henricus Bouwmeester

University of Colorado, Denver

Jack Dongarra

University of Tennessee, Knoxville
Oak Ridge National Laboratory
University of Manchester

5.1   Introduction

Parallel Linear Algebra Software for Multicore Architectures (PLASMA) is a numerical software library for solving problems in dense linear algebra on systems of multicore processors and multisocket systems of multicore processors [1]. PLASMA offers routines for solving a wide range of problems in dense linear algebra such as nonsymmetric, symmetric, and symmetric positive definite systems of linear equations, least square problems, singular value problems, and eigenvalue problems (currently only symmetric eigenvalue problems). PLASMA solves these problems in real and complex arithmetic and in single and double precision. PLASMA is designed to give high efficiency on homogeneous multicore processors and multisocket systems of multicore processors. As of today, the majority of such systems are on-chip symmetric multiprocessors with classic *super-scalar* processors as their building blocks (x86 and alike) augmented with short-vector SIMD extensions (SSE and alike). PLASMA has been designed to supercede LAPACK [2], principally by restructuring the software to achieve much greater efficiency on modern computers based on multicore processors.

The interesting part of PLASMA from the mutithreading perspective is the variety of scheduling mechanism utilized by PLASMA. In the next subsection, the main design principles of PLASMA are introduced. Section 5.2 discusses the different multithreading mechanisms employed by PLASMA. Section 5.3 introduces PLASMA's most powerfull mechanism of dynamic runtime task scheduling with the QUARK scheduler and briefly iterates through QUARK's extensions essential to the implementation of a production-quality numerical software. Section 5.4 highlights the advantages of dynamic scheduling for parallel-task composition by showing how PLASMA implements explicit matrix inversion using the Cholesky factorization. Section 5.5 covers the concept of task aggregation when large clusters of tasks are offloaded to a GPU accelerator. Finally, section 5.6 discusses the very important case of using QUARK for exploiting nested parallelism.

5.1.1   PLASMA Design Principles

The main motivation behind the PLASMA project is performance shortcomings of LAPACK [2] and ScaLAPACK [3] on shared memory systems, specifically systems consisting of multiple sockets of multicore processors. The three crucial elements that allow PLASMA to achieve performance greatly exceeding that of LAPACK and ScaLAPACK are the implementation of *tile algorithms*, the application of *tile data layout*, and the use of *dynamic scheduling*. Although some performance benefits can be delivered by each one of these techniques on its own, it is only the combination of all of them that delivers maximum performance and highest hardware utilization.

Tile algorithms are based on the idea that we process the matrix by square tiles of relatively small size, such that a tile fits entirely in one of the cache levels associated with one core. This way, a tile can be loaded to the cache and processed completely before being evicted back to the main memory. Of the three types of cache misses, *compulsory*, *capacity*, and *conflict*, the use of tile algorithms minimizes the number of capacity misses since each operation loads the amount of data that do not "overflow" the cache.

Tile layout is based on the idea of storing the matrix by square tiles of relatively small size, such that each tile occupies a continuous memory region. This way, a tile can be loaded to the cache memory efficiently, and the risk of evicting it from the cache memory before it is completely processed is minimized. Of the three types of cache misses, *compulsory*, *capacity*, and *conflict*, the use of tile layout minimizes the number of conflict misses since a continuous region of memory will completely fill out a set-associative cache memory before an eviction can happen. Also, from the standpoint of multithreaded execution, the probability of *false sharing* is minimized. It can only affect the cache lines containing the beginning and the ending of a tile.

Dynamic scheduling is the idea that we assign work to cores based on the availability of data for processing at any given point in time and is also referred to as *data-driven* scheduling. The concept is related closely to the idea of expressing computation through a task graph, often referred to as the DAG (*Direct Acyclic Graph*), and the flexibility of exploring the DAG at runtime. Thus, to a large extent, dynamic scheduling is synonymous with *runtime scheduling*. An important concept here is the *critical path*, which defines the upper bound on the achievable parallelism and needs to be pursued at the maximum speed. This is in direct opposition to the *fork-and-join* or *data-parallel* programming models, where artificial synchronization points expose serial sections of the code, where multiple cores are idle, while sequential processing takes place.

5.1.2 PLASMA Software Stack

Starting from the PLASMA Version 2.2 released in July 2010, the library is built on top of standard software components, all of which are either available as open source or are standard OS facilities. Some of them can be replaced by packages provided by hardware vendors for efficiency reasons. Figure 5.1 presents the current structure of PLASMA's software stack. Following is a brief bottom-up description of individual components.

Basic Linear Algebra Subprograms (BLAS) [4] is a de facto standard set of basic linear algebra operations, such as vector and matrix multiplication. CBLAS is the C language interface to BLAS [5]. Most commercial and academic implementations of BLAS also provide CBLAS. (LAPACK) [2] is a software library for numerical linear algebra, a direct predecessor of PLASMA, providing routines for solving linear systems of equations, linear least square problems, eigenvalue problems, and singular value problems. CLAPACK [6] is

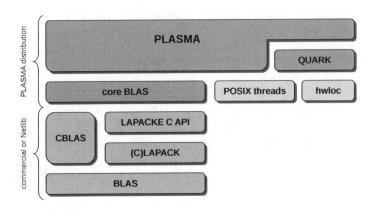

FIGURE 5.1: The software stack of PLASMA, Version 2.3.

a version of LAPACK available from Netlib, created by automatically trans-
lating FORTRAN LAPACK to C with the help of the F2C [7] utility. It
provides the same FORTRAN calling convention as the "original" LAPACK.
LAPACKE C API [8] is a proper C language interface to LAPACK (or CLA-
PACK).

"core BLAS" is a set of serial kernels, the building blocks for PLASMA's
parallel algorithms. PLASMA scheduling mechanisms coordinate the execution of these kernels in parallel on multiple cores. PLASMA relies on POSIX
threads for access to the system's multithreading capabilities and on the
hwloc [9] library for the control of thread affinity. PLASMA employs *static
scheduling*, where threads have their work statically assigned and coordinate
execution through progress tables, but can also rely on the QUARK [10] scheduler for dynamic (runtime) scheduling of work to threads.

5.1.3   PLASMA Scheduling

By now, multicore processors are ubiquitous in both low-end consumer
electronics and high-end servers and supercomputer installations. This has led
to the emergence of a myriad of multithreading frameworks, both academic
and commercial, embracing the idea of task scheduling: Cilk [11], OpenMP
(tasking features) [12], Intel Threading Building Blocks [13], just to name a
few prominent examples. One especially important category is multithreading systems based on dataflow principles, which represent the computation as
a *Direct Acyctlic Graph* (DAG) and schedule tasks at runtime through resolution of data hazards: *Read after Write* (RaW), *Write after Read* (WaR),
and *Write after Write* (WaW). PLASMA's scheduler QUARK is an example
of such a system. Two other, very similar, academic projects are also available: StarSs [14] from Barcelona Supercomputer Center and StarPU [15] from

INRIA Bordeaux. While all three systems have their strengths and weaknesses, QUARK has vital extensions for use in a numerical library.

5.2 Multithreading in PLASMA

PLASMA has to be intialized with the call to the `PLASMA_Init` function before any calls to its computational routines can be made. When the user's (master) thread calls the `PLASMA_Init` function and specifies N as the number of cores to use, PLASMA launches $N-1$ additional (worker) threads and puts them in a waiting state. In the master thread, control returns to the user. When the user calls PLASMA's computational routine, the worker threads are woken up and all the threads including the user's master thread enter the routine. Upon completion, worker threads return back to the waiting state, while the master thread returns from the call. A call to the `PLASMA_Finalize` function terminates the worker threads.

PLASMA is *thread-safe*. Multiple PLASMA instances, referred to as *contexts*, can be active at the same time without conflicts with the restriction that one user thread can be associated with one context only, i.e., one thread can only call `PLASMA_Init` once before it calls `PLASMA_Finalize`. (Currently, contexts are managed implicitly. Context handle is not provided to the user.) The typical usage scenario is to have a serial code (with a single thread) and launch one instance of PLASMA to use all the cores in the system. However, it is also possible for the user to spawn, e.g., four threads, each of which creates a four-thread instance of PLASMA, in order to exploit 16-way parallelism. In such a case, each PLASMA instance synchronizes its four threads while the user has to manage synchronization among the four PLASMA instances. PLASMA provides mechanisms for managing *thread affinity*, i.e., controlling the placement of threads on the physical cores.

PLASMA's statically scheduled routines follow the *Single Program Multiple Data* (SPMD) programming paradigm. Each thread knows its thread ID (`PLASMA_RANK`) and the total number of threads within the context (`PLASMA_SIZE`) and follows a specific execution path based on that information. Synchronization is implemented though shared progress tables and busy waiting. If a thread cannot progress, it yields the core to the OS (`sched_yield`). PLASMA's dynamically scheduled routines follow the *super-scalar* programming paradigm, where the code is written sequentially and parallelized at runtime through the analysis of the dataflow between different tasks. When a static routine is called, all the threads simply enter the SPMD code of the routine. When a dynamic routine is called, the master thread enters the super-scalar routine and begins queueing tasks using QUARK's task queueing calls. At the same time, all the worker threads enter QUARK's *worker loop*, where they keep executing the queued tasks until notified by the master. The master also participates in the execution of tasks unless overwhelmed with the job of queueing.

PLASMA's computational routines are implemented using either static or dynamic scheduling or both. PLASMA provides a switch, which can be changed at runtime, to decide if static or dynamic scheduling is preferred, for routines with both implementations. If a sequence of user calls invokes a mixed set of routines with static implementations only and dynamic implementation only, PLASMA switches the scheduling mode on the flight. Statically scheduled rouitnes are separated with barriers, and each switch from static scheduling to dynamic scheduling, and vice versa, invokes a barrier. However, any continuous sequence of dynamically scheduled routines is free of barriers.

5.3   Dynamic Scheduling with QUARK

The QUARK scheduler targets multicore, multisocket shared memory systems. The main design principle behind QUARK is implementation of the dataflow model, where scheduling is based on data dependencies between tasks in the task graph. The second principle is constrained use of resources, with bounds on space and time complexity. The dataflow model is implemented through analysis of data hazards, discussed in the following paragraphs. The constrained use of resources is accomplished by exploration of the task graph by a *sliding window*.

Even relatively small problems in dense linear algebra (those that can be handled by a laptop or a desktop computer) can easily generate DAGs with hundreds of thousands or even millions of tasks. Generation and exploration of the entire DAG of such size would not be feasible. Instead, as execution proceeds, tasks are continuously generated and executed. At any given point in time, only a relatively small number of tasks (on the order of one thousand) is stored in the task pool. The size of the sliding window is a tunable parameter, allowing trading of time and space overhead for scheduling flexibility.

Parallelization using QUARK relies on two steps: transforming function calls to task definitions and replacing function calls with task-queueing constructs. Listing 5.1 shows how PLASMA defines QUARK *dgemm* (matrix multiply) task using a call to CBLAS. Similar to Cilk and SMPSs, functions implementing parallel tasks must be side-effect free (cannot use global variables, etc.) The task definition takes QUARK handle as its only parameter, declares all arguments as local variables, fetches them from QUARK using the `quark_unpack_args_` construct, and executes the work (which in this example is just calling the `cblas_dgemm` function).

In order to change a function call to a task invocation, one needs to replace the function call with a call to `QUARK_Insert_Task()`, pass the task name (pointer) as the first parameter, preceede each parameter with its size, and

```
void CORE_dgemm_quark(Quark *quark) {
    int transA, transB;
    int m, n, k;
    double alpha, beta;
    double *A, *B, *C;
    int lda, ldb, ldc;

    quark_unpack_args_13(quark, transA, transB, m, n, k,
                         alpha, A, lda, B, ldb, beta, C,
                         ldc);
    cblas_dgemm( CblasColMajor,
                 (CBLAS_TRANSPOSE)transA,
                 (CBLAS_TRANSPOSE)transB,
                 m, n, k, alpha, A, lda, B, ldb, beta, C,
                 ldc);
}
```

LISTING 5.1: QUARK dgemm (matrix multiply) task definition in PLASMA.

follow with direction. This is shown in Listing 5.2. Array (pointer) arguments are preceeded with the size of memory they occupy, in bytes, and followed by one of the following directions: INPUT, OUTPUT, or INOUT. Scalar arguments are passed by a reference (pointer), preceeded with the size of their datatype, and followed by VALUE in place of the direction. Although scalar arguments are passed by reference, passing of scalars has the *pass by value* semantics. (A copy of each scalar argument is made at the time of call to Insert_Task().)

QUARK schedules tasks at runtime through the resolution of data hazards (dependencies): RaW, WaR, and WaW. The RaW hazard, often referred to as the *true dependency*, is the most common dependency. It defines the relation between a task writing ("creating") the data and the task reading ("consuming") the data. The latter task has to wait until the former task completes.

The WaR hazard is caused by a situation where a task attempts to write (modify) data before a preceding task is finished reading the data. In such case, the writer has to wait until the reader completes. The dependency is not referred to as a true dependency because it can be eliminated by renaming (making a copy) of the data. Although the dependency is unlikely to appear often in dense linear algebra, is has been encountered and has to be handled by the scheduler to ensure correctness.

The WaW hazard is caused by a situation where a task attempts to write data before a preceding task is finished writing the data. The final result is expected to be the output of the latter task, but if the dependency is not preserved (and the former task completes after the latter one), incorrect output will result. This is an important dependency in hardware design of processor pipelines, where resource contention can be caused by a limited

```
QUARK_Insert_Task(quark, CORE_dgemm_quark, task_flags,
    sizeof(PLASMA_enum),    &transA,    VALUE,
    sizeof(PLASMA_enum),    &transB,    VALUE,
    sizeof(int),            &m,         VALUE,
    sizeof(int),            &n,         VALUE,
    sizeof(int),            &k,         VALUE,
    sizeof(double),         &alpha,     VALUE,
    sizeof(double)*nb*nb,   A,                      INPUT,
    sizeof(int),            &lda,       VALUE,
    sizeof(double)*nb*nb,   B,                      INPUT,
    sizeof(int),            &ldb,       VALUE,
    sizeof(double),         &beta,      VALUE,
    sizeof(double)*nb*nb,   C,                      INOUT,
    sizeof(int),            &ldc,       VALUE,
    0);
```

LISTING 5.2: QUARK dgemm (matrix multiply) task invocation (queueing) in PLASMA.

number of registers. The situation is, however, quite unlikely for a software scheduler, where the occurrence of the WaW hazard means that some data are produced and overwritten before they are consumed. The same as the WaR hazard, the WaW hazard can be removed by renaming.

5.4 Parallel Composition

One vital feature of QUARK, or any other super-scalar scheduler, is the *parallel composition*, i.e., the ability to construct larger task graphs from a set of smaller task graphs. The benefit is exposing more parallelism in the combined task graph than each of the components posseses alone. PLASMA's routine for computing an inverse of a symmetric positive definite matrix is a great example [16].

The appropriate direct method to compute the solution of a symmetric positive definite system of linear equations consists of computing the Cholesky factorization of that matrix and then solving the underlying triangular systems. It is not recommended to use the inverse of a matrix in this case. However, some applications need to explicitly form the inverse of the matrix. A canonical example is the computation of the variance-covariance matrix in statistics. Higham [17, p. 260] lists more such applications.

The matrix inversion presented here follows closely the one in LAPACK and ScaLAPACK. The inversion is performed in place, i.e., the data structure initially containing matrix A is gradually overwritten with the result, and eventually A^{-1} replaces A. (No extra storage is used.) The algorithm involves three steps: computing the Cholesky factorization $(A = LL^T)$, inverting the L factor (computing L^{-1}), and finally, computing the inverse matrix $A^{-1} = L^{-1^T} L^{-1}$. In LAPACK the three steps are performed by the functions: POTRF, TRTRI, and LAUUM. In PLASMA the steps are performed by functions that process the matrix by tiles, define the work in terms of tile operations (tile kernels), and use QUARK to queue, schedule, and execute the kernels. Listing 5.3 shows an excerpt of PLASMA implementation of the three functions. Each one includes four loops, the first one with three levels of nesting, the other two with two levels of nesting. All work in expressed through elementary BLAS operations encapsulated in core_blas kernels: POTRF, TRSM, SYRK, GEMM, TRTRI, TRMM, LAUUM.

The scheduler addresses three important problems in parallel software development: complexity, productivity, and performance. A super-scalar scheduler eliminates the complexity of writing parallel software by automatically parallelizing algorithms defined sequentially and guaranteeing parallel correctness of sequentially expressed algorithms. For some workloads in dense linear algebra manual parallelization is relatively straightforward. A good example here is the Cholesky factorization (when considered alone) and its statically scheduled implementation in PLASMA [18]. For other operations it becomes nontrivial. Designing a static schedule for the combined three operations of the matrix inversion would be much harder. Finally, for some operations it becomes prohibitively complex. A good example here are PLASMA routines for band reductions through bulge chasing [19].

Another benefit of the scheduler is productivity. Thanks to the fact that sequential correctness guarantees parallel correctness, the scheduler facilitates very rapid development of numerical software, where manual design of parallel codes is labor intensive and error prone, causing long development times. Yet another benefit is the ability to do rapid prototyping of new algorithms and analysis of their parallel performance without the tedious work of parallelization. Finally, the scheduler also provides for increased performance by identifying parallel scheduling opportunities where a human programmer would miss them. Also, it is resilient to fluctuations in task-execution time, OS jitter, and adverse effects of resource sharing. (Such adverse effects will cause a graceful degradation rather than a catastrophic performance loss).

Figures 5.2 and 5.3 further strengthen those points. Figure 5.2 shows the three DAGs of the three components of the matrix inversion and the aggregate DAG of the entire inversion. The DAG aggregation is a natural behavior of QUARK and happens automatically upon invocation of the three routines on Listing 5.3. The DAG created by stacking the DAGs of the three components on top of each other is taller and thinner, which means its sequential part (*the critical path*) is longer and its parallelism is more confined. The aggregate DAG is shorter and wider, meaning shorter critical path and more parallelism.

```
void plasma_pdpotrf_quark(...) {
    for (k = 0; k < M; k++) {
        QUARK_CORE_dpotrf(...);
        for (m = k+1; m < M; m++) {
            QUARK_CORE_dtrsm(...);
        }
        for (m = k+1; m < M; m++) {
            QUARK_CORE_dsyrk(...);
            for (n = k+1; n < m; n++) {
                QUARK_CORE_dgemm(...);
            }
        }
    }
}
void plasma_pdtrtri_quark(...) {
    for (n = 0; n < N; n++) {
        for (m = n+1; m < M; m++) {
            QUARK_CORE_dtrsm(...);
        }
        for (m = n+1; m < M; m++) {
            for (k = 0; k < n; k++) {
                QUARK_CORE_dgemm(...);
            }
        }
        for (m = 0; m < n; m++) {
            QUARK_CORE_dtrsm(...
        }
        QUARK_CORE_dtrtri(...)
    }
}
void plasma_pdlauum_quark(...) {
    for (m = 0; m < M; m++) {
        for(n = 0; n < m; n++) {
            QUARK_CORE_dsyrk(...);
            for(k = n+1; k < m; k++) {
                QUARK_CORE_dgemm(...);
            }
        }
        for (n = 0; n < m; n++) {
            QUARK_CORE_dtrmm(...);
        }
        QUARK_CORE_dlauum(...);
    }
}
```

LISTING 5.3: PLASMA functions, POTRF, TRTRI, LAUUM, computing the inverse of a symmetric positive-definite matrix using QUARK for super-scalar parallelization.

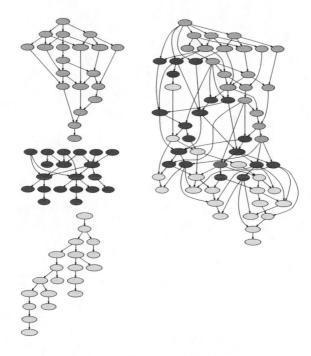

FIGURE 5.2: Three separate DAGs for the components of matrix inversion and their combined DAG.

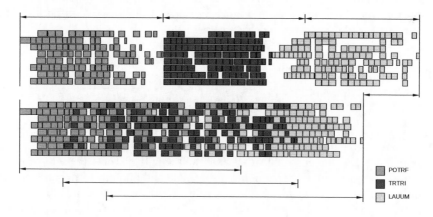

FIGURE 5.3: Trace of a small matrix inversion run on eight cores with and without barriers between phases.

Figure 5.3 shows execution traces of the matrix inversion with and without barriers between phases. The parallelism of each separate phase is limited by its data dependencies, causing gaps in execution. However, when the barriers

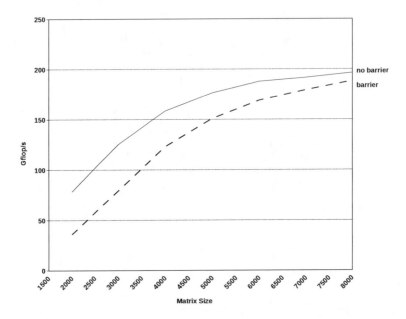

FIGURE 5.4: Performance difference between matrix inversion codes with and without barriers on a 48-core AMD-bases system.

are removed, the three phases fill out each other's gaps, which results in a shorter overall execution time. Figure 5.4 shows the difference in performance when using a large system with 48 cores.

5.5   Task Aggregation

Task aggregation is a feature of QUARK allowing for creation of large tasks representing agglomerates of smaller tasks. Not only do these tasks have a large number of "inherited" dependencies, but also the number of dependencies is not known at compile time and is determined at runtime instead. This feature has two main applications. One is to mitigate scheduling overheads of very fine-granularity tasks by increasing the level of granularity. Another is to offload work to data-parallel devices such as GPUs. Here, the latter case will serve as an example.

The main problem of existing approaches to accelerating dense linear algebra using GPUs is that GPUs are used like monolithic devices, i.e., like another "core" in the system. The massive disproportion of computing power between the GPU and the standard cores creates problems in work scheduling

```
for (k = 0; k < SIZE; k++) {
    QUARK_Insert_Task(CORE_sgeqrt, ...);
    for (m = k+1; m < SIZE; m++)
        QUARK_Insert_Task(CORE_stsqrt, ...);
    for (n = k+1; n < SIZE; n++)
        QUARK_Insert_Task(CORE_sormqr, ...);
    for (m = k+1; m < SIZE; m++)
        for (n = k+1; n < SIZE; n++)
            QUARK_Insert_Task(CORE_stsmqr, ...);
}
```

LISTING 5.4: QUARK code for the QR factorization using CPU cores only.

and load balancing. As an alternative, the GPU can be treated as a set of cores, each of which can efficiently handle work at the same granularity as a standard CPU core. The difficulty here comes from the fact that GPU cores cannot synchronize work in any other way than through a global barrier. In other words, the tasks offloaded to the GPU have to be independent.

The crucial concept here is one of task aggregation. The GPU kernel is an aggregate of many CPU kernels, i.e., one call to the GPU kernel replaces many calls to CPU kernels. Because of the data-parallel nature of the GPU, the CPU calls constituing the aggregate GPU call cannot have dependencies among them. The GPU task call can be pictured in the DAG as a cluster of CPU tasks with arrows coming into the cluster and out of the cluster but no arrows connecting the tasks inside the cluster. In order to support this model, QUARK allows for queueing of tasks with a dynamic range of dependencies. Initially, a task with no dependencies is created through a call to QUARK_Task_Init. Then, any number of dependencies can be added to the task using calls to QUARK_Task_Pack_Arg. Finally, the task can be queued with a call to QUARK_Insert_Task_Packed.

The tile QR factorization will serve as an example here. Listing 5.4 shows a QUARK implementation of the tile QR factorization using CPUs only. This particular example consists of five loops with three levels of nesting. It is built out of four core_blas kernels: GEQRT, TSQRT, ORMQR, and TSMQR. More details can be found in the PLASMA literature. Listing 5.5 shows modifications necessary to offload some of the tasks to the GPU. Here, the last loop nest is split into the CPU part and the GPU part. The split is done along the n dimension. While the CPUs get *lookahead* columns of the matrix to process, the GPU gets $SIZE - lookahead$ columns to process. The cuda_stsmqr kernel implements the GPU work, such that one tile of the matrix is (approximately) processed by one multiprocessor of the GPU. At the same time, all dependencies corresponding to all the tile operations are created inside the double loop nest.

Although GPU acceleration in PLASMA is currently in a prototype stage, Figure 5.5 clearly shows that this approach allows us to efficiently combine

```
for (k = 0; k < SIZE; k++) {
    QUARK_Insert_Task(CORE_sgeqrt, ...);

    for (m = k+1; m < SIZE; m++)
        QUARK_Insert_Task(CORE_stsqrt, ...);
    for (n = k+1; n < SIZE; n++)
            QUARK_Insert_Task(CORE_sormqr, ...);
    for (m = k+1; m < SIZE; m++)
        for (n = k+1; n < k+1+lookahead; n++)
            QUARK_Insert_Task(CORE_stsmqr, ...);
    task = QUARK_Task_Init(cuda_stsmqr, ...);
    for (m = k+1; m < SIZE; m++)
        for (n = k+1+lookahead; n < SIZE; n++) {
            QUARK_Task_Pack_Arg(task, &C1, INOUT);
            QUARK_Task_Pack_Arg(task, &C2, INOUT);
            QUARK_Task_Pack_Arg(task, &V2, INPUT);
            QUARK_Task_Pack_Arg(task, &T, INPUT);
        }
    QUARK_Insert_Task_Packed(task);
}
```

LISTING 5.5: QUARK code for the QR factorization using CPUs and a GPU.

the power of a GPU and a large number of conventional CPU cores. In this particular case, the 14 cores (SMs) of the Fermi GPU combined with 24 conventional AMD cores were capable of delivering performance in excess of half a TeraFlop/s.

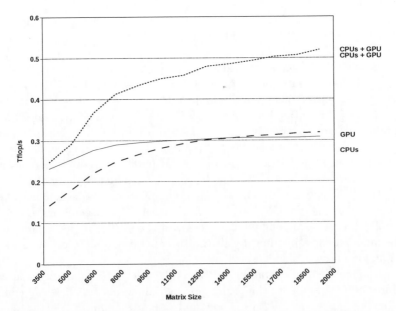

FIGURE 5.5: Performance of the QR factorization using 24 AMD-based cores and an NVIDIA Fermi GPU.

5.6   Nested Parallelism

This section describes two unique contributions, which are the use of nested parallelism in QUARK and fine-grained parallelization of the LU factorization of a matrix panel using the recursive algorithm [20].

5.6.1   The Case of Partial Pivoting

Figure 5.6 shows the initial factorization steps of a matrix subdivided into nine tiles (a 3 × 3 grid of tiles). The first step is a recursive parallel factorization of the first panel consisting of three leftmost tiles. Only when this finishes may the other tasks start executing, which creates an implicit synchronization point. To avoid the negative impact on parallelism, we execute this step on multiple cores (see Section 5.6.2 for further details) to minimize the running time. However, we use the nested parallelism model, as most of the tasks are handled by a single core, and only the panel tasks are assigned to more than one core. Unlike similar implementations [21], we do not use all cores to handle the panel. There are two main reasons for this decision. First, we use dynamic scheduling, which enables us to hide the negative influence of the panel factorization behind more efficient work performed by concurrent tasks. And second, we have clearly observed the effect of diminishing returns

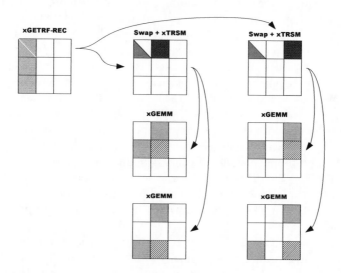

FIGURE 5.6: Execution breakdown for recursive tile LU factorization: factorization of the first panel using the parallel kernel is followed by the corresponding updates to the trailing submatrix.

```
function xGETRFR(M, N, column) {
    if N == 1 {                                    single column, recursion stops
        idx = split_IxAMAX(...)              compute local maximum of modulus
        gidx = combine_IxAMAX(idx)                      combine local results
        split_xSCAL(...)                                    scale local data
    } else {
        xGETRFR(M, N/2, column)              recursive call to factor left half
        xLASWP(...)                                        pivoting forward
        split_xTRSM(...)                                    triangular solve
        split_xGEMM(...)                                   Schur's complement
        xGETRFR(M, N-N/2, column+N/2)       recursive call to factor right half
        xLASWP(...)                                        pivoting backward
    }
}
```

LISTING 5.6: Pseudocode for the recursive panel factorization.

when using too many cores for the panel. Consequently, we do not use them all, and instead, we keep the remaining cores busy with other critical tasks.

The next step is pivoting to the right of the panel that has just been factorized. We combine in this step the triangular update (xTRSM in the BLAS parlance) because there is no advantage of scheduling them separately due to cache locality considerations. Just as the panel factorization locks the panel and has a potential to temporarily stall the computation, the pivot interchange has a similar effect. This is indicated by a rectangular outline encompassing the tile updated by xTRSM of the tiles below it. Even though so many tiles are locked by the triangular update, there is still a potential for parallelism because pivot swaps and the triangular update itself for a single column are independent of other columns. We can then easily split the operations along the tile boundaries and schedule them as independent tasks. This observation is depicted in Listing 5.6 by showing two xTRSM updates for two adjacent tiles in the topmost row of tiles instead of one update for both tiles at once.

The last step shown in Listing 5.6 is an update based on the Schur complement. It is the most computationally intensive operation in the LU factorization and is commonly implemented with a call to a Level 3 BLAS kernel called xGEMM. Instead of a single call that performs the whole update of the trailing submatrix, we use multiple invocations of the routine because we use a tile-based algorithm. In addition to exposing more parallelism and the ability to alleviate the influence of algorithm's synchronization points (such as the panel factorization), by splitting the Schur update operation, we are able to obtain better performance than a single call to a parallelized vendor library [22].

One thing not shown in Listing 5.6 is pivoting to-the-left because it does not occur in the beginning of the factorization. It is necessary for the second and subsequent panels. The swaps originating from different panels have to be ordered correctly but are independent for each column, which is the basis for running them in parallel. The only inhibitor of parallelism, then, is the fact that the swapping operations are inherently memory bound because they do not involve any computation. On the other hand, the memory accesses are done with a single level of indirection, which makes them very irregular in practice. Producing such memory traffic from a single core might not take advantage of the main memory's ability to handle multiple outstanding data requests and the parallelism afforded by NUMA hardware. It is also noteworthy to mention that the tasks performing the pivoting behind the panels are not located on the critical path and therefore, are not essential for the remaining computational steps in the sense that they could be potentially delayed toward the end of the factorization.

5.6.2 Implementation Details of Recursive Panel Factorization

Listing 5.6 shows the pseudocode of our recursive implementation of panel factorization. Even though the panel factorization is a lower-order term—$O(N^2)$—from the computational complexity perspective [23], it still poses a problem in the parallel setting from the theoretical [24] and practical standpoints [21]. To be more precise, the combined panel factorizations' complexity for the entire matrix is $O(N^2NB)$, where N is panel height (and matrix dimension) and NB is panel width. For good performance of BLAS calls, panel width is commonly increased. This creates tension if the panel is a sequential operation because a larger panel width results in a larger Amdahl's fraction [25]. Our own experiments revealed this to be a major obstacle to proper scalability of our implementation of tile LU factorization with partial pivoting, a result consistent with related efforts [21].

Aside from gaining high-level formulation that is free of low-level tuning parameters, recursive formulation allows us to dispense of a higher-level tuning parameter commonly called algorithmic blocking. There is already panel width, a tunable value used for merging multiple panel columns together. Nonrecursive panel factorizations could potentially establish another level of tuning called *inner-blocking* [26, 22]. This is avoided in our implementation.

5.6.3 Data Partitioning

The challenging part of the parallelization is the fact that the recursive formulation suffers from inherent sequential control flow that is characteristic of the column-oriented implementation employed by LAPACK and ScaLA-PACK. As a first step then, we apply a one-dimensional (1D) partitioning technique that has proven successful before [21]. We employed this technique for the recursion-stopping case: single-column factorization. The recursive formulation of the LU algorithm poses another problem, namely, the use of Level 3 BLAS call for triangular solve, xTRSM() and LAPACK's auxiliary routine for swapping named xLASWP(). Both of these calls do not readily lend themselves to the 1D partitioning scheme for two main reasons: (1) each call to these functions occurs with a variable matrix size, and (2) 1D partitioning makes the calls dependent upon each other, thus creating synchronization overhead. The latter problem is fairly easy to see as the pivoting requires data accesses across the entire column, and memory locations may be considered random. Each pivot element swap then requires coordination between the threads that the column is partitioned amongst. The former issue is more subtle in that the overlapping regions of the matrix create a memory hazard that may be at times masked by the synchronization effects occurring in other portions of the factorization. To deal with both issues at once, we chose to use 1D partitioning across the rows and not across the columns as before. This removes the need for extra synchronization and affords us parallel execution, albeit a limited one due to the narrow size of the panel.

The Schur's complement update is commonly implemented by a call to Level 3 BLAS kernel xGEMM(), and this is also a new function that is not present within the panel factorizations from LAPACK and ScaLAPACK. Parallelizing this call is much easier than all the other new components of our panel factorization. We chose to reuse the across-column 1D partitioning to simplify the management of overlapping memory references and to again reduce resulting synchronization points.

To summarize the observations that we have made throughout the preceding text, we consider data partitioning among the threads to be of paramount importance. Unlike the PCA method [21], we do not perform extra data copy to eliminate memory effects that are detrimental to performance, such as TLB misses, false sharing, etc. By choosing the recursive formulation, we rely instead on Level 3 BLAS to perform these optimizations for us. Not surprisingly, this

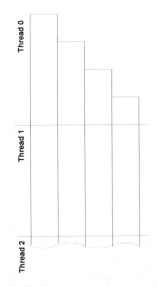

FIGURE 5.7: Fixed partitioning scheme used in the parallel recursive panel factorization.

was also the goal of the original recursive algorithm and its sequential implementation [20]. What is left to do for our code is the introduction of parallelism that is commonly missing from Level 3 BLAS when narrow rectangular matrices are involved.

Instead of low-level memory optimizations, we turn our focus towards avoiding synchronization points and let the computation proceed asynchronously and independently as long as possible until it is absolutely necessary to perform communication between threads. One design decision that stands out in this respect is the fixed partitioning scheme. Regardless of the current column height (within the panel being factored), we always assign the same number of rows to each thread except for the first thread. Figure 5.7 shows that this causes a load imbalance, as the thread number 0 has progressively smaller amounts of work to perform as the panel factorization progresses from the first to the last column. This is counterbalanced by the fact that the panels are relatively tall compared to the number of threads, and the first thread usually has greater responsibility in handling pivot bookkeeping and synchronization tasks.

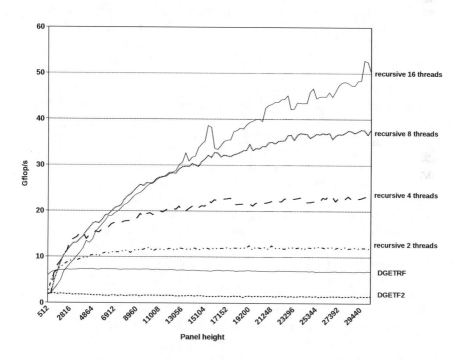

FIGURE 5.8: Scalability study of the recursive parallel panel factorization with various panel width 256.

5.6.4   Scalability Results of the Parallel Recursive Panel Kernel

Figure 5.8 shows a scalability study on a NUMA machine featuring a six-core AMD processor of our parallel recursive panel LU factorization with four different panel widths—32, 64, 128, and 256—against equivalent routines from LAPACK. We limit our parallelism level to 16 cores because our main factorization needs the remaining cores for trailing matrix updates. When compared with the panel factorization routine xGETF2() (mostly Level 2 BLAS), we achieve super-linear speedup for a wide range of panel heights with the maximum achieved efficiency exceeding 550%. In an arguably more relevant comparison against the xGETRF() routine, which could be implemented with mostly Level 3 BLAS, we achieve perfect scaling for 2 and 4 threads and easily exceed 50% efficiency for 8 and 16 threads. This is consistent with the results presented in the related-work section [21].

5.6.5   Further Implementation Details and Optimization Techniques

We exclusively use lockless data structures [27] throughout our code. This choice was dictated by fine-granularity synchronization, which occurs during the pivot selection for every column of the panel and at the branching points of the recursion tree. Synchronization using mutex locks was deemed inappropriate at such frequency as it has a potential of incurring system call overhead.

Together with lockless synchronization, we use *busy waiting* on shared-memory locations to exchange information between threads using a coherency protocol of the memory subsystem. While fast in practice [21], this causes extraneous traffic on the shared-memory interconnect, which we aim to avoid. We do so by changing busy waiting for computations on independent data items. Invariably, this leads to riching the parallel granularity levels that are most likely hampered by spurious memory-coherency traffic due to false sharing. Regardless of the drawback, we feel this is a satisfactory solution, as we are motivated by avoiding busy waiting, which creates even greater demand for inter-core bandwidth because it has no useful work to interleave with the shared-memory polling. We refer to this optimization technique as *delayed waiting*.

Another technique we use to optimize the inter-core communication is what we call *synchronization coalescing*. The essence of this method is to conceptually group unrelated pieces of code that require a synchronization into a single aggregate that synchronizes once. The prime candidate for this optimization is the search and the write of the pivot index. Both of these operations require a synchronization point. The former needs a parallel reduction operation, while the latter requires global barrier. Neither of these are ever considered to be related to each other in the context of sequential parallelization. But with our synchronization coalescing technique, they are deemed related in the communication realm, and consequently, we implemented them in our code as a single operation.

Finally, we introduced a *synchronization avoidance* paradigm whereby we opt for multiple writes to shared-memory locations instead of introducing a memory fence (and potentially a global thread barrier) to ensure global data consistency. Multiple writes are usually considered a hazard and are not guaranteed to occur in a specific order in most of the consistency models for shared-memory systems. We completely sidestep this issue, however, as we guarantee algorithmically that each thread writes exactly the same value to memory. Clearly, this seems as an unnecessary overhead in general, but in our tightly coupled parallel implementation, this is a worthy alternative to either explicit (via inter-core messaging) or implicit (via memory-coherency protocol) synchronization. In short, this technique is another addition to our contention-free design.

Portability, and more precisely, performance portability, was also an important goal in our overall design. In our lock-free synchronization, we heavily rely on shared-memory consistency—a problematic feature from the portability standpoint. To address this issue reliably, we make two basic assumptions about the shared-memory hardware and the software tools, both of which, to our best knowledge, are satisfied on a majority of modern computing platforms. From the hardware perspective, we assume that memory coherency occurs at the cache line granularity. This allows us to rely on global visibility of loads and stores to nearby memory locations. What we need from the compiler toolchain is an appropriate handling of C's volatile keyword. This, combined with the use of primitive data types that are guaranteed to be contained within a single cache line, is sufficient in preventing unintended shared-memory side effects.

Bibliography

[1] E. Agullo, J. Demmel, J. Dongarra, B. Hadri, J. Kurzak, J. Langou, H. Ltaief, P. Luszczek, and S. Tomov. Numerical linear algebra on emerging architectures: The PLASMA and MAGMA projects. *J. Phys.: Conf. Ser.*, 180(1), 2009.

[2] E. Anderson, Z. Bai, C. Bischof, L. S. Blackford, J. W. Demmel, J. J. Dongarra, J. Du Croz, A. Greenbaum, S. Hammarling, A. McKenney, and D. Sorensen. *LAPACK Users' Guide*. SIAM, Philadelphia, PA, 1992.

[3] L. S. Blackford, J. Choi, A. Cleary, E. D'Azevedo, J. Demmel, I. Dhillon, J. J. Dongarra, S. Hammarling, G. Henry, A. Petitet, K. Stanley, D. Walker, and R. C. Whaley. *ScaLAPACK Users' Guide*. SIAM, Philadelphia, PA, 1997.

[4] Basic Linear Algebra Technical Forum. *Basic Linear Algebra Technical Forum Standard*, August 2001.

[5] C interface to the BLAS. http://www.netlib.org/blas/blast-forum/cblas. tgz, 2013.

[6] CLAPACK (f2c'ed version of LAPACK). http://www.netlib.org/ clapack/, 2013.

[7] f2c. http://www.netlib.org/f2c/, 2013.

[8] LAPACK C interface. http://www.netlib.org/lapack/lapacke.tgz, 2013.

[9] Portable hardware locality (hwloc). http://www.open-mpi.org/projects/ hwloc/, 2013.

[10] Asim YarKhan, Jakub Kurzak, and Jack Dongarra. QUARK Users' Guide: Queuing and Runtime for Kernels. Technical report UT-ICL-11-02, University of Tennessee Innovative Computing Laboratory, Knoxville, Tennessee 37996, April 2011.

[11] R. D. Blumofe, C. F. Joerg, B. C. Kuszmaul, C. E. Leiserson, K. H. Randall, and Y. Zhou. Cilk: An efficient multithreaded runtime system. In *Principles and Practice of Parallel Programming, Proceedings of the Fifth ACM SIGPLAN Symposium on Principles and Practice of Parallel Programming, PPOPP'95*, pages 207–216, Santa Barbara, CA, July 19–21 1995. ACM.

[12] OpenMP Architecture Review Board. *OpenMP Application Program Interface, Version 3.0*, 2008.

[13] Intel Threading Building Blocks. http://www.threadingbuildingblocks. org/, 2013.

[14] Barcelona Supercomputing Center. *SMP Superscalar (SMPSs) User's Manual, Version 2.0*, 2008.

[15] C. Augonnet, S. Thibault, R. Namyst, and P. Wacrenier. StarPU: A unified platform for task scheduling on heterogeneous multicore architectures. *Concurrency Computat. Pract. Exper.*, 23(2):187–198, 2010.

[16] E. Agullo, H. Bouwmeester, J. Dongarra, J. Kurzak, J. Langou, and L. Rosenberg. Towards an efficient tile matrix inversion of symmetric positive definite matrices on multicore architectures. In *Proceedings of the 9th International Meeting on High Performance Computing for Computational Science, VECPAR'10*, Berkeley, CA, June 22–25 2011.

[17] Nicholas J. Higham. *Accuracy and Stability of Numerical Algorithms*. Society for Industrial and Applied Mathematics, Philadelphia, PA, second edition, 2002.

[18] J. Kurzak, H. Ltaief, J. J. Dongarra, and R. M. Badia. Scheduling dense linear algebra operations on multicore processors. *Concurrency Computat.: Pract. Exper.*, 21(1):15–44, 2009.

[19] Piotr Luszczek, Hatem Ltaief, and Jack Dongarra. Two-stage tridiagonal reduction for dense symmetric matrices using tile algorithms on multicore architectures. In *Proceedings of IPDPS 2011*, Anchorage, Alaska, May 16–20 2011.

[20] Fred G. Gustavson. Recursion leads to automatic variable blocking for dense linear-algebra algorithms. *IBM Journal of Research and Development*, 41(6):737–755, November 1997.

[21] Anthony M. Castaldo and R. Clint Whaley. Scaling LAPACK panel operations using parallel cache assignment. *Proceedings of the 15th ACM SIGPLAN Symposium on Principles and Practice of Parallel Programming*, pages 223–232, 2010.

[22] Emmanuel Agullo, Bilel Hadri, Hatem Ltaief, and Jack Dongarrra. Comparative study of one-sided factorizations with multiple software packages on multi-core hardware. In *SC '09: Proceedings of the Conference on High Performance Computing Networking, Storage and Analysis*, pages 1–12, New York, NY, 2009. ACM.

[23] E. Anderson and J. Dongarra. Implementation guide for LAPACK. Technical Report UT-CS-90-101, University of Tennessee, Computer Science Department, April 1990. LAPACK Working Note 18.

[24] G. M. Amdahl. Validity of the single-processor approach to achieving large scale computing capabilities. In *AFIPS Conference Proceedings*, volume 30, pages 483–485, Atlantic City, N.J., APR 18–20 1967. AFIPS Press, Reston, VA.

[25] John L. Gustafson. Reevaluating Amdahl's Law. *Communications of ACM*, 31(5):532–533, 1988.

[26] E. Agullo, J. Demmel, J. Dongarra, B. Hadri, J. Kurzak, J. Langou, H. Ltaief, P. Luszczek, and S. Tomov. Numerical linear algebra on emerging architectures: The PLASMA and MAGMA projects. *Journal of Physics: Conference Series*, 180, 2009.

[27] Håkan Sundell. *Efficient and Practical Non-Blocking Data Structures*. Department of computer science, Chalmers University of Technology, Göteborg, Sweden, November 5, 2004. PhD diss.

Chapter 6

Efficient Aho-Corasick String Matching on Emerging Multicore Architectures

Antonino Tumeo

Pacific Northwest National Laboratory

Oreste Villa

NVIDIA

Simone Secchi

Università di Cagliari

Daniel Chavarría-Miranda

Pacific Northwest National Laboratory

6.1 Introduction

String-matching algorithms are critical to several scientific fields. Beside text processing and databases, emerging applications such as DNA protein sequence analysis, data mining, information-security software, antivirus, and machine learning all exploit string matching algorithms [1]. All these applications usually process a large quantity of textual data, require high performance, and/or predictable execution times. Among all the string matching

143

algorithms, one of the most studied, especially for text processing and security applications, is the Aho-Corasick algorithm.

Aho-Corasick is an exact, multipattern string-matching algorithm that performs the search in a time linearly proportional to the length of the input text independent from pattern set size. However, depending on the implementation, when the number of patterns increases, the memory occupation may raise drastically. In turn, this can lead to significant variability in the performance, due to the memory access times and the caching effects. This is a significant concern for many mission-critical applications and modern high-performance architectures. For example, security applications such as Network Intrusion Detection Systems (NIDS), must be able to scan network traffic against very large dictionaries in real time. Modern Ethernet links reach up to 10 Gbps, and malicious threats are already well over one million and exponentially growing [2]. When performing the search, a NIDS should not slow down the network or let network packets pass unchecked. Nevertheless, on the current state-of-the-art cache-based processors, there may be a large performance variability when dealing with big dictionaries and inputs that have different frequencies of matching patterns. In particular, when few patterns are matched and are all in the cache, the procedure is fast. Instead, when they are not in the cache, often because many patterns are matched and the caches are continuously thrashed, they must be retrieved from the system memory, and the procedure is slowed down by the increased latency.

Efficient implementations of string-matching algorithms have been the focus of several works, targeting Field Programmable Gate Arrays [3, 4, 5, 6], highly multithreaded solutions like the Cray XMT [7], multicore processors [8] or heterogeneous processors like the Cell Broadband Engine [9, 10]. Recently, several researchers have also started to investigate the use Graphic Processing Units (GPUs) for string-matching algorithms in security applications [11, 12, 13, 14]. Most of these approaches mainly focus on reaching high peak performance or try to optimize the memory occupation rather than looking at performance stability. However, hardware solutions support only small dictionary sizes due to lack of memory and are difficult to customize, while platforms such as the Cell/B.E. are very complex to program.

The emergence of multicore, multithreaded, and General Purpose Graphic Processing Unit (GPGPU) platforms has made viable software implementations of high-throughput string-matching algorithms. String matching, and in particular the Aho-Corasick algorithm, is a good candidate for execution on these parallel architectures since the search can be easily parallelized by dividing the input text among the different threads or cores. However, obtaining high throughputs still requires adequate mapping, and performance stability may remain a concern.

This chapter presents a study of the behavior of the Aho-Corasick string-matching algorithm on a set of modern multicore and multithreaded architectures. We discuss the implementation and the performance of the algorithm on modern x86 multicores, multithreaded Niagara 2 processors, and GPUs from the previous and current generation. In particular, we perform our experiments on the following platforms:

- A dual-socket Intel Xeon W5590 (Nehalem architecture, four cores per processor, two threads per core)

- A quad-socket AMD Opteron 6176SE (Magny Cours architecture, two dies per processor, six cores per die)

- A dual-socket Niagara 2 (eight cores per processor, eight threads per core)

- One to four Tesla C1060 boards (based on the previous generation Tesla T10 GPU)

- One to four Tesla C2050 boards (based on the current-generation Tesla T20 GPU, codename "Fermi")

We focus our attention on streaming applications with unknown inputs, a situation typical of security systems and often problematic for the search process. We look carefully at how each solution achieves the objectives of supporting large dictionaries (up to 190,000 patterns), obtaining high performance, enabling flexibility and customization, and limiting performance variability.

The remainder of this chapter is organized as follows. Section 6.2 discusses related work. Section 6.3 briefly discusses the Aho-Corasick algorithm and details the systems used in our study. Section 6.4 presents the implementation of the algorithm on the various platforms, detailing the specific optimizations for the GPU code. Section 6.5 discusses our experimental results, presenting an analysis of the effects of the optimizations on the two GPUs used in this work and the comparison among all the platforms. Finally, Section 6.6 concludes the chapter.

6.2   Related Work

String matching is a critical kernel for many applications. Thus, researchers have proposed a large number of approaches to accelerate its execution. These include faster algorithms, novel hardware implementations, and optimized implementations exploiting hardware accelerators; multicore, multithreaded, and heterogeneous processors; and GPUs.

There are many different algorithm formulations. One of their main differences is the ability to recognize a single or multiple patterns in each execution. Knuth-Morris-Pratt [15] (KMP) and Boyer-Moore [16] (BM) are the most well known single-pattern matching algorithms. KMP constructs a partial-match table by preprocessing a pattern and is able to skip characters of the input text when a mismatch occurs. BM, instead, uses a skipping heuristic that can achieve sublinear execution time if the suffix of the pattern appears infrequently in the input text. Among the most know multiple-pattern matching algorithms, we can find Aho-Corasick [17] (AC), Commentz-Walter [18] (CW), and Wu-Manber [19] (WM). AC constructs a finite state machine (FSM) that enables the detection of multiple patterns in a single pass. CW combines BM with AC and is faster than AC for small pattern sets with long patterns. WM adopts techniques from BM. It has the highest average case performance and the smallest memory footprint for small or medium numbers of patterns, but it does not scale well when using tens of thousands of patterns. SFK Search [11] is an AC modification used in SNORT, a famous open-source tool for implementing network intrusion-detection systems, that uses prefix tree (trie) data structures to reduce memory requirements. Single-pattern matching formulations may be extended to recognize multiple patterns, but they generally require naive and less-efficient multipass implementations.

In this chapter, we focus our attention on the AC algorithm. AC is common in text processing and security applications when large inputs and large dictionaries are involved. It is also used for bioinformatic applications, although current software tools for largescale systems, such as such as ScalaBlast [20], or for GPUs, such as MUMmerGPU [21], generally resort to inexact alignment methods due to the errors in the genome sequences. AC has received a lot of interest from researchers because the independence of its time complexity with respect to the size of the pattern set makes it a good candidate for time-sensitive applications. However, since the memory occupation increases significantly while increasing the number of patterns, the algorithm poses various challenges in terms of space complexity and in performance stability due to the variability of memory and cache access times.

Several hardware designs for pattern-matching accelerators on Field Programmable Gate Arrays (FPGAs) have been proposed. Many of them focus on AC, sometimes extending the automaton to support regular expressions [22, 23, 24]. In general, they try to exploit parallelism by implementing multiple comparators for each pattern [3, 4, 6, 25]. The main limit for these approaches is memory occupation. To cope with this limit, several optimizations to hardware implementations of the AC algorithm have been proposed. The optimizations include bitmap node and patch compression [26], Suffix Based Traversing (SBT) [27], the splitting of the automaton in smaller automata that handle different bits of the input symbols [24], Bloom filters [28], and the combined use of a parallel Binary Search Tree (pBST) and a Tail Acceleration Finite Automaton (TAFA) [29]. However, except for [29], which can reach up to 90,000 patterns, all these solutions cannot manage more than 10,000 patterns. Furthermore, they all are difficult to customize. Thus, real-world solutions often include an FPGA front end that executes a nonexact matching (on regular expressions) and triggers a slower software back end on exact patterns such as in [5]. This may cause further complexities for managing the communication between the FPGA and the software back end.

The interest in software-based solutions for multicore and multithreaded processors has considerably risen in the past few years due to the increased performance and memory sizes of new architectures. Software implementations exploiting heterogeneous architectures such as the Cell/B.E. have recently appeared [9, 10, 30]. Villa et al. [9] and Scarpazza et al. [10] focus on AC, while [30] presents a solution that integrates a compression mechanism through the B-FSM [31] algorithm. All these works show that a significant programming effort is required to obtain high performance from the Cell/B.E.; therefore, it is difficult to integrate and modify the algorithms to target different application areas.

GPUs have recently become an interesting architecture for implementing pattern-matching algorithms. The first approaches such as [12] or [32], respectively, employing Cg and standard graphic APIs, were not successful and did not show any significant performance improvement with respect to more common general-purpose processors. On the other hand, solutions based on NVIDIA CUDA [13, 14, 33, 34] demonstrate more interesting results. Gnort [13] exploits CUDA for implementing SNORT's pattern-matching part, while GrAVity [14] integrates a CUDA-based matching engine in the open-source antivirus ClamAV. Both focus on the Deterministic Finite Automaton (DFA) representation of the AC algorithm. Smith et al. [33] examine both a standard and an extendend DFA formulation for recognizing regular expressions, while Naghmouchi et al. [34] explore small-rule-set expression matching on GPUs, describing how to implement and optimize the FLEX kernel on up to four NVIDIA GPUs.

Some works have started to explore high-performance solutions on multicores, on Shared-memory Multi-Processors (SMPs), and on scalable systems composed of multiple nodes. Pasetto et al. [8] looked at the efficient matching of regular expressions on multicore processors such as the Intel Xeon, the AMD Opteron, the IBM Power 6 and the Sun Niagara. Park and Demirdag [35], instead discuss a Java-based solution for distributed, network-connected systems. Finally, Michailidis and Margaritis [36] analyze a very basic MPI implementation.

In this chapter, we start from these foundations and analyze the performance trade-offs of the AC algorithm on a range of emerging multicore and multithreaded architectures. We consider SMP systems with multiple processors and multiple GPUs. Different from other works, our analysis focuses on very large dictionaries (from 20,000 to 190,000 patterns). Dictionaries of such sizes determine several issues in performance and storage and are uncommon in literature.

6.3   Preliminaries

In this section, we initially describe the Aho-Corasick string-matching algorithm and then present the details of the systems that we use for our study.

6.3.1   Aho-Corasick String-Matching Algorithm

The AC string-matching algorithm [17] scans an input text T of length m and detects any exact occurrence of each of the patterns of a given *dictionary*, including partially and completely overlapping occurrences.

A *pattern* is a finite sequence of symbols from an alphabet, while a *dictionary* is a set of patterns $P = \{p_1, p_2, \ldots, p_k\}$. The AC algorithm works by constructing an automaton starting from the dictionary. The automaton takes as input a given text and analyses it symbol by symbol. Each final state of the automaton directly corresponds to a matching pattern. There are two variants of the algorithm. The first uses a Nondeterministic Finite Automaton with ϵ moves (NFA-ϵ), while the second uses a Deterministic Finite State Automaton (DFA). We call the first variant *AC-fail* and the second one *AC-opt*. We initially introduce AC-fail and then discuss AC-opt as an improvement.

The AC-fail automaton is obtained from the keyword tree (also known as a trie) of the dictionary. For each pattern, the trie has a path that starts from the root node. The edges of the path are labeled as the symbols in the pattern, and distinct edges leaving a node have distinct labels. When deriving the automaton from the trie, nodes and edges, respectively, become states and transitions of the automaton. The root node becomes the initial state, and the nodes associated with identifiers become final states. Transitions from the root node to itself are added for each symbol in the alphabet that does not have edges leaving the root node. Finally, for each state, failure transitions are added. Failure transitions are used to move the automaton from one state to another when the current state does not have a regular transition associated with the current input symbol. Failure transitions reuse information associated with the last input symbols (suffix), moving the automaton to a state representing patterns that begin with such a suffix (i.e., they have that suffix as a prefix). To perform the move, they do not consider the current input.

Figure 6.1(a) shows an example of the AC-fail automaton with the dictionary $P = \{TEA, THE, THEME, SET\}$. Note that the pattern THE is included in the pattern $THEME$, so final state 2 is on the same path that is followed to reach final state 3. As an example, we consider the input text $SETHEME$, which matches the patterns SET, THE, and $THEME$ in a partially overlapping way. The automaton moves from initial state 0 to the final state 4 through regular transitions as the symbols S, E, T are progressively analyzed. When the automaton gets into final state 4, the pattern SET

is recognized. Then, the symbol H is analyzed, but it does not allow any regular transition. Consequently, a failure transition is invoked. Since the longest suffix that corresponds to a prefix in the automaton is T, the failure function selects state a to retry the input symbol H. After the ϵ move to a is done (dashed line *f1* in Figure 6.1(a)) a regular transition for H is found. The automaton thus continues to operate and, when E is encountered, final state 2 is reached, and the pattern THE is matched. Finally, M and E are analyzed, and they move the automaton to final state 3, recognizing also the pattern $THEME$. If instead of $SETHEME$, the input string was $SETAHEME$, after the first invocation of the failure transition at symbol A, setting the automaton to state a, a second failure transition would have been called, setting the automaton to initial state 0 (dashed line *f2* in Figure 6.1(a)). The possibility to have multiple state transitions on a failure is a potential performance drawback as the size of the dictionary grows [37].

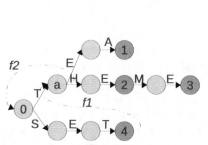

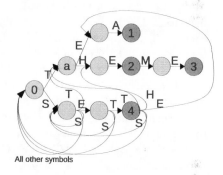

(a) AC-fail. Failure transitions from final state 4 to state a and from state a to state 0 are represented.

(b) AC-opt. For the sake of clarity, we only represent the transitions from the states of pattern SET. Note, however, that transitions are specified for all the patterns and associated states.

FIGURE 6.1: Sample AC-fail and AC-opt automata with dictionary $\{TEA,$ $THE, THEME, SET\}$. Solid lines represent regular transitions, dashed lines represent failure transitions, darker grey nodes represent final states.

AC-opt solves this problem by replacing all the failure transitions with regular ones. This is obtained at the cost of increased memory requirements, since for each state it is necessary to define transitions for each possible symbol input, whereas the failure transition is only associated with the suffix and thus to a state. The result is a Deterministic Finite-state Automaton (DFA), which has exactly one transition per state and input symbol. In Figure 6.1(b), we show the possible state transitions from the states that allow recognition of the pattern SET. Now, for each state that is traversed to reach final state 4, regular transitions are defined for all the possible input symbols. In particular, we can see that in place of failure transition $f1$, there are now regular transitions for input symbols T, H, E, S, and for all other symbols (which bring the automaton back to initial state 0). Since the objective of this paper is to discuss multicore and multithreaded architectures from the point of view of performance and stability, we consider the AC-opt variant of the algorithm. On one hand, since each input symbol for each state corresponds to exactly one transition, this approach should guarantee predictable performance. On the other hand, the higher memory requirements may lead to performance variability due to the memory and cache behavior.

6.3.2   Systems

For the x86 platforms, we consider two systems. The first has 2 Intel Xeon W5590 processors, while the second integrates 4 AMD Opteron 6176SE processors. The Xeon W5590 is a quadcore processor based on the Nehalem architecture, able to support up to two threads per core through Simultaneous MultiThreading (SMT), dubbed as HyperTreading by Intel. Each core has private 32-KB L1 caches for instructions and data and a private 256-KB L2 cache. All the cores share an 8-MB L3 cache. The processor has an operating frequency at 3.33 GHz and integrates a memory controller with three channels, allowing a bandwidth of up to 32 GB/s. The processors are connected together and with the rest of the system through two QPI links. Our system has a total of 32 GB of DDR3 memory at 1333 MHz. The Opteron 6176SE, codename "Magny Cours" is a dual die processor that integrates 12 cores (6 per die). Each core has private instruction and data L1 caches of 64 KB each and a private L2 of 512 KB. Each die has a shared cache of 6 MB and two DDR3 channels. The two dies are connected through a HyperTransport coherent connection. The processor operates at a frequency of 2.3 GHz. The theoretical memory bandwidth for a 12-core processor is 42.7 GB/s, but the northbridge in each die only runs at 1.8 GHz, limiting the bandwidth to a maximum of 28.8 GB/s. The four processors in our system are connected through HyperTransport links, and the overall system has 256 GB of DDR3 memory at 1333 MHz.

For the Niagara 2 system, we consider an UltraSparc T5240 blade, which is composed of two processors at 1165 MHz, sharing 32 GB of DDR2 memory. Niagara 2 has eight SPARC cores, and each core supports the concurrent execution of eight threads, for a total of 64 threads per chip. The cores implement the 64-bit SPARC V9 instruction set, and has one load/store unit, two execution units, one floating-point unit, and a cryptographic/stream processing unit. Furthermore, the cores have an 8-way 16-KB instruction cache and a 4-way 8-KB data cache. At each clock cycle, two threads among the eight available can be selected for execution. The eight cores access a shared 4-MB Level 2 16-way set associative cache, which is divided into eight banks of 512 KB each. Niagara 2 has four memory controllers on chip, each controlling two FBDIMM channels, clocked by the DRAM clock at 400 MHz. The read bandwidth is 51.2 GB/s, while the write bandwidth is 25.6 GB/s (writes execute at half the rate of the reads).

For the GPU-based systems, we consider boards integrating two different generations of GPUs. The first is the Tesla C1060, which uses the T10 GPU. The T10 has 30 Streaming Multiprocessors (SMs), SIMD units composed of eight Streaming Processors (SPs), two Super Function Units (SFUs), a Dual Precision Streaming Processor (DP), a scratch-pad memory of 16 KB (known as *shared memory*), the instruction unit (I-Unit), the instruction cache (I-Cache), and a constant cache (C-Cache). A group of 32 threads, called *warp* is executed by an SM in SIMD style over multiple cycles (a minimum of 4 for simple instructions). An SM can allocate up to 1024 threads (32 warps). On the other hand, a thread block is composed of a maximum of 16 warps, allocated on a single SM, with exclusive access to a portion of the shared memory. Multiple SMs are combined in a Texture Processor Cluster (TPC), which provides a Texture Unit and a Texture Cache. In the T10, three SMs form a single TPC. Double-precision peak performance is one-eighth of the single precision. All SMs access the board memory, known as *global memory*. The SMs communicate with the global memory through a crossbar connected to several 64-bit memory controllers. The host CPU also accesses the global memory through the PCI Express bus. Each SM issues memory operations for a half-warp (16 threads) at a time, and operations should be correctly aligned to coalesce in a single memory transaction. Memory operations on the global memory have a latency of 400–600 cycles, which is hidden by switching warps on the SMs. In the Tesla C1060, the SMs are clocked at 1.3 GHz. Other parts of the GPU are clocked at 602 MHz. The board has 4 GB of GDDR3 memory at 800 MHz, determining a peak bandwidth of 102 GB/s.

The second board considered for this chapter is the Tesla C2050, which uses the T20 GPU, codenamed "Fermi." The T20 is a radical departure from the T10 architecture. An SM is composed of 32 SPs, four SFUs, 64 KB of on-chip memory configurable either as 48 KB of Level 1 cache and 16 KB of shared memory or as 16 KB of L1 cache and 48 KB of shared memory, two I-Units, the I-Cache, and the C-Cache. The I-Units fetch and issue instructions from two different warps, which are then executed simultaneously on the two groups of 16 SPs. A simple instruction (e.g., MUL, ADD, or MAD) from a warp takes a minimum of two clock cycles to be executed. In the T20, double precision computations are performed by the SPs. However, when double precision is used, only one warp can issue, thus reducing the speed to half with respect to single-precision computations. Each SM has 32,768 registers and can keep up to 1536 threads active. The memory subsystem has been completely overhauled with respect to the T10. There is no longer the concept of a TPC. Instead, the 16 Load/Store units are directly integrated in the SMs. Four texture units and the texture cache are also part of an SM. L1 caches are not coherent across SMs. All the SMs interface to a second-level cache of 768 KB. Differently from the T10, this L2 cache is used not only for textures, but for all data. The L2 cache can significantly increase the speed of atomic operations. The global memory, composed of GDDR5 RAM modules, is again accessed through 64 bits memory controllers. In the T20, there are two Direct Memory Access (DMA) engines connected to the PCI-Express bus, so inbound and outbound data transfers from the host can proceed simultaneously. T20 also supports Error Correcting Code for all the memory hierarchy. The board used in this work has 14 SMs (448 SPs), working at the frequency of 1.15 GHz, and 3 GB of memory at 1.5 GHz.

For the multi-GPU benchmarks, we use machines with up to four Tesla C1060 and Tesla C2050 boards. The Tesla C1060 boards are hosted on a system integrating an AMD Phenom II X4940 (4 GHz) at 3 GHz and 16 GB of RAM. The boards connect to the NVIDIA GeForce 8200 chipset (an nForce chipset with integrated graphic) through PCI Express 2.0 x8 links (4 GB/s peak). The Tesla C2050 boards are hosted on a system with 2 Intel Xeon X5560 (Nehalem architecture, six cores each at 2.67 GHz) and connected to the Intel 5500 chipset with PCI Express 2.0 links all configured at x16 (8 GB/s peak). The system has 24 GB of memory.

6.4   Algorithm Design

In this section, we discuss the implementation of our algorithm for multi-core and multithreaded processors and GPUs. We initially present the basic design adopted for the multicore and the multithreaded machines and then discuss how it has been modified to fit the architectural constraints of NVIDIA GPUs. Starting from the same principles, we designed shared-memory implementations using `pthreads` for the dual Xeon, the quad Opteron, and the dual Niagara 2. For the GPU kernel, instead, we used NVIDIA CUDA. For all the implementations, our algorithm design is based on the following cornerstones:

- minimize the number of memory references

- reduce memory contention

Algorithm 6.1 shows the basic steps of the AC string-matching algorithm. For each symbol, a minimum of three loads are performed. The first, in Step 2, loads the symbol itself. The second, in Step 3, loads the `next_node` in the automaton. As previously discussed with AC fail, this step may involve multiple loads due to the execution of multiple failure transitions. With AC opt instead, which executes exactly one transition per symbol, it is possible to always execute a single load. Finally, the third load, in Step 5, checks if the `current_node` is final, leading to the matching of a pattern. However, in our implementation, we remove the necessity to execute the load for final-state checking. Furthermore, if the ASCII alphabet is used, symbols are only 8 bits long. Since symbols of the input text are contiguous, the frequency of the load in Step 2 can be reduced to one for every four or eight symbols if 32-bit or 64-bit architectures, respectively, are used. On the other hand, the load for next_node in Step 3 is unpredictable and has no locality at all since it depends on the current input symbol and the current_node. This load is the main cause of performance variability on cache-based architectures. When the input text matches a large number of patterns, the entire graph describing the automaton will be accessed during the matching process. If the graph is big, many unpredictable memory locations are accessed, and cache lines are easily thrashed. When the input text only matches a few patterns, instead, most of the time the algorithm will stay on node 0 or nodes at a few levels from node 0 (normally in the first two levels, considering a breadth-first exploration of the states of the automaton) since the regular transitions will often move the automaton to them. This will determine a large number of cache hits and minimal cache line replacement. This is true even when a limited number of patterns matches with high frequency.

Algorithm 6.1 Basic steps of the DFA-based AC string-matching algorithm

1. **Load** node "0" (or root) from main memory in current_node.
2. **Load** one input symbol from main memory in symbol.
3. **Load** the next_node from main memory (following the link from the current_node, labeled by symbol).
4. **Assign** next_node to current_node.
5. **Check if** current_node is final (if it is, the last symbols are a matching pattern).
6. **Repeat** starting from Step 2 **until** there are no more input symbols.

In our implementation, we represent the DFA of AC-opt in a State Transition Table (STT). The STT is a table composed of as many rows as there are nodes in the DFA and as many colums as there are symbols of the alphabet. Each STT line represents a *node*. Each cell in an STT line, which is indexed by the current_node and the current input symbol, represents a transition of the DFA and stores the address of the beginning of the STT line corresponding to the next_node for that transition.

Figure 6.2 shows the STT (with a reduced number of transitions) of the example dictionary ($\{TEA, THE, THEME, SET\}$) used for Figure 6.1. The STT is composed of 11 lines (the root node 0 plus the 10 nodes in the original DFA) and 256 columns (256 symbols of the ASCII alphabet).

For the implementation on the multicore and multithreaded processors, the STT lines are aligned at 256-byte such that the least significant byte of the address is equal to zero. This allows using the least significant bit of each STT cell to store the boolean information indicating if that transition is in a final state or not, eliminating the need for the additional load in Step 5. In Figure 6.2, transitions to final states are represented with bold gray lines. From the figure, it is possible to see how each cell is organized (pointer to the next state and bit flagging the transition as a transition to a final state). Since

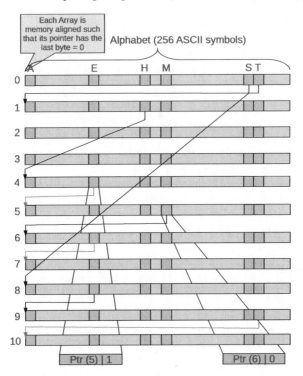

FIGURE 6.2: STT structure of the DFA in Figure 6.1(b). Only a few transitions are represented. Transitions in bold gray are final.

the `next_node` is obtained through the dereferencing of the `current_node +` `symbol` pointer, STT lines must always have the same size. Alphabet symbols not used in the dictionary have to be explicitly represented as transitions to the root node. This representation is expensive in terms of memory utilization but guarantees the execution of a single memory operation for Step 3. On current high-performance multicore and multithreaded systems, memory utilization is becoming only a secondary issue since it is now common to have workstations with 24 or 32 GB of RAM. This is a typical example where memory space can be traded for speed. Furthermore, if the STT and input symbol streams become too big, it is still possible to resort to multipass approaches, such that the same input is alternatively analyzed with different STTs.

Parallel operation is achieved by assigning to each thread a `current_node` and a distinct section of the input text. At runtime the input text is buffered and split into chunks, which are then assigned to each processing element. Since shared-memory machines are used, the same STT is accessed by all the threads. The chunks overlap partially to allow matching of those patterns that cross a boundary. The overlapping is equal to the length of the longest pattern in the dictionary minus one symbol. The inefficiency of the overlapping (replicated work) is measured as (longest pattern $-$ 1)/(size of the chunk).

For the GPU implementation, our algorithm implementation still starts from an STT, but its organization is slightly different, as explained in the following section.

6.4.1   GPU Algorithm Design

The algorithm design for the GPUs starts from principles similar to the multithreaded design, but optimizes several aspects to comply with their architectural constraints. Basically, our approach assigns a small chunk of the input text to each CUDA thread, which can access the same shared STT in the global memory of the GPU. This allows generation of the required large number of threads, while maintaining high utilization of each thread. Patterns that cross chunk boundaries are again supported through partial overlapping of the chunks. For the GPU implementation, we use a fixed chunk size of 2 KB, which is a good trade-off in terms of number of threads generated (for reasonably sized input texts) and utilization of each thread. With this chunk size, the inefficiency due to chunk overlapping is limited to 0.7% in the worst case (patterns of 16 bytes).

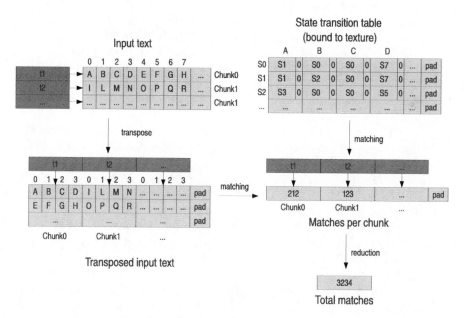

FIGURE 6.3: The implementation of the algorithm on GPUs.

Figure 6.3 shows the main elements of the GPU implementation of the algorithm. The main differences with respect to the `pthreads` implementation reside in the STT organization. Instead of using a table composed of pointers, in CUDA we use a table composed of indices. Rows still represent states and columns symbols, but states are addressed by indices and not by pointers. Each cell of the table, which is 32 bits in size, contains the index of the next state in the first 31 bits, and the flag that tags transitions to final states in the last bit. The code in the kernel opportunely masks the value loaded from each cell to distinguish the two parts. Using 31 bits for representing the next state index allows allocation of up to two billion states, which, considering a reasonably sized alphabet (such as the ASCII alphabet), accounts for well over the 6 GBs of memory currently available on high-performance GPUs.

Current GPUs require following specific alignment rules for accessing data in the board memory to exploit the high memory bandwidth available. We recall that in current NVIDIA GPUs, threads are scheduled for execution in warps (groups of 32 threads). In particular, the Tesla T10 GPU, used on the Tesla C1060 board, issues a single transaction for each memory segment addressed in a half-warp (group of 16 threads) independently from the order in which each thread in the half warp addresses memory words. Obviously, if all the accessed words reside in the same segment, a single transaction per half-warp is issued (i.e., all the memory accesses of a half-warp are *coalesced* in a single transaction). The maximum size for a transaction is 64 bytes. This is in contrast to previous GPUs where to issue a single transaction, words accessed by a half-warp should reside in the same memory segment and also be sequential. The Tesla T20 GPU, instead, used in the Tesla C2050 board, supports single transactions of up to 128 bytes (i.e., the data read by a full-warp when accessing 32-bits words). Furthermore, T20 integrates L1 caches (one for each streaming multiprocessor) and an L2 cache (unified for all the GPUs), which have lines of 128 bytes too. So, subsequent accesses to unaligned data already loaded in one of the caches are serviced at the throughput of the respective cache.

CUDA also allows the use of texture-fetching units to load read-only data. Texture units do not need to respect alignment rules and can take advantage of a texture cache. As previously discussed, during the matching process, the accessed locations of the STT are unpredictable. Thus, even on the T10 and the T20 GPUs, it is not possible to limit the number of memory transactions by coalescing together accesses from different threads. For these reasons, our implementation binds the STT to the texture memory. In the considered GPUs, the texture cache has a working set of 8 KB and is optimized for two-dimensional locality. The benefits of caching for string matching, however, as previously introduced, are limited to those cases where only the first few levels of the STT are accessed. There is also a limit on the size of textures in CUDA: when bound to linear memory, they can allocate up to 2^{27} elements (corresponding to an STT of 512 MB). So, when big STTs are used, we bind them to texture only partially. This allows the retention of the benefits of caching for the majority of cases where it can effectively improve performance without significantly impacting the others cases. Nevertheless, since the T20 GPU already has caches for data, the use of texture memory may bring, depending on the application, limited improvements or even some performance reductions. In the experimental result section, we evaluate this aspect further.

Using the shared memories (i.e., the scratchpad memories integrated in each Streaming Multiprocessor) for caching STT lines is not effective for big dictionaries. Shared memory has very small dimensions (only 16 KB on the T10, configurable at 16 or 48 KB on the T20, with the L1 caches, respectively, set at 48 or 16 KB) and can only store a limited number of states with respect to the overall number of states of the dictionaries considered in this chapter (20,000 minimum). This is not enough memory to cover even a single level of the STT. Furthermore, its effective exploitation also requires the implementation of a replacement algorithm.

On the T10 GPU, another optimization for the STT is required (independent of the use of the texture memory) to remove the possibility of some hardware-related memory hotspots. In fact, the memory controllers of the T10 GPU manage consecutive 256-byte-wide memory partitions. Depending on the number of memory controllers implemented on the GPU, on the alphabet, on the input streams, and on the number of matches, the accesses may concentrate only on a partition, thus saturating the related memory controller. This problem is known in the GPGPU community as partition camping [38]. On the GPU of the Tesla C1060 board, there are eight memory controllers (and, consequently, eight partitions); thus, with rows of 256 cells (ASCII alphabet) of 32 bits (1024 bytes in total), a partition has the cells for the same 64 symbols on odd or even rows. So, with inputs that have particular symbol frequencies, the accesses may concentrate only on few partitions. To solve this problem, we added to each line of the STT a padding corresponding to the size of a partition, allowing a better distributed access pattern. On the T20, this problem is not present anymore since the address space managed by the different memory controllers is not sequential but instead scrambled.

Due to the memory-alignment requirements, another important optimization is performed on the input text. In applications that stream data from a single source, the input text is sequentially buffered in the host memory. If the input text is moved as is to the global memory of the graphic card, each thread will start loading symbols with a stride corresponding to the chunk size. Consequently, reads cannot be coalesced, and a transaction for each symbol is executed to load the data. To solve this problem, as Figure 6.3 shows, our CUDA implementation applies a transposition after copying the input text in the global memory of the GPU. The input data structure can be thought of as a matrix where chunks correspond to the rows. After the transposition, chunks are disposed on the columns. This disposition allows the CUDA threads executing in parallel to access, with each load, data that are physically contiguous, permitting the memory controllers to coalesce loads for the same half-warp (in the T10) or for the same warp (in the T20). However, to further reduce the number of memory transactions, the input text is transposed into groups of four symbols. In fact, each transaction in the T10 can contain up to 64 bytes, corresponding to 32 bits for each thread of a half-warp. On the

T20, instead, each transaction contains up to 128 bytes, but it pertains to a full-warp. Since for this implementation we consider the ASCII alphabet, each symbol has a size of 8 bits. Thus, the chunks of input text are transposed in blocks of four symbols, which are then read with a single (coalesced) load in the main loop of the pattern-matching algorithm. When the number of chunks is not an integer multiple of the number of threads in a half-warp or a warp, our implementation also adds padding so that each row starts aligned with a memory segment. The transposition is performed directly on the GPU with a fast and optimized kernel, and it is transparent to the rest of the system.

The matching results are collected per chunk. A reduction is applied to gather all the results and send them back to the host through a single memory-copy operation. Again, the reduction operation is performed on the GPU with an optimized kernel.

6.5   Experimental Results

In this section, we present the results of our study. We initially introduce the experimental setup, discussing the input sets and the dictionaries used. We then discuss the effects of the optimizations on the CUDA kernels, for both the Tesla C1060 (T10 based) and the Tesla C2050 (T20 "Fermi" based) boards. Finally, we show the scaling of the algorithm on the various machines, discussing the trade-offs in peak throughput, performance variability, and size of dictionaries.

6.5.1   Experimental Setup

We performed our experiments on the systems presented in Section 6.3.2. All of them have the common characteristic of using processors that can be considered multicores, even the GPUs. We also look at how the AC string-matching algorithm scales on multisocket and on multi-GPU systems.

We have implemented various versions of the AC algorithm as described in Section 6.4 for the GPU systems, the x86 SMPs, and the Niagara 2. The implementations have two main phases: building the State Transition Table (STT) and executing string matching against the built STT. The STT building phase is performed offline and stored in a file representation. We focus our experiments on the string-matching phase since this is the critical portion in the realistic use of this algorithm for real-world applications.

For our experiments, we use three different dictionaries: **English**, a 20,000-pattern data set with the most common words from the English language and an average length of 8.5 bytes; **Dictionary 1**, an ~190,000-pattern data set with mostly text entries with an average length of 16 bytes; **Random**, a 50,000-pattern data set with entries generated at random from the ASCII alphabet with an uniform distribution and an average length of 8 bytes. Dictionaries with more text-like entries have higher frequencies of alphabetical ASCII symbols. We also use four different input streams for each dictionary: **Text**, the English text of the King James Bible; **TCP**, captured TCP/IP traffic; **Random**, a random sample of characters from the ASCII alphabet; and **Itself**, which corresponds to feeding the dictionary itself as input text. Using the dictionary itself as an input will exhibit the "heaviest" matching behavior, forcing the algorithm to thoroughly explore the STT and influencing significantly the performance of the algorithm. All the results are reported in Gigabits per second (Gbps).

6.5.2   GPU Optimizations

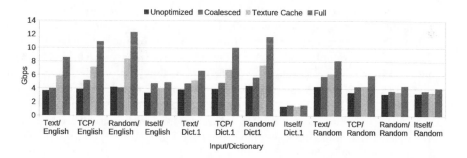

FIGURE 6.4: Effects of the CUDA optimizations on the Tesla C1060.

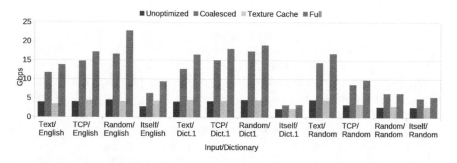

FIGURE 6.5: Effects of the CUDA optimizations on the Tesla C2050.

Figure 6.4 and Figure 6.5 show the effects of the two main optimizations of the GPU kernel on the Tesla C1060 and the Tesla C2050, respectively. The figures report the throughputs of the previously discussed combinations of inputs and dictionaries with an unoptimized kernel (*Unoptimized*), the kernel executing the transposition to coalesce the input text (*Coalesced*), the kernel storing the STTs in the Texture Memory (*Texture Cache*), and the kernel integrating both the optimizations (*Full*). The throughputs also take into account the transfer of the input text from the host to the GPU and, for the Coalesced and Fully optimized kernels, the execution of the transposition to coalesce the accesses to the input text. For the Tesla C1060 (Figure 6.4), we can see that in light matching conditions, the use of the texture memory gives the biggest benefits. The reason is that texture memory allows the use of the texture cache. In low matching conditions, the texture cache has sufficient size to cache the first levels of the STTs, and its lines rarely get replaced since other states are rarely accessed. Also, allocating the STT in the texture memory (even only partially), removes a large number of uncoalesced memory accesses. However, in heavy matching situations (i.e., dictionaries matched with themselves), the transposition of the input text allows better performance than the use of the texture memory. With all the optimizations, the maximum performance reached is a little over 12 Gbps for light matching conditions, while it remains around 1.5 Gbps for the worst case. For the Tesla C2050 (Figure 6.5) in the standard configuration for the caches (i.e., 16 KB of L1 cache enabled per SM, L2 always active), instead, the benefits of coalesced memory accesses for the input text are much more significant than only allocating the STT in the texture nemory. The reason is that the benefits due to the STT caching in texture memory are already obtained with the L1 and the L2 caches, especially in low matching conditions where the data have much reuse, and thus the uncoalesced reads only affect the first accesses. On the other hand, the input text has only spatial locality but no temporal locality and thus no reuse for data previously loaded. Since the Fermi architecture always tries to perform 128 byte transactions, with the chosen chunk sizes of 2 KB, coalescing the inputs maximizes the utilization of each memory transaction.

We conducted further experiments using only the L2 cache (disabling L1 through a specific compiler directive) and configuring the L1 caches to 48 KB (through intrinsics in the code). It is not possible to disable the L2 caches since the memory staging area is tightly connected to it. We executed tests with these parameters for all the versions of the code (Unoptimized, only Coalesced, only Texture Cache, and Full optimization), but we only report in Figure 6.6 the tests with the STT allocated in texture memory and uncoalesced input streams and in Figure 6.7 the results with the fully optimized code since they are the most significant. From Figure 6.6 we can see that completely bypassing the L1s and using only the L2 brings better performance when the texture caches are also used. Using larger L1s sometimes even reduces the performance. The reason is that the texture caches are optimized mainly for spatial locality, while the L1 caches are optimized for temporal locality; thus, they are somewhat conflicting. This is particularly evident under heavy matching when the STT is thoroughly explored and has almost no temporal locality. When using all the optimizations and changing the configuration of the L1 caches, as Figure 6.7 shows, the results are mixed. However, it appears that when there is

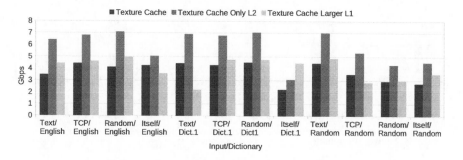

FIGURE 6.6: Effects of the cache hierarchy of the Tesla C2050 on the code with only texture optimizations.

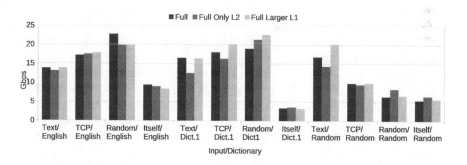

FIGURE 6.7: Effects of the cache hierarchy of the Tesla C2050 on the fully optimized code.

heavy matching (dictionaries matched with themselves) bypassing the L1 can slightly raise the throughput because the relatively small L1s do not influence the performance with misses. In many light matching cases (but not in all), increasing the size of the L1 increases the throughput because of the higher temporal locality. In particular, for the Random input and English dictionary combination, the fully optimized kernel with standard setting for the caches gets better performance. Also, in the Text and Dictionary 1 combination, the fully optimized kernel running with the standard cache settings has performance similar to the same kernel executed with larger L1 caches. The reasons are that the English dictionary fits completely in the texture memory, so partial allocation is not used, and all the accesses to the STT go through the textures units and caches, while for Dictionary 1, the states that are allocated in the texture memory mainly coincide with those of the textual patterns.

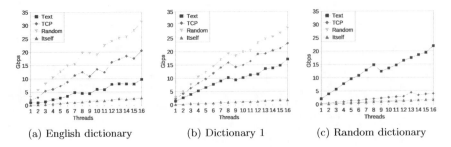

(a) English dictionary   (b) Dictionary 1   (c) Random dictionary

FIGURE 6.8: Scaling on the dual Xeon W5590.

6.5.3   Scaling

Figure 6.8 shows how the performance of the algorithm scales when increasing the number of threads on the dual Xeon W5590. For all the combinations of inputs and dictionaries, the throughput appears to increase almost linearly up to the number of physical cores of the system. The threads are pinned so as to progressively use all the physical cores before starting to exploit Hyper-Threading. The reasons for the small variance in performance with respect to the linear trend when using fewer threads than cores are mainly due to the static partitioning of the chunks processed by each thread. In fact, since the size of the chunks changes, the number of matching patterns in each chunk may also change, thus influencing the overall performance of the parallel algorithm. All the graphs show that in light matching conditions, when beginning to use more than one thread per core, the performance lowers before starting to increase again. However, it does increase with a linear trend. In heavy matching conditions, instead, the curve is smoother since the cache is constantly thrashed and the only performance limit is the sustained memory bandwidth. There is a significant variance among light and heavy matching conditions.

Figure 6.9 presents the throughputs obtained with the various combinations of inputs and dictionaries on the quad Opteron 6176SE. We executed benchmarks by increasing the number of threads one by one, but we only

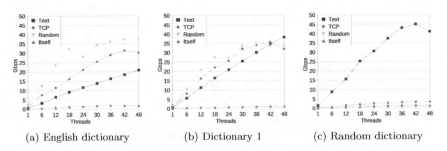

(a) English dictionary   (b) Dictionary 1   (c) Random dictionary

FIGURE 6.9: Scaling on the quad Opteron 6176SE.

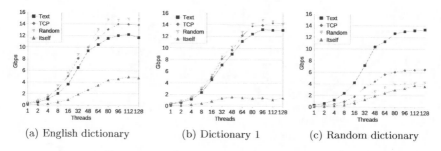

(a) English dictionary   (b) Dictionary 1   (c) Random dictionary

FIGURE 6.10: Scaling on the dual Niagara 2.

report the steps corresponding to filling a die in the plots. The results are more varied. In general, it seems that with the text input, on the dictionaries that are mostly based on textual patterns, the increase in throughput remains almost linear. With inputs that have different symbol frequencies, however, there is much more variability, which depends on the number of patterns that match in each chunk and, consequently, on the cache behavior of the different dies. When using the text input with the random dictionary, instead, the performance has an almost linear trend up to 36 processors, but it starts lowering with a high number of threads (i.e., when adding the two dies of another socket), even if the peak performance reached is the highest with respect to all the other combinations. The reason is that the system starts experiencing conflicts for the accesses on the same cells of the STT, even if there is a limited number of matches. In fact, the text input uses a limited range of symbols of the ASCII alphabet, thus requiring the system to always access the same columns of the STT. The performance variability between heavy matching and light matching cases shown by the quad Opteron systems are the highest in our study.

The scaling of the throughput of the dual Niagara 2 system, reported in Figure 6.10, is particularly interesting, especially when compared to the trends shown by the x86 architectures. For Niagara 2, we executed the `pthreads` implementation of the algorithm, progressively increasing the number of threads from 1 to 128 (64 per processor). However, we report in the figure only the relevant runs. Niagara 2 is able to obtain significant speedups with up to 80 threads. With light and medium matching data sets, the speedups are almost linear up to 64 threads. At 80 threads, the speedups start reducing and become marginal over 96 threads. In these conditions, the performance is quite stable (i.e., similar results for different dictionaries and input streams). In fact, multithreading allows hiding the memory access latency efficiently. However, with heavy matching data sets (i.e., dictionaries matched with themselves or with input streams containing a large number of recognized patterns), the performance is significantly lower. With these data sets, the level of multithreading supported by Niagara 2 is not sufficient to hide all the latency,

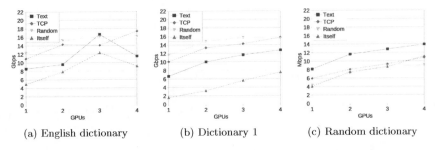

(a) English dictionary   (b) Dictionary 1   (c) Random dictionary

FIGURE 6.11: Scaling with multiple Tesla C1060 boards.

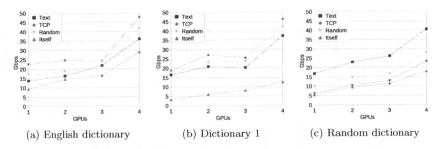

(a) English dictionary   (b) Dictionary 1   (c) Random dictionary

FIGURE 6.12: Scaling with multiple Tesla C2050 (Fermi-based) boards.

and the performance is reduced due to the thrashing of the small second-level cache, memory host-spots, and inter-socket communication.

Figures 6.11 and 6.12 present the scaling obtained with multiple GPU boards. The parallelization among multiple GPUs is obtained by further partitioning the input set. The Tesla C1060s reach a peak performance slightly over 16 Gbps. The Tesla C2050s, instead, reach a peak performance near 48 Gbps. With the Tesla C1060s and the Random and Dictionary 1 dictionaries, adding more GPUs generally increases the performance. However, the increases in throughput are not linear, and with the Random input set, the throughput flattens when moving from three to four GPUs. With the English dictionary, the throughputs are more varied. In heavy matching conditions (Text input and dictionary matched with itself), the performance even decreases when moving from three to four GPUs due to the different chunking of the input set and the saturation of the PCI Express bus (we measure performance end to end). In light matching conditions, instead, performance flattens (Random input) or increases (TCP input) when moving from three to four GPUs. With the Tesla C2050s, increasing the number of GPUs increases the throughput in all cases. However, increasing the number of GPUs from three to four increases the performance more than when moving from two to three GPUs. This behavior is mainly due to the PCI Express (PCI-E) switch in the

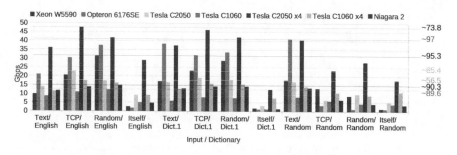

FIGURE 6.13: Comparison of all the architectures discussed in the chapter with variability percentage

Intel 5500 chip set of the machine used for the tests, which performs better when four end points are used.

Figure 6.13 shows a comparison among all the architectures discussed in the chapter. The figure also reports the performance variability for each architecture between the lowest and the highest throughputs obtained with the various combinations of Input and Dictionaries. The highest performance is reached with four Tesla C2050s (48 Gbps). The 4-socket Opteron 6176SE is near, reaching almost 45 Gbps but on different data sets. Nevertheless, the difference among the highest throughput and the lowest throughput for the four Tesla C2050s is only 73.8%, while for the Opteron the variability is 97%, the highest in our evaluation. The 2-socket Xeon W5590 reaches slightly over 30 Gbps with a variability of 95.3%. A single Tesla C2050 outperforms four Tesla C1060s, with a peak of 24 Gbps versus 16 Gbps. The variability for a single Tesla C2050 is 85.4%, while for the four Tesla C1060s is 56.5%, the lowest in our evaluation. For a single Tesla C1060, the variability is 89.6%, only slightly lower than the 2-socket Niagara 2 (90.3%). On the other hand, peak performance for the Niagara 2 is slightly higher than for the Tesla C1060 (14 Gbps versus 12 Gbps). In general, multi-GPU configurations have lower variability. In fact, the scaling on multiple GPUs is mainly limited by the PCI-E bandwidth, which in turn limits peak performances for light matching benchmarks. Heavy matching benchmarks, instead, get a considerable speedup by using multiple GPUs since they are not PCI bandwidth limited and have similar matching behavior even if the input sets are further chunked for execution on different GPUs. Even if multithreaded architectures such as the Niagara 2 and the GPUs somehow limit the variability due to the higher latency tolerance with respect to x86 systems, they still employ cache hierarchies (private L1s and a small unified L2 for the Niagara 2, texture caches for the Tesla C1060, texture caches, private L1s and small L2 for the Tesla C2050). Thus, their throughputs still vary in a range from 85 to 90%. The x86 architectures, instead, which employ complex cache architectures with up to three levels of cache (even on different dies inside the same socket for the Opteron) exhibit the highest performance variability, 97% and 95%, respectively, for the Opteron and the Xeon system.

6.6   Conclusions

In this chapter, we presented a study of the behavior of the Aho-Corasick string-matching algorithm on a set of modern multicore and multithreaded architectures. The Aho-Corasick algorithm is able to perform the search in a time linearly proportional to the length of the input text independently from the pattern set size. However, in practical implementations, when using big dictionaries, it may present significant performance variability (i.e., an "irregular" behavior) due to the memory access times and the cache effects.

We presented the implementation of the algorithm on a 2-socket Xeon W5590 ("Nehalem" architecture) system, a 4-socket 6176SE Opteron ("Magny Cours" architecture) system, a 2-socket Niagara 2 system and on systems integrating from one to four Tesla C1060s (T10 GPU) or Tesla C2050s (T20 "Fermi" GPU). We discussed several optimizations of the algorithm for the CUDA-based GPUs, and evaluated their effects on the two different architectures. We analyzed the performance trade-offs due to the new cache hierarchies in Fermi. We presented a full evaluation of the scaling of the algorithm on the various platforms, discussing trade-offs in terms of peak performance, performance variability, and data set size.

We believe that this work represents an important case study for a better understanding of the behavior of irregular algorithms on modern architectures.

Bibliography

[1] Dechang Chen and Xiuzhen Cheng (Eds.). *Pattern Recognition and String Matching.* Springer, 2003.

[2] Symantec Corporation. Symantec Global Internet Security Threat Report. White paper, Apr. 2008.

[3] Y. H. Cho and W. H. Mangione-Smith. Deep packet filter with dedicated logic and read only memories. In *FCCM '04: 12th Annual IEEE Symposium on Field-Programmable Custom Computing Machines*, pages 125–134, 2004.

[4] Ioannis Sourdis and Dionisios Pnevmatikatos. Fast, Large-Scale String Match for a 10Gbps FPGA-Based Network Intrusion. In *FPL '03: 13th*

Conference on Field Programmable Logic and Applications, pages 880–889, 2003.

[5] Abhishek Mitra, Walid Najjar, and Laxmi Bhuyan. Compiling PCRE to FPGA for accelerating SNORT IDS. In *ANCS '07: The 3rd ACM/IEEE Symposium on Architecture for Networking and Communications Systems*, pages 127–136, 2007.

[6] C. R. Clark and D. E. Schimmel. Scalable pattern matching for high speed networks. In *FCCM '04: 12th Annual IEEE Symposium on Field-Programmable Custom Computing Machines*, pages 249–257, 2004.

[7] Oreste Villa, Daniel Chavarria-Miranda, and Kristyn Maschhoff. Input-independent, scalable and fast string matching on the Cray XMT. In *IPDPS 2009: 23rd IEEE International Symposium on Parallel & Distributed Processing*, pages 1–12, 2009.

[8] Davide Pasetto, Fabrizio Petrini, and Virat Agarwal. Tools for Very Fast Regular Expression Matching. *Computer*, 43:50–58, 2010.

[9] Oreste Villa, Daniele Paolo Scarpazza, and Fabrizio Petrini. Accelerating Real-Time String Searching with Multicore Processors. *Computer*, 41(4):42–50, 2008.

[10] Daniele Paolo Scarpazza, Oreste Villa, and Fabrizio Petrini. Exact multi-pattern string matching on the Cell/B.E. processor. In *CF '08: The 2008 Conference on Computing Frontiers*, pages 33–42, 2008.

[11] Martin Roesch. Snort: Lightweight Intrusion Detection for Networks. In *LISA*, pages 229–238, 1999.

[12] Nigel Jacob and Carla Brodley. Offloading IDS Computation to the GPU. In *ACSAC '06: 22nd Annual Computer Security Applications Conference*, pages 371–380, 2006.

[13] Giorgos Vasiliadis, Spiros Antonatos, Michalis Polychronakis, Evangelos P. Markatos, and Sotiris Ioannidis. Gnort: High Performance Network Intrusion Detection Using Graphics Processors. In *RAID '08: 11th International Symposium on Recent Advances in Intrusion Detection*, pages 116–134, 2008.

[14] Giorgos Vasiliadis and Sotiris Ioannidis. Gravity: A massively parallel antivirus engine. In Somesh Jha, Robin Sommer, and Christian Kreibich, editors, *Recent Advances in Intrusion Detection*, volume 6307 of *Lecture Notes in Computer Science*, pages 79–96, Berlin, 2010. Springer.

[15] Donald E. Knuth, Jr, and Vaughan R. Pratt. Fast pattern matching in strings. *SIAM Journal on Computing*, 6(2):323–350, 1977.

[16] Robert S. Boyer and J. Strother Moore. A fast string searching algorithm. *Communications of the ACM*, 20(10):62–72, 1977.

[17] Alfred V. Aho and Margaret J. Corasick. Efficient string matching: An aid to bibliographic search. *Communications of the ACM*, 18(6):333–340, 1975.

[18] Beate Commentz-Walter. A String Matching Algorithm Fast on the Average. In *Proceedings of the 6th Colloquium, on Automata, Languages and Programming*, pages 118–132, 1979.

[19] S. Wu and U. Manber. A fast algorithm for multi-pattern searching. Technical Report TR-94-17, 1994.

[20] Christopher Oehmen and Jarek Nieplocha. ScalaBLAST: A scalable implementation of BLAST for high-performance data-intensive bioinformatics analysis. *IEEE Transactions on Parallel and Distributed Systems*, 17:740–749, 2006.

[21] Cole Trapnell and Michael C. Schatz. Optimizing data intensive GPGPU computations for DNA sequence alignment. *Parallel Computing*, 35(8-9):429 – 440, 2009.

[22] Janghaeng Lee, Sung Ho Hwang, Neungsoo Park, Seong-Won Lee, Sunglk Jun, and Young Soo Kim. A high performance NIDS using FPGA-based regular expression matching. In *SAC '07: the 2007 ACM symposium on Applied computing*, pages 1187–1191, 2007.

[23] R. Sidhu and V. Prasanna. Fast regular expression matching using FPGAs. In *FCCM '01: 9th Annual IEEE Symposium on Field-Programmable Custom Computing Machines*, pages 227–238, 2001.

[24] Hong-Jip Jung, Z.K. Baker, and V.K. Prasanna. Performance of FPGA implementation of bit-split architecture for intrusion detection systems. In *IPDPS 2006: 20th IEEE International Symposium on Parallel & Distributed Processing*, page 8, 2006.

[25] Ioannis Sourdis and Dionisios Pnevmatikatos. Pre-Decoded CAMs for Efficient and High-Speed NIDS Pattern Matching. In *FCCM '04: The 12th Annual IEEE Symposium on Field-Programmable Custom Computing Machines*, pages 258–267, 2004.

[26] N. Tuck, T. Sherwood, B. Calder, and G. Varghese. Deterministic memory-efficient string matching algorithms for intrusion detection. In *INFOCOM 2004: 23rd IEEE International Conference on Computer Communications*, volume 4, pages 2628–2639, 2004.

[27] Yutaka Sugawara, Mary Inaba, and Kei Hiraki. Over 10Gbps string matching mechanism for multi-stream packet scanning systems. In *FPL*

'04: 14th Conference on Field Programmable Logic and Applications, pages 484–493, 2004.

[28] Sarang Dharmapurikar and John Lockwood. Fast and scalable pattern matching for content filtering. In *ANCS '05: ACM Symposium on Architecture for Networking and Communications Systems*, pages 183–192, 2005.

[29] Y.-H.E. Yang, Hoang Le, and V.K. Prasanna. High performance dictionary-based string matching for deep packet inspection. In *INFOCOM 2010: 29th IEEE International Conference on Computer Communications*, pages 1–5, 2010.

[30] Francesco Iorio and Jan Van Lunteren. Fast pattern matching on the CELL Broadband Engine. In *WCSA: Workshop on Cell Systems and Applications, affiliated with ISCA08*, 2008.

[31] Jan van Lunteren. High-performance pattern-matching for intrusion detection. In *INFOCOM 2006: 25th IEEE International Conference on Computer Communications*, pages 1–13, 2006.

[32] Nen-Fu Huang, Hsien-Wei Hung, Sheng-Hung Lai, Yen-Ming Chu, and Wen-Yen Tsai. A GPU-based multiple-pattern matching algorithm for network intrusion detection systems. In *AINAW 2008: 22nd International Conference on Advanced Information Networking and Applications*, pages 62–67, 2008.

[33] R. Smith, N. Goyal, J. Ormont, K. Sankaralingam, and C. Estan. Evaluating GPUs for network packet signature matching. In *ISPASS 2009: International Symposium on Performance Analysis of Systems and Software*, pages 175–184, 2009.

[34] Jamin Naghmouchi, Daniele Paolo Scarpazza, and Mladen Berekovic. Small-ruleset regular expression matching on GPGPUs: Quantitative performance analysis and optimization. In *ICS '10: 24th International Conference on Supercomputing*, pages 337–348, 2010.

[35] Jin Hwan Park and Bernard A. Demirdag. High performance pattern matching with dynamic load balancing on heterogeneous systems. In *PDP '06: 14th Euromicro International Conference on Parallel, Distributed, and Network-Based Processing*, pages 285–290, 2006.

[36] Panagiotis D. Michailidis and Konstantinos G. Margaritis. String matching problem on a cluster of personal computers: Experimental results. In *15th International Conference Systems for Automation of Engineering and Research*, pages 71–75, 2001.

[37] Bruce W. Watson. The performance of single-keyword and multiple-keyword pattern matching algorithms. Technical Report 19, Eindhoven University of Technology, 1994.

[38] Greg Ruetsch and Paulius Micikevicius. Optimizing Matrix Transpose in CUDA. NVIDIA white paper, Jan. 2000.

Chapter 7

Sorting on a Graphics Processing Unit (GPU)

Shibdas Bandyopadhyay

University of Florida

Sartaj Sahni

University of Florida

7.1   Graphics Processing Units

Contemporary graphics processing units (GPUs) are massively parallel many-core processors. NVIDIA's Tesla GPUs, for example, have 240 scalar processing cores (SPs) per chip [22]. These cores are partitioned into 30 streaming multiprocessors (SMs) with each SM comprised of eight SPs. Each SM shares a 16-KB local memory (called shared memory) and has a total of 16,384 32-bit registers that may be utilized by the threads running on this SM. Besides registers and shared memory, on-chip memory shared by the cores in an SM also includes constant and texture caches. The 240 on-chip cores also share a 4-GB off-chip global (or device) memory. Figure 7.1 shows a schematic of the Tesla architecture. With the introduction of CUDA (Compute Unified Driver Architecture) [35], it has become possible to program GPUs using C. This has resulted in an explosion of research directed toward expanding the applicability of GPUs from their native computer graphics applications to a wide variety of high-performance computing applications.

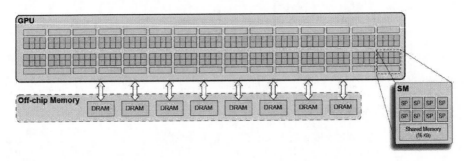

FIGURE 7.1: NVIDIA's Tesla GPU [26].

GPUs operate under the master-slave computing model (see [25]) in which there is a host or master processor to which is attached a collection of slave processors. A possible configuration would have a GPU card attached to the bus of a PC. The PC CPU would be the host or master, and the GPU processors would be the slaves. The CUDA programming model requires the user to write a program that runs on the host processor. At present, CUDA supports host programs written in C and C++ only, though there are plans to expand the set of available languages [35]. The host program may invoke *kernels*, which are C functions, that run on the GPU slaves. A kernel may be instantiated in synchronous (the CPU waits for the kernel to complete before proceeding with other tasks) or asynchronous (the CPU continues with other tasks following the spawning of a kernel) mode. A kernel specifies the computation to be done by a thread. When a kernel is invoked by the host program, the host program specifies the number of threads that are to be created. Each thread is assigned a unique ID, and CUDA provides C-language extensions to enable a kernel to determine which thread it is executing. The host program groups threads into blocks by specifying a block size at the time a kernel is invoked. Figure 7.2 shows the organization of threads used by CUDA.

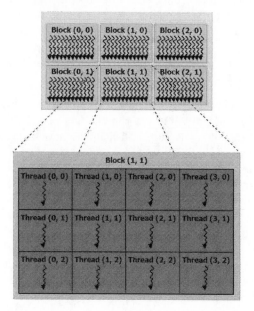

FIGURE 7.2: Cuda programming model [35].

The GPU schedules the threads so that a block of threads runs on the cores of an SM. At any given time, an SM executes the threads of a single block, and the threads of a block can execute only on a single SM. Once a block begins to execute on an SM, that SM executes the block to completion. Each SM schedules the threads in its assigned block in groups of 32 threads called a *warp*. The partitioning into warps is fairly intuitive with the first 32 threads forming the first warp, the next 32 threads form the next warp, and so on. A *half-warp* is a group of 16 threads. The first 16 threads in a warp form the first half-warp for the warp, and the remaining 16 threads form the second half-warp. When an SM is ready to execute the next instruction, it selects a warp that is ready (i.e., its threads are not waiting for a memory transaction to complete) and executes the next instruction of every thread in the selected warp. Common instructions are executed in parallel using the eight SPs in the SM. Noncommon instructions are serialized. So it is important, for performance, to avoid thread divergence within a warp. Some of the other factors important for performance are as follows:

1. Since access to global memory is about two orders of magnitude more expensive than access to registers and shared memory, data that are to be used several times should be read once from global memory and stored in registers or shared memory for reuse.

2. When the threads of a half-warp access global memory, this access is accomplished via a series of memory transactions. The number of memory transactions equals the number of different 32-byte (64-byte, 128-byte, 128-byte) memory segments that the words to be accessed lie in when each thread accesses an 8-bit (16-bit, 32-bit, 64-bit) word. Given the cost of a global memory transaction, it pays to organize the computation so that the number of global memory transactions made by each half-warp is minimized.

3. Shared memory is divided into 16 banks in round-robin fashion using words of size 32 bits. When the threads of a half-warp access shared memory, the access is accomplished as a series of one or more memory transactions. Let S denoted the set of addresses to be accessed. Each transaction is built by selecting one of the addresses in S to define the broadcast word. All addresses in S that are included in the broadcast word are removed from S. Next, up to one address from each of the remaining banks is removed from S. The set of removed addresses is serviced by a single memory transaction. Since the user has no way to specify the broadcast word, for maximum parallelism, the computation should be organized so that, at any given time, the threads in a half warp access either words in different banks of shared memory, or they access the same word of shared memory.

4. Volkov et al. [30] have observed greater throughput using operands in registers than operands in shared memory. So, data that are to be used often should be stored in registers rather than in shared memory.

5. Loop unrolling often improves performance. However, the `#pragma` unroll statement unrolls loops only under certain restrictive conditions. Manual loop unrolling by replicating code and changing the loop stride can be employed to overcome these limitations.

6. Arrays declared as register arrays get assigned to global memory when the CUDA compiler is unable to determine at compile time what the value of an array index is. This is typically the case when an array is indexed using a loop variable. Manually unrolling the loop so that all references to the array use indexes known at compile time ensures that the register array is, in fact, stored in registers.

7.2 Sorting Numbers on GPUs

One of the very first GPU sorting algorithms, an adaptation of bitonic sort, was developed by Govindraju et al. [12]. Since this algorithm was developed before the advent of CUDA, the algorithm was implemented using GPU pixel shaders. Zachmann et al. [13] improved on this sort algorithm by using Bitonic Trees to reduce the number of comparisons while merging the bitonic sequences. Cederman et al. [7] have adapted quick sort for GPUs. Their adaptation first partitions the sequence to be sorted into subsequences, sorts these subsequences in parallel, and then merges the sorted subsequences in parallel. A hybrid sort algorithm that splits the data using bucket sort and then merges the data using a vectorized version of merge sort is proposed by Sintron et al. [28]. Satish et al. [26] have developed an even faster merge sort. In this merge sort, two sorted sequences A and B are merged by a thread block to produce the sequence C when A and B have less than 256 elements each. Each thread reads an element of A and then does a binary search on the sequence B with that element to determine where it should be placed in the merged sequence C. When the number of elements in a sequence is more than 256, A and B are divided into a set of subsequences using a set of splitters. The splitters are chosen from the two sequences in such a way that the interval between successive splitters is small enough to be merged by a thread block. The fastest GPU merge-sort algorithm known at this time is Warpsort [31]. Warpsort first creates sorted sequences using bitonic sort, each sorted sequence being created by a thread warp. The sorted sequences are merged in pairs until only a small number of sequences remain. The remaining sequences are partitioned into subsequences that can be pairwise merged independently, and finally this pairwise merging is done with each warp merging a pair of subsequences. Experimental results reported in [31] indicate that Warpsort is about 30% faster than the merge-sort algorithm of [26]. Another comparison-based sort for GPUs—GPU sample sort—was developed by Leischner et al. [20]. Sample sort is reported to be about 30% faster than the merge sort of [26], on average, when the keys are 32-bit integers. This would make sample sort competitive with Warpsort for 32-bit keys. For 64-bit keys, sample sort is twice as fast, on average, as the merge sort of [26].

Several sources [27, 34, 19, 26, 23, 1] have adapted radix sort to GPUs. Radix sort works in phases where each phase sorts on a digit of the key using, typically, either a count sort or a bucket sort. The counting to be done in each phase may be carried out using a prefix sum or scan [5] operation that is quite efficiently done on a GPU [27]. Harris et al.'s [34] adaptation of radix sort to GPUs uses the radix 2 (i.e., each phase sorts on a bit of the key) and uses the bit split technique of [5] in each phase of the radix sort to reorder numbers by the bit being considered in that phase. This implementation of radix sort is available in the CUDA Data Parallel Primitive (CUDPP) library [34]. For

32-bit keys, this implementation of radix sort requires 32 phases. In each phase, expensive scatter operations to/from the global memory are made. Le Grand et al. [19] reduce the number of phases and hence the number of expensive scatters to global memory by using a larger radix, 2^b, for $b > 0$. A radix of 16, for example, reduces the number of phases from 32 to 8. The sort in each phase is done by first computing the histogram of the 2^b possible values that a digit with radix 2^b may have. Satish et al. [26] further improve the 2^b-radix sort of Le Grand et al. [19] by sorting blocks of data in shared memory before writing to global memory. This reduces the randomness of the scatter to global memory, which, in turn, improves performance. The radix sort implementation of Satish et al. [26] is included in NVIDIA's CUDA SDK 3.0. Bandyopadhyay and Sahni [1] developed the radix sort algorithm, GRS, which is suitable for sorting records with many fields. GRS outperforms the SDK radix sort algorithm while sorting numbers by reducing the number of steps and using an additional storage in the global memory. Merrill and Grimshaw [23] have developed an alternative radix sort, SRTS, for GPUs that is based on a highly optimized algorithm, developed by them, for the scan operation and comingling of several logical steps of a radix sort so as to reduce accesses to device/global memory. Presently, SRTS is the fastest GPU radix sort algorithm for 32-bit integers.

The results of [20, 31] indicate that the radix sort algorithm of [26] outperforms both Warpsort [31] and sample sort [20] on 32-bit keys. These results together with those of [23] imply that the radix sort of [23] is the fastest GPU sort algorithm for 32-bit integer keys.

Most of the GPU sorting routines perform some functions repetitively during their operations. These primitives are designed to be easily parallelized on a many-core architecture like a GPU. Two such important primitives that are frequently used in GPU sorting are `scan` and `reduce`. Scan is a function that takes a list of n elements $[\ x_0...x_{n-1}\ ]$ and a binary operator $\oplus$ as input. The output is also a list of n elements $[y_0....y_{n-1}]$. There are two variants of scan, `exclusive` and `inclusive`. For exclusive scan, $y_i := x_0 \oplus x_1 \oplus x_2 \oplus ... \oplus x_{i-1}$ for $i > 0$, and $y_0 = \oplus_0$ ($\oplus_0$ is the identity element defined for operator $\oplus$); while in the case of inclusive scan, $y_i := x_0 \oplus x_1 \oplus x_2 \oplus ... \oplus x_i$. The reduce operation, on the other hand, produces a single value $y = x_0 \oplus x_1 \oplus x_2 \oplus ... \oplus x_{n-1}$. Many useful operations can be designed using these primitives. For example, prefix sum is an exclusive scan with + as the binary operator. Finding the maximum of a sequence is a reduce with $\oplus$ as maximum operator that returns the larger of two elements.

7.2.1   SDK Radix Sort Algorithm

The SDK radix sort algorithm [26] uses a radix of 2^b with $b = 4$ ($b = 4$ was determined, experimentally, to give best results). With $b = 4$ and 32-bit keys, the radix sort runs in eight phases with each phase sorting on 4 bits of the key. Each phase of the radix sort is accomplished in 4 steps as below.

These steps assume the data are partitioned (implicitly) into tiles of size 1024 numbers each. There is a separate kernel for each of these four steps.

Step 1: Sort each tile on the b bits being considered in this phase using the bit-split algorithm of [5].

Step 2: Compute the histogram of each tile. Note that the histogram has 2^b entries with entry i giving the number of elements in the tile for which the b bits considered in this phase equal i.

Step 3: Compute the prefix sum of the histograms of all tiles.

Step 4: Use the prefix sums computed in Step 3 to rearrange the numbers into sorted order of the b bits being considered.

7.2.1.1 Step 1: Sorting tiles

For Step 1, each SM inputs a tile into shared memory using 256 threads (eight warps). So, each thread reads four consecutive numbers facilitating a coalesced read from the global memory. Each thread inputs the four numbers using a variable of type int4, which is 128 bits long. The input data are stored in registers. Next, the 256 threads collaborate to do b rounds of the bit-split scheme of [5]. In each round, all the numbers having 0 in that bit place are moved forward followed by the ones having 1 in the bit position being considered as shown in Figure 7.3.

0011	0000	1100	0111	0011	0100	1101	0011	*input*

0000	1100	0100	0011	0111	0011	1101	0011	*bit 0*

0000	1100	0100	1101	0011	0111	0011	0011	*bit 1*

0000	0011	0011	0011	1100	0100	1101	0111	*bit 2*

0000	0011	0011	0011	0100	0111	1101	1100	*bit 3*

FIGURE 7.3: Bit-split scheme for sorting numbers on 4 bits [5].

To move the numbers having 0s ahead of the ones having 1s, the position of each number in the output is determined, and then the number is written into the shared memory in that position. For the next round, threads read the numbers in the sorted order of the previous bit from the shared memory. A prefix scan-based algorithm called warp-scan is executed by each thread to find out the rank of the four numbers it has read. First, each thread calculates the number of 0s in the current bit position of the four numbers read by it. It then writes the value in a particular location in the shared memory depending on

```
Algorithm warp-scan() \\
{ // Compute the prefix sum of count vector
// warpId is the warp ID = tid >> 5 and warpSize is 32
idx = 2 * warpId * warpSize + tid;
shared[idx] = 0 // Zero out the location
idx += warpSize;
shared[idx] = val; // Copy the count here
shared[idx] += shared[idx - 1];
shared[idx] += shared[idx - 2];
shared[idx] += shared[idx - 4];
shared[idx] += shared[idx - 8];
shared[idx] += shared[idx - 16];
// Convert inclusive to exclusive
return (shared[idx - val);
}
```

LISTING 7.1: Divergence-free warp-scan algorithm.

its thread ID (`tid`), constructing a count vector per warp. Additional storage is allocated in the shared memory adjacent to the count vector and zeored out before the calculation begins. An exclusive prefix scan of the count vector is performed by the 32 threads in a warp in $\log(32) = 5$ steps. During step i ($i = 0...4$), each thread adds its own count with the count which is 2^i away as shown in the algorithm of Listing 7.1 and corresponding schematic in Figure 7.4. This algorithm has no divergence within a warp. However, spurious additions are performed by the threads during each step other than the first step when each addition is necessary.

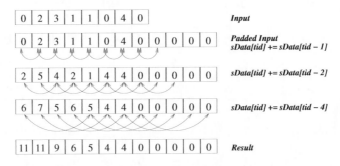

FIGURE 7.4: Warp scan of eight numbers.

A final scan is performed over the last values of eight warps to find out the final position of the four numbers read by a thread. Each thread then writes the numbers out to the shared memory according to the determined position. Following these b rounds of the bit-split algorithm, the tile is in sorted order (of the b bits being considered) in registers. The sorted tile is then written out to global memory.

7.2.1.2 Step 2: Calculating histogram

In this step, the numbers for each half-tile are input (in sorted order) using 256 threads per SM, and the histogram of the b bits being considered is computed for the half-tile input. Note that one can instead input the entire tile and the remaining steps would work fine, but empirically, working with half-tiles from this point onward performs better. For this computation, each thread inputs two numbers from global memory and writes these to shared memory. The threads then determine up to 15 places in the input half-tile where the 4 bits being considered change. For this, each thread simply checks its own number with the next one. If the bits differ, it records the position in a shared-memory array. Once all the threads are done, all positions in the half-tile where the bits change are recorded in the array as shown by the broken arrows in Figure 7.5

FIGURE 7.5: Calculating histogram offsets.

Once the positions are determined, the size of each bucket (i.e., number of elements having same bit values) is determined by finding out the intervals between the positions by the same threads which found the positions during the previous step. The histogram is written out to global memory in column major order, i.e., bucket 0 of all half-tiles followed by bucket 1 of all half-tiles all the way up to bucket 15 of all half-tiles. The positions in a half-tile where the bits differ are also written out to global memory, as this information is needed in the final step.

7.2.1.3 Step 3: Prefix sum of histogram

In Step 3, the prefix sum of the half-tile histograms is computed using the prefix-sum code available in CUDPP [34] and is written to global memory as shown in the Figure 7.6.

7.2.1.4 Step 4: Rearrangement

In Step 4, each SM inputs a half-tile of data, which is sorted on the b bits being considered and uses the computed prefix-sum histogram for this half-tile to write the numbers in the half-tile to their correct position in global memory. Each thread in the SM reads two numbers as before, corresponding to its ID. The tile offsets calculated in Step 2 and the prefix-sum histogram from Step 3 are also read. As the half-tiles are already sorted following Step 1, the thread ID is the same as the position of the number read in the tile. If the

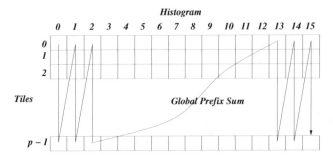

FIGURE 7.6: Column major histogram.

value of the b bits being considered for the current number is r, the prefix-sum histogram gives how many numbers there are with b bit value less than r and also how many numbers in the tiles before it have bit values exactly equal to r. However, one also needs to find out how many of the numbers in the current tile have a bit value exactly equal to r and occur before the number in the sorted order. As per Step 2, the tile offset for bit value $(r-1)$ read from the global memory indicates the place where the bit value change occurs from $(r-1)$ to r, i.e., the place from where numbers with bit value r starts. Hence, there are $(\texttt{tid} - \texttt{tileOffset}[r-1])$ numbers before the current number having bit values exactly equal to r. Hence, the final position of the number will be the sum of this local offset and the prefix-sum histogram. As the numbers are already in sorted order when read by threads, consecutive numbers in the tile read by consecutive threads in a half-warp are written to consecutive locations in global memory. This reduces the random scattering by a significant amount as most of the writes produce coalesced memory transactions. Following this, all numbers are in sorted order in global memory with respect to the b bits being considered in this phase of the radix sort.

7.2.2   GPU Radix Sort (GRS)

Like the SDK radix sort algorithm of [26], GRS accomplishes a radix sort using a radix of 2^b with $b = 4$. Each phase of the radix sort is done in three steps with each step using a different kernel. For purposes of discussion, we assume a tile size of t ($t = 1024$ in the SDK implementation). We define the *rank* of number i in a tile to be the number of integers in the tile that precede number i and have the same value as number i. Since we compute ranks in each phase of the radix sort, number equality (for rank purposes) translates to equality of the b bits of the number being considered in a particular phase. Note that when the tile size is 1024, ranks lie in the range 0 through 1023. The three steps in each phase of GRS are the following:

Step 1: Compute the histogram for each tile as well as the rank of each number in the tile. This histogram is the same as that computed in Step 2 of the SDK radix sort.

Step 2: Compute the prefix sums of the histograms of all tiles.

Step 3: Use the ranks computed in Step 1 to sort the data within a tile. Next, use the prefix sums computed in Step 2 to rearrange the numbers into sorted order of the b bits being considered.

Step 1 requires us to read the numbers in a tile from global memory, compute the histogram and ranks, and then write the computed histogram and ranks to global memory. Step 2 is identical to Step 3 of the SDK algorithm. In Step 3, numbers are again read from global memory. The numbers in a tile are first reordered in shared memory to get the sorted arrangement of Step 1 of the SDK algorithm and then written to global memory so as to obtain the sorted order following Step 4 of the SDK algorithm. This writing of numbers from shared memory to global memory is identical to that done in Step 4 of the SDK algorithm. The following subsections provide implementation details for the three steps of GRS. Note that the sorting done by the SDK radix sort during the very first step is now done using the ranks calculated in the first step, and the sorting is merged into the last step of GRS where the numbers are finally put into their sorted order. As the rank calculation is done without any overhead while calculating the histogram, GRS is doing less work compared to SDK radix sort. However, GRS uses an additional storage for storing the ranks.

7.2.2.1 Step 1: Histogram and ranks

An SM computes the histograms and ranks for 64 tiles at a time employing 64 threads. Listing 7.2 gives a high-level description of the algorithm used by us for this purpose. Our algorithm processes 32 numbers from each tile in an iteration of the for loop. So the number of for loop iterations is the tile size (t) divided by 32. In each iteration of the for loop, the 64 threads cooperate to read 32 numbers from each of the 64 tiles. This is done in such a way (described later) that global memory transactions are 128 bytes each. The data that are read are written to shared memory. Next, each thread reads the 32 numbers of a particular tile from shared memory and updates the tile histogram, which itself resides in shared memory. Although we have enough registers to accommodate the 64 histograms, CUDA relegates a register array to global memory unless it is able to determine, at compile time, the value of the array index. To maintain the histograms in registers, we need an elaborate histogram update scheme whose computation time exceeds the time saved over making accesses to random points in an array stored in shared memory. When a number is processed by a thread, the thread extracts the b bits in use for this phase of the radix sort. Suppose the extracted b bits have the value 12, then the current histogram value for 12 is the rank of the number. The new

```
Algorithm HR()
{ // Compute the histograms and ranks for 64 tiles
  itrs = t / 32; // t = tile size
  for(i = 0; i < itrs; i++)
{
Read 32 numbers from each of the 64 tiles;
  Determine the ranks and update the histograms;
Write the ranks to global memory;
}
  Write the histograms to global memory;
}
```

LISTING 7.2: Algorithm to compute the histograms and ranks of 64 tiles.

histogram value for 12 is one more than the current value. The determined rank is written to shared memory using the same location used by the number (i.e., the rank overwrites the number). Note that once a number is processed, it is no longer needed by Algorithm HR. Once the ranks for the current batch of 32 numbers per tile have been computed, these are written to global memory and we proceed to the next batch of 32 numbers. To write the ranks to global memory, those 64 threads cooperate ensuring that each transaction to global memory is 128 bytes. When the `for` loop terminates, we have successfully computed the histograms for the 64 tiles, and these are written to global memory.

To ensure 128-byte read transactions in global memory, we use an array that is declared as

`shared int4 sKeys4[512];`

Each element of `sKeys4` is comprised of four 4-byte integers, and the entire array is assigned to shared memory. A thread reads in one element of `sKeys4` at a time from global memory and in so doing, four numbers are input. It takes eight threads to cooperate so as to read in 32 numbers (or 128 bytes) from a tile. The 16 threads in a half-warp read 32 numbers from each of two tiles. This read takes two 128-byte memory transactions. With a single read of this type, the 64 threads together are able to read 32 numbers from a total of eight tiles. So, each thread needs to repeat this read eight times (each time targeting a different tile) in order for the 64 threads to input 32 numbers from each of 64 tiles. Besides maximizing bandwidth utilization from global memory, we also need to be concerned about avoiding shared-memory-bank conflicts when the threads begin to process the numbers for their assigned tile. Since shared memory is divided into 16 banks of 4-byte words, storing the numbers in a natural way results in the first number of each tile residing in the same bank. Since in the next step, each thread processes its tile's numbers in the same order, we will have shared-memory conflicts that will cause the reads of

```
// tid is the thread ID and bid is the block ID
 // Determine the first tile handled by this thread
 startTile = (bid * 64 + tid / 8) * (t / 4);
 // Starting number position in the tile
 // keyOffset is the offset for current 32 keys
 keyPos = keyOffset + tid % 8;
 // Shared memory position to write the keys with circular shift
 sKeyPos = (tid / 8) * 8 + (((tid / 8) % 8) + (tid % 8)) % 8;
 // Some constants
 tileSize8 = 8 * (t / 4); tid4 = tid * 4;
 // Initialize the histogram counters
 for(i = 0; i < 16; i++)
{
sHist[tid * 16 + i] = 0;
}
// Wait for all threads to finish
syncthreads();
 curTileId = startTileId;
 for(i = 0; i < 8; i++)
{
 sKeys4[sKeyPos + i * 64] = keysIn4[keyPos + startTile];
 curTileId += tileSize8;
}
 syncthreads();
```

LISTING 7.3: Reading the numbers from global memory.

numbers to be serialized within each half warp. To avoid this serialization, we use a circular shift pattern to map numbers to the array sKeys4. The CUDA kernel code to do this read is given in Listing 7.3.

As stated earlier, to compute the histograms and ranks, each thread works on the numbers of a single tile. A thread inputs one element (four numbers) of sKeys4, updates the histogram using these four numbers, and writes back the rank of these four numbers (as noted earlier, this equals the histogram value just before it is updated). Listing 7.4 gives the kernel code to update the histogram and compute the ranks for the four numbers in one element of sKeys4.

Once the ranks have been computed, they are written to global memory using a process similar to that used to read in the numbers. Listing 7.5 gives the kernel code. Here, ranks4[] is an array in global memory; its data type is int4.

When an SM completes the histogram computation for the 64 tiles assigned to it, it writes the computed 64 histograms to the array counters in global memory. If we view the 64 histograms as forming a 16×64 array, then this array is mapped to the one-dimensional array in column-major order. Listing 7.6 gives the kernel code for this.

```
// Update the histograms and calculate the rank
// startbit is the starting bit position for
// this phase
int4 p4, r4;
for(i = 0; i < 8; i ++)
{
p4 = sKeys4[(tid4) + (i + tid) % 8];
r4.x = sHist[((p4.x >> startbit) & 0xF) * 64 + tid]++;
r4.y = sHist[((p4.y >> startbit) & 0xF) * 64 + tid]++;
r4.z = sHist[((p4.z >> startbit) & 0xF) * 64 + tid]++;
r4.w = sHist[((p4.w >> startbit) & 0xF) * 64 + tid]++;
sKeys4[tid4 + (i + tid) % 8] = r4;
}
syncthreads();
}
```

LISTING 7.4: Processing an element of sHist4[] .

```
curTileId = startTileId;
for(i = 0; i < 8; i++)
{
 ranks4[keyOffset + keyPos + startTileId] = sKeys4[sKeyPos
                                                 + i * 64] ;
 curTileId += tileSize8;
}
syncthreads();
}
```

LISTING 7.5: Writing the ranks to global memory.

```
// Calculate ID of threads amongst all threads
// nTiles is total number of tiles
globalTid = bid * 64 + tid;
for(i = 0; i < 16; i++)
{
counters[i * nTiles + globalTid] = sHist[i * 64 + tid];
}
```

LISTING 7.6: Writing the histograms to global memory.

7.2.2.2   Step 2: Prefix sum of tile histograms

As in Step 3 of the SDK algorithm, the prefix sum of the tile histograms is computed with a CUDPP function call. The SDK radix sort computes the prefix sum of half-tiles while we do this for full tiles. Assuming both algorithms use the same tile size, the prefix sum in Step 2 of GRS involves half as many histograms as does the prefix sum in Step 3 of the SDK algorithm. This difference, however, results in a negligible reduction in run time for Step 2 of GRS versus Step 3 of SDK.

7.2.2.3 Step 3: Positioning numbers in a tile

To move the numbers in a tile to their correct overall sorted position with respect to the b bits being considered in a particular phase, we need to determine the correct position for each number in a tile. The correct position is obtained by first computing the prefix sum of the tile histogram. This prefix sum may be computed using the warp scan algorithm used in the SDK radix sort code corresponding to the algorithm of [26]. The correct position of a number is its rank plus the histogram prefix sum corresponding to the b bits of the number being considered in this phase. As noted earlier, moving numbers directly from their current position to their final correct positions is expensive because of the large number of memory transactions to global memory. Better performance is obtained when we first rearrange the numbers in a tile into sorted order of the b bits being considered and then move the numbers to their correct position in the overall sorted order. Listing 7.7 gives the kernel code. One SM reorders the numbers in a single tile. Since each thread handles four numbers, the number of threads used by an SM for this purpose (equivalently, the number of threads per thread block) is $t/4$.

7.2.3 SRTS Radix Sort

SRTS employs a highly optimized version of the scan kernel developed by Merrill and Grimshaw [23] to perform radix sort. The scan process consists of three kernels: bottom-level reduce, top-level scan, and bottom-level scan.

Bottom-Level Reduce: In this step, the threads in a thread block read inputs and produce a final output per thread block. The scan does not use the traditional method of assigning a unique thread to each input element. Instead, it fixes the number C of thread blocks performing the reduction. Merrill and Grimshaw [23] experimented with different configurations of empty kernels just reading the numbers (one, two, or four at a time) from the global memory and then writing out a single value to the global memory to find the optimum C and the number of threads in a block. Threads in a block loop over the input elements in batches to perform the reduction. The dependencies between different batches are carried over in the SM registers or local shared memory. Threads in a block operate in three different phases while reducing the entire input assigned to it. Firstly, threads read multiple input elements from the current batch and do a serial reduce using the registers. The next phase is the serial reduction in the shared memory by groups of threads until the number of outputs is small enough to be reduced by a single warp of threads. In the last phase, the warp reduces using the strategy of

```
kernel reorderData(keysOut, keysIn4, counters, countersSum,
                   ranks4)
{
// Read the numbers from keysIn4 and put them in
// sorted order in keysOut
// sTileCnt stores tile histogram
// sGOffset stores global prefix-summed histogram
shared sTileCnt[16], sGOffset[16];
// Storage for numbers
shared int sKeys[t];
int4 k4, r4;
// Read the histograms from the global memory
if(tid < 16)
{
sTileCnt[tid] = counters[tid * nTiles + bid];
sGOffset[tid] = countersSum[tid* nTiles + bid];
}
syncthreads();
// Perform a warp scan on the tile histogram
sTileCnt = warp_scan(sTileCnt);
syncthreads();
// Read the numbers and their ranks
k4 = keysIn4[bid * (t / 4) + tid];
r4 = ranks4[bid * (t / 4) + tid];
// Find the correct position and write to the shared mem
r4.x = r4.x + sTileCnt[(k4.x >> startbit) & 0xF];
sKeys[r4.x] = k4.x;
// Similar code for y, z, and w components of
// r4 and k4 comes here
syncthreads();
// Determine the global rank
// Each thread places four numbers at positions
// tid, (tid + t/4), (tid + t/2), and (tid + 3t/2)
radix = (sKeys[tid] >> startbit) & 0xF;
globalOffset.x = sGOffset[radix] + tid - sTileCnt[radix];
keysOut[globalOffset.x] = sKeys[tid];
}
```

LISTING 7.7: Rearranging data.

Listing 7.1 to produce the final value for the current batch. The value is also stored in the shared memory to be used when reducing the elements from the next batch. Using this strategy for reducing the input increases the number of arithmetic operations performed by the threads and hence increases the number of instructions per memory transaction. This helps in offsetting the idle time of an SM that is waiting for data from global memory for a predominantly memory-bound operation like reduction. Figure 7.7 gives a schematic of the entire process. The circles indicate a thread working on two elements.

At the end of bottom-level reduce, there are C partial reductions.

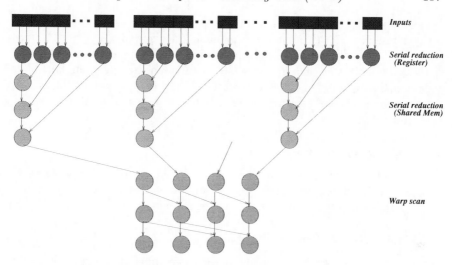

FIGURE 7.7: Bottom level reduction.

Top-Level Scan: A single thread block scans the C values from the previous phase along with some residual input elements, if any, that are not reduced in case the number of input elements is not a perfect multiple of the number of blocks. As only a few elements are scanned, this phase does not contribute much to the overall runtime.

Bottom-Level Scan: This is essentially the same as the first phase, but the scans are seeded with the values obtained from top-level scan. Although similar steps are carried out, the operation is scan instead of reduce. The number of inputs and outputs are the same for the scan operation. Merrill and Grimshaw [23] experimented with empty kernels that read and write values to and from the global memory to determine the optimal values of C and number of threads in a block.

SRTS radix sort starts with the three-step scan process. The scan and the reduce kernels use the summation operation, calculating the prefix sum and sum of all elements, respectively. SRTS augments the kernels with the operations required to radix sort a number of integers, and hence only three kernel launches are needed to do a radix sort with a fewer number of intermediate values passed around compared to SDK radix sort [26]. As with the other radix sort strategies, SRTS progressively radix sorts on 4 bits per phase. Hence, SRTS requires eight phases to completely sort 32-bit integers. Each phase, as mentioned earlier, consists of three steps.

7.2.3.1 Step 1: Bottom-level reduce

This step adds an extract phase before the bottom-level reduce operation. In the extract phase, $r = 16$ flag vectors corresponding to every possible value of 4 bits constructed from the input. For example, if a thread block inputs 64 elements at a time, 16 64-element bit vectors are constructed from the input elements. For the ith input element having a value of j ($0 \leq j \leq 15$) in the current 4 bits being considered, the ith bit of the jth vector is set to 1 while the ith bit of other vectors are set to 0. In the next phase, the bottom-level prefix sum is done on these 16 vectors serially one at a time. The total for each digit is carried over for the next batch of inputs to be processed by the thread block.

7.2.3.2   Step 2: Top-level scan

In this phase, a single block of threads operates over the prefix sums to compute the global prefix sum. The scan is modified to handle the scanning of 16 different sets of partial sums one at a time.

7.2.3.3   Step 3: Bottom-level scan

Bottom-level scan is augmented with two different operations, one performed before the scan and one after the scan. The first one is reading the input and extracting the bits as done in the first step. During the second phase, a bottom-level scan is performed seeded with the values from the top-level scan of Step 2. As in the first phase, 16 bottom-level scans are performed each for 16 possible digit values for 4 bits. After the scan, the sorted position of the input elements with the current batch is known. This position is used in the next phase, when the elements are placed into their sorted order in the final output. First, depending on the prefix sum positions obtained, a shared memory exchange is performed to put the elements within the batch in sorted order. After this, the threads read consecutive elements from the shared memory and write them to global memory depending on the global offset and the local position. As with other radix sort methods described earlier, this strategy generates larger global memory transactions as consecutive elements within the shared memory are written to nearby positions in the global memory. As with the first phase, the radix counters are accumulated and are carried over to the next batch of inputs processed by this block of threads.

Figure 7.8 depicts the three stream kernels augmented with extract and scatter operations.

7.2.4   GPU Sample Sort

Sample sort [20] is a multi-way divide-and-conquer sorting algorithm that performs better when the memory bandwidth is an issue because the data transferred to and from the global memory are less than in a two-way approach. The serial version of sample sort works by first choosing a set of splitters randomly from the input data. The splitters are then sorted and arranged in increasing order of their values. The input data set is divided into buckets delimited by successive splitters. The elements in a particular bucket have values that are bounded by the guarding splitters. Each bucket is sorted

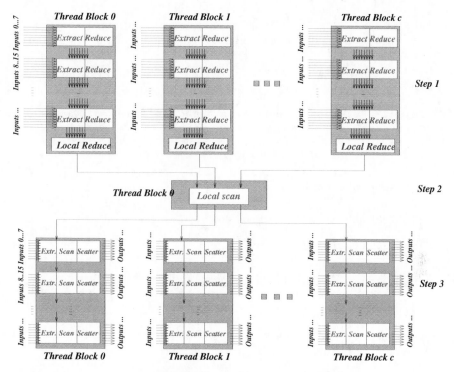

FIGURE 7.8: SRTS steps.

by a recursive application of sample sort. This recursion continues until the size of the bucket becomes less than a certain threshold. At this point, a base sorting algorithm is used to sort the small bucket. Listing 7.8 shows the steps of serial sample sort.

To obtain an efficient parallel version of sample sort, it is necessary to balance the size of buckets assigned to thread blocks. This is done by choosing the splitters from a large randomly selected sample from the input. Once the splitters are selected, the numbers are partitioned into buckets by first dividing the data into equal sized tiles with each tile being assigned to a block of threads. A thread block examines its tile of data and assigns numbers in this tile to buckets whose boundaries are the previously chosen splitters. Finally, the buckets produced for the tiles are combined to obtain global buckets. The steps in a particular iteration of GPU sample sort of [20] are described below.

7.2.4.1   Step 1: Splitter selection

During this step, the splitters are chosen. First, random samples are taken out of the elements in the buckets. A set of splitters is then chosen from these random samples. Finally, splitters are sorted using odd-even merge sort [4] in

```
SampleSort(a[])
 if sizeof(a) ≤ M // A threshold
 {
Sort(a); // Use a sorting method to sort a
 return;
 }
 Select k elements randomly from a and put them in samples[];
 Sort(samples[]);
 for(each element e in a[])
 {
 find i such that samples[i] ≤ e ≤ samples[i + 1];
 Put e in bucket b[i];
 }
 for(each bucket b[i])
 {
SampleSort(b[i]);
 }
```

LISTING 7.8: Serial sample sort.

shared memory, and a binary search tree of splitters is created to facilitate the process of finding the bucket for an element.

7.2.4.2   Step 2: Finding buckets

Each thread block is assigned a part of the input data. Threads in a block load the binary search tree into shared memory and then calculate the bucket index for each element in the tile. At the end, threads store the number of elements in each bucket as a k-entry histogram in the global memory.

7.2.4.3   Step 3: Prefix sum

The per block k-entry histograms are prefix-summed to obtain the global offset for each bucket.

7.2.4.4   Step 4: Placing elements into buckets

Each thread block in this phase again calculates bucket index for all keys in its tile. They also calculate the local offset within the buckets. The local offsets are added to the global offsets from the previous phase to get the final position of the numbers.

Figure 7.9 depicts the steps in an iteration of the GPU sample sort.

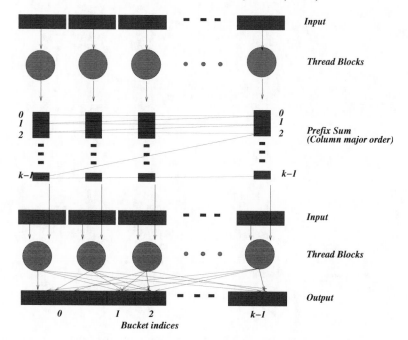

FIGURE 7.9: An iteration of GPU sample sort.

7.2.5 Warpsort

Warpsort [31] is another comparison-based sorting algorithm along the lines of merge sort. It uses bitonic merge to sort the input in a number of stages. Warpsort consists of the following four steps.

7.2.5.1 Step 1: Bitonic sort by warps

The input is first divided into a set of tiles, and each tile is sorted by a thread warp using bitonic merge sort. A bitonic network for n elements is comprised of $\log(n)$ phases with each phase $i(0 \le i \le \log(n) - 1)$ having $(i+1)$ stages. Figure 7.10 shows the bitonic sorting network for eight elements with each arrow indicating a compare-exchange operation where two elements are compared and swapped if the first one is greater (or less depending on the direction of the arrow) than the second.

All the compare-exchanges of a stage need to be performed before moving to the next stage. This requires a global synchronization among the threads. So, bitonic sort by a thread block will require a `syncthreads()` function call after each stage. In warpsort, as the bitonic sort is done by a warp of threads, there is no need for synchronization because threads in a warp are executed in lock-step fashion. However, in each stage, the threads in a warp will be doing compare-exchanges in ascending or descending directions. This will cause a divergence among the threads in a warp and hence will lead

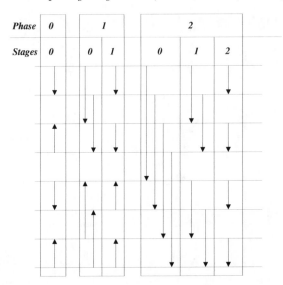

FIGURE 7.10: Bitonic merge sort of eight elements.

to serial execution of threads. It can be seen that for an n-element bitonic network, only $n/2$ compare-exchanges are performed, and half of them are to form ascending pairs while the other half are to form descending pairs (except during the last stage when all the compare-exchanges are in the same order). Warpsort uses this fact to set the number of elements handled by a 32-thread warp to be 128. Each thread then performs two operations during each stage of bitonic sort, and operations performed by the all threads in a warp are for either an ascending pair or a descending pair. Thread divergence is thus avoided by each thread performing compare-exchanges to generate results in a particular sorted order during each step of its execution.

7.2.5.2   Step 2: Bitonic merge by warps

The sorted sequences are merged in this step by a warp using bitonic merge until the number of sequences falls below a threshold. As the threads in a warp only merge a fixed number of elements t in a step, sequences are divided into buffers of size $t/2$. To merge two sorted sequences A and B, $t/2$ elements from each of the sequences are fetched into the buffers in shared memory and then they are merged using the bitonic merge algorithm (bitonic merge is the same as the last step of bitonic sort outlined in Figure 7.10). The smallest $t/2$ elements of the merged sequence are output, and the remaining $t/2$ elements are kept in the buffer. Then, the last elements of the $t/2$ element buffers of A and B are checked to find out from which sequence the next buffer load of $t/2$ elements should be fetched.

7.2.5.3   Step 3: Splitting long sequences

As the number of the sequences decreases geometrically during merging, after some point there are not enough sequences to be merged on different SMs of the GPU. In this step, the long sequences are split into small sequences, which can be merged using all the SMs in the GPU. To split l sequences of size n into subsequences of size s, a random sample of $s * k$ elements are chosen from the sequences where k is a constant whose value depends on the trade-off between choosing more random samples for a good set of splitters and the time needed for choosing those random samples. The set of random samples is then sorted, and the elements in positions that are a multiple of k form the global set of splitters. These s splitters are then used to partition each of these l sequences into s subsequences.

7.2.5.4   Step 4: Final merge by warps

This step is essentially the same as Step 2. The Smaller sequences produced in Step 3 can now be merged independently by thread warps to produce the final output.

Figure 7.11 gives all the steps in Warpsort.

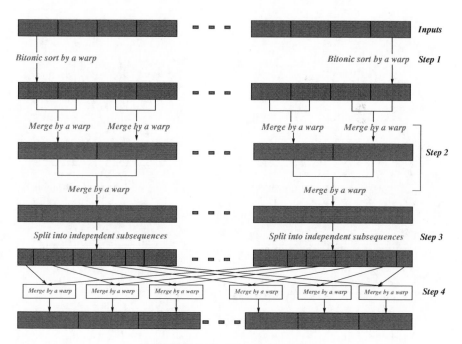

FIGURE 7.11: Warpsort steps.

7.2.6   Comparison of Number-Sorting Algorithms

According to the results reported in [20] and [31], radix sort–based algorithms perform better than sample sort and warpsort. Hence, here, we compare SDK radix sort, GRS, and SRTS. These algorithms were coded and run using NVIDIA CUDA SDK 3.0 on a Tesla C1060 GPU. For each of the experiments, the runtime reported is the average runtime for 100 randomly generated sequences. While comparing SDK, GRS, and SRTS the same random sequences are used for the three algorithms.

For the first set of comparisons, SDK, GRS, and SRTS are used to sort from one to ten million numbers. As shown in Figure 7.12(a), SDK runs 20% to 7% faster than GRS for one to three million numbers, respectively. However, GRS outperforms SDK for sorting four million numbers and more. It runs 11% to 21% faster than SDK for four to ten million numbers, respectively. SRTS is the best performing algorithm among the three by running 53% to 57% faster than GRS for one to ten million numbers. This performance differential is also observed while sorting even larger sets of numbers. Figure 7.12(b) shows the runtimes of SDK, GRS, and SRTS starting from ten million numbers with an increment of ten million. For 100 million numbers, GRS runs 21% faster than SDK and 53% slower than SRTS.

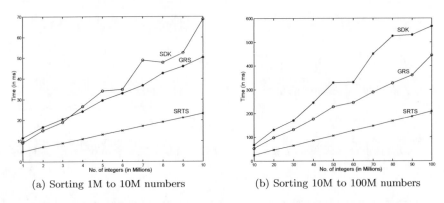

(a) Sorting 1M to 10M numbers   (b) Sorting 10M to 100M numbers

FIGURE 7.12: Sorting numbers using radix sorts.

7.3   Sorting Records on GPUs

This section refers to results reported in [33].

7.3.1   Record Layouts

A record R is comprised of a key k and m other fields $f_1, f_2, \cdots, f_m$. For simplicity, we assume that the key and each other field occupies 32 bits.

Let k_i be the key of record R_i, and let f_{ij}, $1 \leq j \leq m$ be this record's other fields. With our simplifying assumption of uniform-size fields, we may view the n records to be sorted as a two-dimensional array `fieldsArray[][]` with `fieldsArray[i][0]` $= k_i$ and `fieldsArray[i][j]` $= f_{ij}$, $1 \leq j \leq m$, $1 \leq i \leq n$. When this array is mapped to memory in column-major order, we get the `ByField` layout of [2]. This layout was used also for the AA-sort algorithm developed for the Cell Broadband Engine in [16] and is essentially the same as that used by the GPU radix sort algorithm of [26]. When the fields array is mapped to memory in row-major order, we get the `ByRecord` layout of [2]. A third layout, `Hybrid`, is employed in [23]. This is a hybrid between the `ByField` and `ByRecord` layouts. The keys are stored in an array, and the remaining fields are stored using the `ByRecord` layout. Essentially then, in the `Hybrid` layout, we have two arrays. Each element of one array is a key, and each element of the other array is a structure that contains all fields associated with an individual record. In this chapter, we limit ourselves to the `ByField` and `Hybrid` layouts. We do not consider the `ByRecord` layout because it appears that the most effective way to sort in this layout is to first extract the keys, sort (`key`, `index`) pairs, and then reorder the records into the obtained sorted permutation. The last two steps are identical to the steps in an optimal sort for the `Hybrid` layout, so we expect that good strategies to sort in the `Hybrid` layout will also be good for the `ByRecord` layout. When the sort begins with data in a particular layout format, the result of the sort must also be in that layout format.

7.3.2 High-Level Strategies for Sorting Records

At a high level, there are two very distinct approaches to sort multifield records. In the first, we construct a set of tuples (k_i, i), where k_i is the key of the ith record. Then, these tuples are sorted by extending a number sort algorithm so that whenever the number sort algorithm moves a key, the extended version moves a tuple. Once the tuples are sorted, the original records are rearranged by copying records from the `fieldsArray` to a new array, placing the records into their sorted positions in the new array or in place using a cycle-chasing algorithm as described for a table sort in [15]. The second strategy is to extend a number sort so as to move an entire record every time its key is moved by the number sort. We call the first strategy an indirect and the second strategy a direct strategy for sorting multifield records. There are advantages and disadvantages to each strategy. The indirect strategy seems to perform much less work than the direct during sorting as the satellite data that need to be moved with the key are only integer indexes, while in the direct strategy, they are the entire record. On the flip side, the indirect strategy has a very costly random global-memory-access phase at the end when records are moved to their sorted positions, whereas the direct strategy does not have this phase.

7.3.3   Sample Sort for Sorting Records

Sample sort [20] for sorting numbers is described in section 7.2.4. It is a multi-way sorting strategy that reduces the number of times data need to be moved to and from global memory. When sample sorting records using the direct strategy outlined in Section 7.3, you need to move the records only during the fourth phase because in all other phases, only the keys are required. This distribution of the records from the large bucket to small buckets is repeated multiple times until the size of the bucket is below a specified threshold. Finally, quicksort is done on the records when the bucket size is small. Records are also moved during the partitioning phase of the quicksort within a small bucket. The fourth phase and the quicksort part of sample sort can be extended to handle records in ByField and Hybrid formats. In ByField layout, while moving fieldsArray[i] to outfieldsArray[j], threads can move the corresponding fields as shown in Listing 7.9.

```
outKey[j] = key[i];
// Move the fields
for(f = 1; f <= m; f++) {
 outfieldsArray[j][p] = fieldsArray[i][p];
}
```
LISTING 7.9: Moving records in ByField layout.

Similarly, in the Hybrid (ByRecord) format, fields can be moved by a thread while moving the keys. Listing 7.10 shows the code to move records assuming that fieldsArray[i] and outfieldsArray[j] are structures that contain the fields.

```
outKey[j] = key[i];
// Move the fields
outfieldsArray[j] = fieldsArray[i];
```
LISTING 7.10: Moving records in Hybrid layout.

We observe that in the ByField layout, threads in a warp access adjacent elements in global memory, resulting in a memory access coalescing. While in the Hybrid layout, threads in a warp access words in the global memory that are potentially far apart, generating global-memory transactions of size at most 16 bytes. We employ a strategy of grouping the threads together so that we can generate larger memory transactions. Rather than a single thread reading and writing the entire record, we employ a group of threads to read and write the records into the global memory cooperatively. Then, this same group of threads iterates to read and write other records cooperatively. This ensures larger global-memory transactions. As an example, lets say the record is of 64 bytes in length and as each thread can read in 16 bytes of data

```
// Determine the number of threads required to
// read the entire record
numThrdsInGrp = sizeof(Rec) / 16;
// Total number of records to be read = number of
// threads in the group
numItrs = numThrdsInGrp;
// Number of records read in a single iteration
// by all threads
nRecsPerItr = numThrds / numThrdsInGrp;
// Convert Record arrays to int4 arrays
recInt4 = (int4 *)rec; outRecInt4 = (int4 *)outRec;
// Determine the starting record and position in the group
// for this thread
startRec = startOffset + threadId / numThrdsInGrp;
posInGrp = threadId % numThrdsInGrp;
for(i = 0; i < numItrs; i++)
{
outRecInt4[mapInOut(startRec) + posInGrp] =
recInt4[startRec + posInGrp];
startRec += numThrdsInGrp;
}
```

LISTING 7.11: Optimized version of moving records in `Hybrid` layout.

using an `int4` data type, we can group four threads together so that they can read the entire record together. Then, this thread group iterates over to read other records until all the records assigned to the thread group have been read. Let `numThreads` denote the number of threads in a block, and suppose that each thread is to read in one record and put it into the proper place in the output array. Assume that records from `startOffset` to (`startOffset + numThreads`) are processed by this thread block. For the sake of clarity of the pseudocode, we assume that there is a map `mapInToOut` that determines the proper position in the output array. In the case of sample sort, it is the binary search tree constructed out of the splitters that determines the position of a particular record in the output array. Listing 7.11 outlines the optimized version of moving records using coalesced read and write.

7.3.4 SRTS for Sorting Records

SRTS [23] is the fastest radix sort algorithm for sorting numbers. As noted in Section 7.2.3, it is a three-step radix sorting process where highly optimized scan kernels are augmented to perform tasks pertaining to the sorting process. The final scatter of input elements happens during the very last phase. Only keys of the records are required during other phases, so the fields of the record can also be moved during the third phase while scattering the keys. We can use the strategies outlined in Figures 7.9 and 7.10 to scatter the fields in `ByField` and `Hybrid` layouts, respectively. However, due to the way SRTS is implemented using generic programming, it is difficult to use the optimized version of record moving (Figure 7.11) in Hybrid format. The third phase of record scattering occurs only eight times for 32-bit keys during the entire sorting process and does not depend on the number of records being sorted. This indicates there is a possibility, for SRTS, that a direct strategy of moving records while sorting might actually perform better than the indirect strategy of sorting (key, index) pairs followed by rearrangement.

7.3.5   GRS for Sorting Records

GRS [1], which was described in Section 7.3.5, was developed specifically for sorting records along the lines of the SDK radix sort [26] (see Section 7.2.1), but the focus is to reduce the number of times a record is read from or written into the global memory. During Step 1 of the SDK algorithm, all records have to read from the global memory as they are sorted based on the key and moved around the shared memory using the bit-split algorithm described in Section 7.2.1, while in the case of GRS, only keys are read from the global memory. The records are read only once during the very last stage when they have to be put into their sorted order in the global memory. So, GRS reduces the number of times non-key fields of records are read from global memory and written to global memory by 50% (i.e., from 2 to 1). More precisely, suppose n records that have a 4-byte key and m 4-byte fields each are being sorted. We shall ignore global memory I/O of the histograms in our analysis, as for reasonable tile sizes the histograms represent a relatively small amount of the total global memory I/O. The SDK algorithm reads $4mn + 4n$ bytes of data and writes as much data in Step 1 (exclusive of the histogram I/O). In Step 2, only the keys are input from global memory, so in this step, $4n$ bytes of data are read. Step 3 does I/O on histograms only. In Step 4, $4mn + 4n$ bytes are read and written. So, in all, SDK reads $8mn + 12n$ bytes and writes $8mn + 8n$ bytes. The GRS algorithm, on the other hand, reads $4n$ bytes (the n keys) in Step 1 and writes $4n$ bytes (the ranks). Since the ranks require only 2 bytes each (2 bytes are sufficient as long as the tile size is no more than 2^{16}), the writes could be reduced to $2n$ bytes. The Step 2 I/O involves only histograms. In Step 3, all keys, fields, and ranks are read, but only keys and fields are written back to the global memory, so in Step 4, we read $4mn + 8n$ bytes and write $4mn + 4n$ bytes. GRS reads a total of $4mn + 12n$ bytes and writes a total of $4mn + 8n$ bytes. We see that SDK reads (writes) $4mn$ bytes of data more than does GRS. Although the analysis for SDK sort is based on the version for records with multiple fields, it applies also for the number-sort version (i.e., $m = 0$). In the end, GRS does the same amount of global-memory I/O when we are sorting numbers but reads $4mn$ fewer bytes and writes $4mn$ fewer bytes when we are sorting records that have m 4-byte fields in addition to the key field.

Much like SRTS, the final scatter of the records is done in the last phase. As the last phase caters to a very simple implementation, we can efficiently read and write records in this phase. This simplicity enables us to use the algorithms of Listings 7.9 and 7.11 to move records in ByField and Hybrid layouts, respectively. As with SRTS, we only move records eight times during the sorting of records with 32-bit keys. Hence, GRS also has a fair chance of outperforming the first strategy of sorting records by using (key, index) pairs.

7.3.6 Comparison of Record-Sorting Algorithms

The record-sorting algorithms mentioned in the previous sections were implemented using Nvidia CUDA SDK 3.2. Specifically, two versions of sample sort, GRS and SRTS, each corresponding to the direct and indirect strategies for sorting records mentioned in Section 7.3, are evaluated.

1. SampleSort-direct... samplesort algorithm of [20] extended for sorting records.

2. SampleSort-indirect... (key, index) pairs for the records are formed, and then these are sorted using samplesort. Finally, a rearrangement is done to put the entire record in the sorted order.

3. SRTS-direct... SRTS algorithm [23] extended for sorting records.

4. SRTS-indirect... (key, index) pairs for the records are formed, and these are sorted using SRTS. Finally, a rearrangement is done to put the entire record in the sorted order.

5. GRS-direct... GRS algorithm for sorting records [1].

6. GRS-indirect... (key, index) pairs for the records are formed, and these are sorted using GRS. Finally, a rearrangement is done to put the entire record in the sorted order.

The sorting algorithms were run on an Nvidia Tesla C1060, which has 240 cores and 4 GB of global memory. The algorithms were evaluated using randomly generated input sequences. In these experiments, the number of 32-bit fields per record is varied from 2 to 20 (in addition to the key field), and the number of records was 10 million. Also, the algorithms were evaluated for both ByField and Hybrid layouts. For each combination of fields and layout type, the time to sort 10 random sequences was obtained. The standard deviation in the observed runtimes was small, and only the average times are reported.

7.3.7 Runtimes for ByField Layout

Figures 7.13(a)–(c) show the comparison of SampleSort, SRTS, and GRS using direct and indirect strategies for sorting 10 million records with two to nine fields in the ByField layout. During each run, we used the same set of records while comparing these algorithms. SampleSort-indirect runs 36% faster than SampleSort-direct when sorting 10 million records with two fields while it runs 66% faster for records with nine fields. SRTS-indirect runs 37% slower than SRTS-direct sorting 10 million records with two fields while it runs 27% slower for records with nine fields. GRS-indirect runs 27% slower than GRS-direct when sorting 10 million records with two fields while it runs 33% slower for records with nine fields.

Figure 7.13(d) shows the comparison among the faster version of each of these three algorithms. SRTS-direct is the fastest algorithm to sort records in the `ByField` layout when records have between 2 to 11 fields. GRS-direct is the fastest algorithm for sorting records with more than 11 fields. SRTS-direct runs 35% faster than GRS-direct when sorting 10 million records with 2 fields while GRS-direct runs 38% faster than SRTS-direct when records have 20 fields. SampleSort-indirect is the slowest, running 63% slower than GRS when sorting records with 2 fields and 48% slower when sorting records with 20 fields.

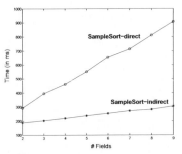

(a) SampleSort-direct and SampleSort-indirect for 10M records (`ByField`)

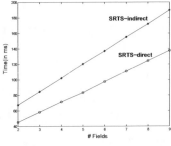

(b) SRTS-direct and SRTS-indirect for 10M records (`ByField`)

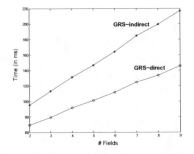

(c) GRS-direct and GRS-indirect for 10M records (`ByField`)

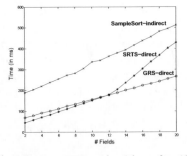

(d) Different sorting algorithms for 10M records (`ByField`)

FIGURE 7.13: Sorting records in `ByField` format.

7.3.8   Runtimes for Hybrid Layout

Figures 7.14(a)–(c) show the comparison of SampleSort, SRTS, and GRS using the direct and indirect strategies for sorting 10 million records with two to nine fields in the Hybrid layout. SampleSort-indirect runs 12% faster than SampleSort-direct when sorting 10 million records with two fields while it runs 72% faster for records with nine fields. SRTS-indirect runs 38% faster than SRTS-direct sorting 10 million records with two fields while it runs 79% faster for records with nine fields. GRS-indirect runs 13% slower than GRS-direct sorting 10 million records with two fields while it runs 41% faster for records with nine fields.

Figure 7.14(d) shows the comparison between the faster version of each of these three algorithms. SRTS-indirect is the fastest algorithm to sort records in the Hybrid layout. SRTS-indirect runs 27% faster than GRS-indirect and 71% faster than SampleSort-indirect when sorting records with two fields while it runs 3% faster than GRS-indirect and 16% faster than SampleSort-indirect when sorting records with 20 fields.

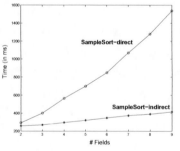

(a) SampleSort-direct and SampleSort-indirect for 10M records (Hybrid)

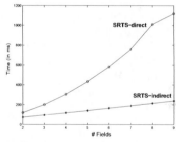

(b) SRTS-direct and SRTS-indirect for 10M records (Hybrid)

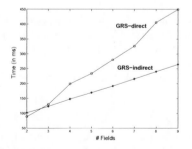

(c) GRS-direct and GRS-indirect for 10M records (Hybrid)

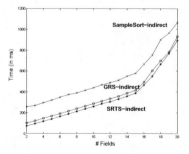

(d) Different sorting algorithms for 10M records (Hybrid)

FIGURE 7.14: Sorting records in Hybrid format.

Acknowledgements

This research was supported, in part, by the National Science Foundation under grants CNS-0963812 and CNS-1115184. The authors acknowledge the University of Florida High-Performance Computing Center for providing computational resources and support that have contributed to the results reported in this chapter.

Bibliography

[1] S. Bandyopadhyay and S. Sahni. GRS—GPU radix sort for large multi-field records, *International Conference on High Performance Computing* (HiPC), 2010.

[2] S. Bandyopadhyay and S. Sahni. Sorting large records on a cell broadband engine, *IEEE International Symposium on Computers and Communications* (ISCC), 2010.

[3] S. Bandyopadhyay and S. Sahni. Sorting on a cell broadband engine SPU, *IEEE International Symposium on Computers and Communications* (ISCC), 2009.

[4] K.E. Batcher. Sorting networks and their applications, *Proc. AFIPS Spring Joint Computing Conference*, vol. 32, 307–314, 1968.

[5] G.E. Blelloch. Vector models for data-parallel computing. Cambridge, MA; MIT Press, 1990.

[6] R. Box and S. Lacey. A fast, easy sort. *Byte*, vol. 4, 315–318. 1991.

[7] D. Cederman and P. Tsigas. GPU-quicksort: A practical quicksort algorithm for graphics processors, *ACM Journal of Experimental Algorithmics* (JEA), 14, 4, 2009.

[8] W. Dobosiewicz. An efficient variation of bubble sort. *Information Processing Letters*, vol. 11, 5–6, 1980.

[9] A. C. Chow, G. C. Fossum, and D. A. Brokenshire. A programming example: Large FFT on the cell broadband engine, IBM White paper, 2005.

[10] A. Drozdek. Worst case for Comb Sort. *Informatyka Teoretyczna i Stosowana*, vol. 9, 23–27, 2005.

[11] B. Gedik, R. Bordawekar, and P. Yu. CellSort: High performance sorting on the Cell processor. *VLDB*, 1286–1297, 2007.

[12] N. Govindaraju, J. Gray, R. Kumar, and D. Manocha. Gputerasort: High performance graphics coprocessor sorting for large database management, *ACM SIGMOD International Conference on Management of Data*, 2006.

[13] A. Greb and G. Zachmann. GPU-ABiSort: Optimal parallel sorting on stream architectures, *IEEE International Parallel and Distributed Processing Symposium* (IPDPS), 2006.

[14] H. Hofstee. Power efficient processor architecture and the cell processor, In *Proceedings of 11th International Symposium on High Performance Computer Architecture*, 2005.

[15] E. Horowitz, S. Sahni, and D. Mehta. *Fundamentals of data structures in C++*, Second Edition, Silicon Press, 2007.

[16] H. Inoue, T. Moriyama, H. Komatsu, and T. Nakatani. AA-sort: A new parallel algorithm for multi-core SIMD processors, *16th International Conference on Parallel Architecture and Compilation (PACT)*, 2007.

[17] D. Knuth. *The Art of Computer Programming: Sorting and Searching Volume 3, Second Edition*, Reading, MA: Addition Wesley, 1998.

[18] S. Lai and S. Sahni. Anomalies in parallel branch-and-bound algorithms, *Comm. of the ACM*, vol. 6, 594–602, 1984.

[19] S. Le Grand. Broad-phase collision detection with CUDA. In *GPU Gems 3*. Reading, MA: Addison-Wesley Professional, 2007.

[20] N. Leischner, V. Osipov, and P. Sanders. GPU sample sort, *IEEE International Parallel and Distributed Processing Symposium (IPDPS)*, 2010.

[21] P. Lemke. The performance of randomized Shellsort-like network sorting algorithms, *SCAMP working paper P18/94*, Institute for Defense Analysis, Princeton, NJ, 1994.

[22] E. Lindholm, J. Nickolls, S. Oberman, and J. Montrym. NVIDIA Tesla: A unified graphics and computing architecture, *IEEE Micro*, 28, 3955, 2008.

[23] D. Merrill and A. Grimshaw. Revisiting sorting for GPGPU stream architectures, University of Virginia, Department of Computer Science, Technical Report CS2010-03, 2010.

[24] R. Sedgewick. Analysis of Shellsort and related algorithms, *4th European Symposium on Algorithms*, 1996.

[25] S. Sahni. Scheduling master-slave multiprocessor systems, *IEEE Trans. on Computers*, 45, 10, 1195–1199, 1996.

[26] N. Satish, M. Harris, and M. Garland. Designing efficient sorting algorithms for manycore GPUs, *IEEE International Parallel and Distributed Processing Symposium* (IPDPS), 2009.

[27] S. Sengupta, M. Harris, Y. Zhang, and J.D. Owens. Scan primitives for GPU computing, *Graphics Hardware 2007*, 97–106, 2007.

[28] E. Sintorn and U. Assarsson. Fast parallel GPU-sorting using a hybrid algorithm, *Journal of Parallel and Distributed Computing*, 10, 1381–1388, 2008.

[29] D. Sharma, V. Thapar, R. Ammar, S. Rajasekaran, and M. Ahmed. Efficient sorting algorithms for the Cell Broadband Engine, *IEEE International Symposium on Computers and Communications (ISCC)*, 2008.

[30] V. Volkov and J.W. Demmel. Benchmarking GPUs to tune dense linear algebra. *ACM/IEEE Conference on Supercomputing*, 2008.

[31] X. Ye, D. Fan, W. Lin, N. Yuan, and P. Ienne. High performance comparison-based sorting algorithm on many-core GPUs, *IEEE International Parallel and Distributed Processing Symposium (IPDPS)*, 2010.

[32] Y. Won and S. Sahni. Hypercube-to-host sorting, *Journal of Supercomputing*, vol. 3, 41–61, 1989.

[33] S. Bandyopadhyay and S. Sahni. Sorting large multifield records on a GPU, *IEEE International Conference on Parallel and Distributed Systems (ICPADS)*, 2011.

[34] CUDPP: CUDA Data-Parallel Primitives Library, http://www.gpgpu.org/developer/cudpp/, 2009.

[35] NVIDIA CUDA Programming Guide, *NVIDIA Corporation*, version 3.0, Feb. 2010.

[36] Cell Broadband Engine, http://www-01.ibm.com/chips/techlib/techlib.nsf/products/Cell_Broadband_Engine, 2013.

Chapter 8

Scheduling DAG-Structured Computations

Yinglong Xia

IBM T.J. Watson Research Center

Viktor K. Prasanna

University of Southern California, Los Angeles

Many computational solutions can be expressed as directed acyclic graphs (DAGs) with weighted nodes. In parallel computing, scheduling such DAGs onto multi/many-core processors remains a fundamental challenge. In this chapter, we first introduce a modularized scheduling method for DAG-structured computations on multicore processors, where various heuristics can be utilized for each module within the scheduling framework. This modular design enables the scheduler to be adaptive to various underlying architectures. In particular, we develop lock-free data structures for reducing the overhead due to coordination. Second, we extend the scheduling method to many-core processors. Considering the architectural characteristics of many-core processors, we propose a hierarchical scheduler with dynamic thread grouping. Such a scheduler dynamically adjusts the number of threads used for scheduling and task execution. Therefore, it can adapt to the input task graph to improve performance. We evaluate the proposed method using various data sets. We also use exact inference in junction trees, a real-life problem in machine learning, to evaluate the performance of the proposed schedulers on multi/many-core processors.

8.1   Introduction

Given a program, we can represent the program as a *directed acyclic graph* (DAG) with weighted nodes, in which the nodes represent code segments, and there is an edge from node v to node $\tilde{v}$ if the output from the code segment performed at v is an input to the code segment at $\tilde{v}$. The weight of a node represents the (estimated) execution time of the corresponding code segment. Such a DAG is called a *task-dependency graph*. The computations that can be represented as task-dependency graphs are called *DAG-structured computations* [1, 2].

The objective of scheduling for DAG-structured computations on multi/many-core processors is to minimize the overall execution time by allocating the tasks to the cores, while preserving the precedence constraints [2]. Prior work has shown that system performance is sensitive to scheduling [3, 4, 5]. Efficient scheduling on typical multicore processors such as the AMD Opteron and Intel Xeon requires balanced workload allocation and minimum scheduling overhead [6]. The *collaborative scheduling* is a type of work sharing based on distributed scheduling methods, where we distribute the scheduling activities across the cores and let the schedulers dynamically collaborate with each other to balance the workload. Collaboration requires locks to prevent concurrent write to shared variables. However, locks can result in significant synchronization overhead. The traditional lock-free data structures are based on hardware-specific atomic primitives, such as CAS2, which are not available in some modern systems [7]. Thus, we utilize a software-lock-free data structure to improve the performance of the collaborative scheduler. Unlike general-purpose multicore processors, many-core processors are more interested in how many tasks from a DAG can be completed efficiently over a period of time rather than how quickly an individual task can be completed. Examples of existing many-core processors include the Sun UltraSPARC T1 (Niagara) and T2 (Niagara 2), which support up to 32 and 64 concurrent threads, respectively [8]. The Nvidia Tesla and Tilera TILE64 are also available. Scheduling DAG-structured computations on many-core processors remains a fundamental challenge in parallel computing.

Our contributions in this chapter include the following: (a) We propose a lightweight collaborative scheduling method and a hierarchical scheduling method for DAG structured computations on multicore and many-core processors, respectively. (b) We develop lock-free local task lists to reduce coordination overhead during scheduling. (c) We propose a dynamic thread-grouping technique to merge or partition the thread groups at runtime. (d) We experimentally show the efficiency of the proposed methods compared with various baseline methods.

The rest of this chapter is organized as follows: In Section 8.2, we provide the background of task scheduling for DAG-structured computations. Section 8.3 introduces related work on task scheduling and lock-free data structures. Sections 8.4 and 8.5 discuss the collaborative scheduling method and hierarchical scheduling method, respectively. We summarize the scheduling methods for DAG-structured computations on multicore and many-core processors in Section 8.6.

8.2 Background

In our context, the input to task scheduling is a directed acyclic graph (DAG), where each node represents a task, and the edges correspond to precedence constraints among the tasks. Each task in the DAG is associated with a *weight*, which is the estimated execution time of the task. A task can begin execution only if all its predecessors have completed execution [9]. The task-scheduling problem is to map the tasks in a given DAG onto the cores so as to minimize the overall execution time on a parallel computing system. Scheduling DAG-structured computation is in general an *NP-complete* problem [10, 11].

Scheduling DAG-structured computations can be *static* or *dynamic*. Unlike static scheduling, which requires complete and accurate information on the DAG and platform to determine the schedule of tasks before execution begins, dynamic scheduling decides the mapping and scheduling of tasks on the fly. In addition, dynamic scheduling can tolerate error in estimated task weights [2]. We focus on dynamic scheduling in this chapter. Dynamic scheduling methods can be *centralized* or *distributed*. Centralized scheduling dedicates a thread to execute a scheduler and uses the remaining threads to execute the tasks allocated by the scheduler. Distributed scheduling, however, integrates a scheduler into each thread. These schedulers cooperate with each other to achieve load balance across the threads. We consider distributed scheduling on homogeneous multi/many-core processors.

Work sharing [12] and *work stealing* [13] are two generic approaches for task scheduling. In work sharing, when a task becomes ready, it is allocated to a thread to satisfy some objective function, for example, balancing the workload across the threads. By workload of a thread, we mean the total weight of the tasks allocated to the thread before the tasks have been executed. In work stealing, however, every thread allocates new tasks to itself and the underutilized threads attempt to steal tasks from others [12]. Randomization is used for work stealing. This also reduces coordination overhead. When the task weights are not available and the input DAG has sufficient parallelism, work stealing shows encouraging performance [13]. When the task weights are known and the input DAGs are not limited to nested parallelism, work sharing can lead to improved load balance across the threads [12]. The main downside of the work-sharing approach is that a thread must know the others' workloads when allocating new tasks. This can increase coordination overhead in lock-based implementations. Our proposed method is a type of work-sharing scheduler. We use software-lock-free data structures to reduce the overhead due to synchronization.

To design an efficient scheduler, we must take into account the architectural characteristics of processors. General-purpose multicore processors consist of several homogeneous cores with separate or shared caches, which play a crucial role in synchronization across the cores. Therefore, efficient schedulers on such processors must evenly distribute the workload and reduce overheads due to synchronization. In contrast, almost all the existing many-core processors have relatively simple cores compared with general-purpose multicore processors, e.g., AMD Opteron and Intel Xeon. For example, the pipeline of the UltraSPARC T2 does not support out-of-order (OoO) execution and therefore results in a longer delay. However, the fast context switch of such processors overlaps such delays with the execution of another thread. For this reason, the UltraSPARC generally shows higher throughput when enough parallel tasks are available [8].

Directly utilizing traditional scheduling methods such as centralized or distributed scheduling can degrade the performance of DAG-structured computations on multi/many-core processors. For example, many centralized scheduling methods dedicate a thread to allocate tasks, which brings challenges in terms of load balance across the threads. In addition, the single thread in charge of scheduling may not be able to serve the rest of the threads in time on many-core processors, especially when the tasks are completed quickly. On the other hand, distributed scheduling requires multiple threads to schedule tasks. This limits the resources for task execution. Many schedulers accessing shared variables can also result in costly synchronization overhead. Therefore, an efficient scheduling method must be able to adapt itself to input-task dependency graphs.

8.3   Related Work

The scheduling problem has been extensively studied for several decades. Early algorithms optimized scheduling with respect to specific task-dependency graphs, such as the fork-join graph [14], albeit general programs come in a variety of structures. Beaumount et al. [15] compared a centralized scheduler to a distributed scheduler. The comparison is based on multiple bag-of-task applications, which are different from DAG-structured computations. Squillante and Nelson studied work-sharing and -stealing approaches in task scheduling and pointed out that task migration in shared-memory multiprocessors increased contention for the bus and the shared memory itself [12]. This implies the importance of a lock-free data structure for schedulers. Dongarra et al. [16] explored dynamic data-driven execution of tasks in dependency graphs with compute-intensive tasks interconnected with dense and complex dependencies, such as the tiled Cholesky, the tiled QR, and the tiled LU factorizations, where the workload is divisible. In [17], Zhao discussed scheduling techniques for heterogeneous resources, unlike the platforms considered in this chapter. Scheduling techniques have been utilized by several programming systems such as Cilk [13], Intel Threading Building Blocks (TBB) [18], and OpenMP [19]. All these systems rely on a set of extensions to common imperative programming languages, and involve a compilation stage and various runtime systems. TBB and Cilk employ work stealing for scheduling; they do not consider task weights. In contrast with these systems, we focus on scheduling methods for DAGs on general-purpose multicore processors. Almost all the existing lock-free data structures are based on atomic primitives. However, some atomic primitives used in these structures require special hardware support, e.g., Double-Word Compare-And-Swap (CAS2), which is not available in modern computer systems [20]. Unlike [21], we propose a software lock-free data structure without using atomic primitives. In our approach, we duplicate/partition shared data to ensure exclusive access to shared variables on general-purpose multicore processors. In addition, to the best of our knowledge, the above systems are not optimized for many-core processors.

8.4   Lock-Free Collaborative Scheduling

We propose a lightweight scheduling framework for DAG-structured computations on general-purpose multicore processors. We distribute the scheduling activities across the cores and let the schedulers collaborate with each other to balance the workload. The scheduling framework consists of multiple modules, where various heuristics can be employed for each module to improve the performance. In addition, we develop a lock-free local task list for the scheduler to reduce the scheduling overhead on multicore processors.

8.4.1   Components

We modularize the collaborative scheduling method and show the components in Figure 8.1, where each thread hosts a scheduler consisting of several modules. The input-task dependency graph is shared among all threads. The task-dependency graph is represented by a list called the *global task list* (GL) on the left-hand side of Figure 8.1, which is accessed by all the threads shown on the right-hand side. Unlike the global task list, the other modules are hosted by each thread.

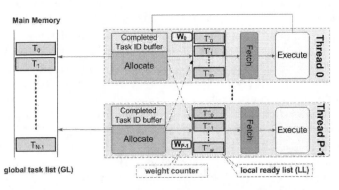

FIGURE 8.1: Components of the collaborative scheduler.

The GL represents the given task-dependency graph. Figure 8.2(a) shows a portion of the task-dependency graph. Figure 8.2(b) shows the corresponding part of the GL. As shown in Figure 8.2(c), each element in the GL consists of task ID, dependency degree, task weight, successors, and the task metadata (e.g., application-specific parameters). The *task ID* is the unique identity of a task. The *dependency degree* of a task is initially set as the number of incoming edges of the task. During the scheduling process, we decrease the dependency degree of a task once a predecessor of the task is processed. The *task weight* is the estimated execution time of the task. We keep the task IDs of the *successors* along with each task to preserve the precedence constraints of the task-dependency graph. When we process a task T_i, we can locate its successors directly using the successor IDs instead of transversing the global task list. In each element, there is *task metadata*, such as the task type and pointers to the data buffer of the task, etc. The GL is shared by all the threads, and we use locks to protect dependency degree d_i, $0 \leq i < N$.

Every thread has a *Completed Task ID buffer* and an *Allocate module* (Figure 8.1). The Completed Task ID buffer in each thread stores the IDs of tasks that are processed by the thread. Initially, all the Completed Task ID buffers are empty. During scheduling, when a thread completes the execution

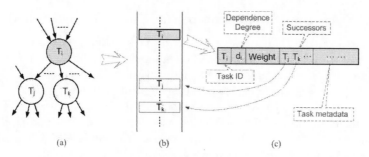

FIGURE 8.2: (a) A portion of a task-dependency graph. (b) The corresponding representation of the global task list (GL). (c) The data of element T_i in the GL.

of an assigned task, the ID of the task is inserted into the Completed Task ID buffer in that thread. For each task in the Completed Task ID buffer, the Allocate module decrements the task-dependency degree of the successors of the task. For example, if the ID of T_i in Figure 8.2 was fetched by Thread l from the Completed Task ID buffer, then Thread l locates the successors of T_i and decreases their task-dependency degrees. If the task-dependency degree of a successor T_j, becomes 0, the Allocate module allocates T_j to a thread. Heuristics can be used in task allocation to balance the workload across the threads. Our framework permits plug-and-play insertion of such modules. In this section, we allocate a task to the thread with the smallest workload at the time of completion of the task execution.

The *local task list* (LL) in each thread stores the tasks allocated to the thread. For load balance, the Allocate module of a thread can allocate tasks to any thread. Thus, the LLs are actually *shared* by all threads. Each LL has a *weight counter* associated with it to record the total workload of the tasks currently in the LL. Once a task is inserted into (or fetched from) the LL, the weight counter is updated.

The *Fetch module* takes a task from the LL and sends it to the *Execute module* in the same thread for execution. Heuristics can be used by the Fetch module to select tasks from the LL. For example, tasks with more children can have higher priority for execution [2]. In this chapter, we use a straightforward method, where the task at the head of the LL is selected by the Fetch module. Once the execution of the task is completed, the Execute module sends the ID of the task to the Completed Task ID buffer, so that the Allocate module can accordingly decrease the dependency degree of the successors of the task.

We emphasize that Figure 8.1 shows the *framework* of collaborative scheduling, where various heuristics can be used for the components. For example, genetic algorithms or randomization techniques can be used for the Allocate and Fetch modules [2]. In this chapter, we focus on the reduction of overhead of collaborative scheduling with respect to straightforward implementations of the components.

8.4.2   An Implementation of the Collaborative Scheduler

Based on the framework of collaborative scheduling in Section 8.4.1, we present a sample implementation of collaborative scheduling in Algorithm 8.1. We use the following notations in the algorithm: GL denotes the global task list; LL_i denotes the local task list in Thread i, $0 \leq i < P$; d_T and w_T denote the dependency degree and the weight of task T, respectively; W_i is the weight counter of Thread i, i.e., the total weight (estimated execution time) of the tasks currently in LL_i; δ_M is a given threshold. The statements in boxes involve shared variable accesses.

Algorithm 8.1 A Sample Implementation of Collaborative Scheduling

---

    {Initialization}
1: Let $S = \{T_i | d_i = 0\}, 0 \leq i < P$
2: Evenly distribute tasks in S to $LL_i, 0 \leq i < P$

    {Scheduling}
3: **for** Thread i $(i = 0 \cdots P - 1)$ **pardo**
4:    **while** $\boxed{\text{GL} \cup LL_i \neq \phi}$ **do**

        {Completed Task ID buffer & Allocate module}
5:       **if** |Completed Task ID buffer| $> \delta_M$ or $LL_i = \phi$ **then**
6:          **for all** $T \in$ {successors of tasks in the Completed Task ID buffer of Thread i} **do**
7:             $\boxed{d_T = d_T - 1}$
8:             **if** $d_T = 0$ **then**
9:                $\boxed{\text{fetch } T \text{ from GL and append it to } LL_j,}$
                   $\boxed{\text{where } j = \arg \min_{t=1 \cdots P}(W_t)}$
10:                $\boxed{W_j = W_j + w_T}$
11:             **end if**
12:         **end for**
13:         Completed Task ID buffer $= \phi$
14:       **end if**

        {Fetch module}
15:       $\boxed{\text{fetch a task } T' \text{ from the head of } LL_i}$
16:       $\boxed{W_i = W_i - w_{T'}}$

        {Execute module}
17:       execute T' and place the task ID of T' into the Completed Task ID buffer of Thread i
18:    **end while**
19: **end for**

Lines 1 and 2 in Algorithm 8.1 initialize the local task lists. In Lines 3–19, the algorithm performs task scheduling iteratively until all the tasks are processed. Lines 5–14 correspond to the Allocate module, where the algorithm decreases the dependency degree of the successors of tasks in the Completed Task ID buffer (Line 7), then allocates the successors with dependency degree equal to 0 into a target thread (Line 9). Line 5 determines when the Allocate module should work. When the number of tasks in the Completed Task ID buffer is greater than the threshold δ_M, the Allocate module fetches tasks from the GL. The motivation is that accessing the shared GL less frequently can reduce the lock overhead. We choose the target thread as the one with the smallest workload (Line 9), although alternative heuristics can be used. Line 15 corresponds to the Fetch module, where we fetch a task from the head of the local task list. Line 16 updates W_i according to the fetched task. Finally, in Line 17, T' is executed in the Execute module, and its ID is placed into the Completed Task ID buffer.

Due to the collaboration among threads (Line 9), all local task lists and weight counters are **shared** by all threads. Thus, in addition to the GL, we must avoid concurrent write to LL_i and W_i, $0 \leq i < P$, in Lines 9, 10, 15, and 16.

8.4.3   Lock-Free Data Structures

In Algorithm 8.1, collaboration among the threads requires locks for the weight counters and local task lists to avoid concurrent writes. For example, without locks, multiple threads can concurrently insert tasks into local task list LL_j, for some j, $0 \leq j < P$ (Line 9, Algorithm 8.1). In such a scenario, the tasks inserted by a thread can be overwritten by those inserted by another thread. Similarly, without locks, data race can occur to weight counter W_j (Line 10, Algorithm 8.1). Therefore, locks must be used for weight counters and local task lists. Locks serialize the execution and incur increasing overheads as P increases. Thus, we focus on eliminating the locks for local task lists and weight counters.

We propose a lock-free organization for the local task lists and weight counters. We substitute P weight counters for *each* original weight counter and P circular lists for *each* local task list, so that each weight counter or circular list is updated *exclusively* during scheduling. Thus, there are P^2 lock-free weight counters and P^2 lock-free circular lists. The lock-free organization is shown in Figure 8.3, where the dashed box on the left-hand side shows the organization of local task list LL_i and weight counter W_i in Thread i for some $i, 0 \leq i < P$; the dashed circles on the right-hand side represent the P threads. Although there are more queues and counters than shown in Figure 8.1, this organization does not stress the memory. Each queue can be maintained by three variables only: head, tail, and counter. If each variable is an integer, then the total size of these variables is less than 1 KB, which is negligible compared with the cache size of almost all multicore platforms.

The lock-free local task list in Thread i, $0 \leq i < P$, denoted by LL_i, consists of P circular lists $Q_i^0, Q_i^1, \cdots, Q_i^{P-1}$. The jth circular list Q_i^j, $0 \leq$

$j < P$, corresponds to the jth portion of LL_i. Each circular list Q_i^j has two pointers, $head_i^j$ and $tail_i^j$, pointing to the first and last task, respectively. Tasks are fetched from (inserted into) Q_i^j at the location pointed to by $head_i^j$ $(tail_i^j)$.

The solid arrows in Figure 8.3 connect the heads of the circular lists in LL_i to Thread i, which shows that Thread i can fetch tasks from the head of *any* circular list in LL_i. Corresponding to Line 15 of Algorithm 8.1, we let Thread i fetch tasks from the circular lists in LL_i in a round-robin fashion. The dash-dotted arrows connecting Thread j, $0 \le j < P$, to the tail of Q_i^j imply that Thread j allocates tasks to LL_i by inserting the tasks at the tail of circular list Q_i^j. If Q_i^j is full, Thread j inserts tasks into Q_k^j, $0 \le k < P$ and $k \ne i$, in a round-robin fashion. According to Figure 8.3, each head or tail is updated by one thread only.

We use P weight counters in Thread i to track W_i, the workload of LL_i. The P weight counters are denoted W_i^0, W_i^1, $\cdots$, W_i^{P-1} in Figure 8.3, and W_i is given by $W_i = \sum_{j=0}^{P-1} W_i^j$. Note that all the solid arrows pass through weight counter W_i^i in Figure 8.3, which means that Thread i must update W_i^i when it fetches a task from *any* circular list. The dash-dotted arrow passing through W_i^j for some j, $0 \le j < P$, indicates that Thread j must update W_i^j when it inserts a task into Q_i^j. Note that the value of W_i^j is *not* the total weight of tasks in Q_i^j. Therefore, each weight counter is updated by one thread only during scheduling.

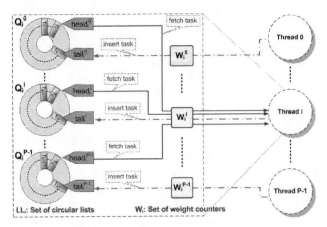

FIGURE 8.3: Lock-free organization of local task list (LL_i) and a set of weight counters in Thread i and their interaction with respect to each thread.

8.4.4 Correctness

When a thread is writing data to shared variables, e.g., the weight counters, the head and tail of the circular lists, another thread may read stale values of these shared variables if we eliminate locks. However, such a stale value does not impact the *correctness*. By correctness, we mean that all the precedence constraints are satisfied, and each of the tasks in the DAG is scheduled onto some thread.

All the precedence constraints of a given DAG-structured computation are satisfied because of the locks for the GL. For each task T in the GL, the task-dependency degree d_T is protected by a lock to avoid concurrent write. Thus, according to Lines 7 and 8 in Algorithm 8.1, T can be allocated only if all the precedence tasks of T are processed. Therefore, the precedence constraints are satisfied regardless of potentially stale values.

When a stale value of a weight counter W_i^j is read by a thread, an error is introduced into the workload estimate for LL$_i$, i.e., $W_i = \sum_{j=0}^{P-1} W_i^j$. Due to the inaccurate workload estimate, the thread may allocate a new task to a different thread, compared with the case for which the exact workload is known. This does not affect the correctness but may lead to an unbalanced load. However, since the weight counters are checked in every iteration of scheduling (Lines 4–18, Algorithm 8.1), the updated value is taken into consideration in the next iteration of some thread. Therefore, except for temporal influence on load balance, the stale values of the weight counters have no impact on correctness.

We analyze the impact of the stale value of $head_j^i$, $0 \le i, j < P$, by considering Line 9 in Algorithm 8.1, where Thread i allocates task T to Thread j by appending T to LL$_j$. To append T to the lock-free local task list LL$_j$, Thread i first checks if Q_j^i is full. If not, T is appended at $tail_j^i$ in Q_j^i. A circular list Q_j^i is full if and only if $head_j^i = tail_j^i + 1$. Consider the following scenario: assume Q_j^i is full, and Thread j fetches a task from Q_j^i. Then, Thread j updates $head_j^i$, and Thread i allocates task T to Thread j. In such a scenario, T can be appended to Q_j^i, since Thread j has taken a task away. However, if Thread i reads the stale value of $head_j^i$, then Q_j^i still appears to be full. Thus, Thread i finds another circular list Q_k^i, $k \ne j$, in a round-robin fashion for task allocation where Q_k^i is not full. Therefore, the stale value of $head_j^i$ does not affect the correctness, but may lead to an unbalanced load.

Similarly, we analyze the impact of the stale value of $tail_i^j$, $0 \le i, j < P$, by considering Line 15 in Algorithm 8.1, where Thread i fetches task T' from LL$_i$. To fetch a task from a lock-free local task list LL$_i$, Thread i first checks if Q_i^j is empty, where j is chosen in a round-robin fashion. If not empty, the task at $head_i^j$ in Q_i^j is fetched. Otherwise, Thread i fetches a task from the next nonempty circular list in LL$_i$. Consider the following scenario: Assume Q_i^j is empty and Thread j appends task T' to Q_i^j. Then, Thread j updates $tail_i^j$ and Thread i fetches a task from Q_i^j. In such a scenario, Thread i can fetch a task from Q_i^j, since Thread j has inserted T' into Q_i^j. However, if Thread i reads the stale value of $tail_i^j$, then Q_i^j still appears to be empty. Thus, Thread j fetches a task from another circular list. The updated value of $tail_i^j$ can be read in the next iteration of scheduling. Then Thread j can fetch T' for execution. Therefore, the stale value of $tail_i^j$ does not affect the correctness.

8.4.5   Experiments

We conducted experiments on three state-of-the-art homogeneous multicore systems: (1) The dual Intel Xeon 5335 (Clovertown) quadcore platform containing two Intel Xeon x86_64 E5335 processors, each having four cores. The processors ran at 2.00 GHz with a 4-MB L2 cache each and a 16-GB DDR2 shared memory. The operating system was Red Hat Enterprise Linux WS Release 4 (Nahant Update 7). We installed a GCC Version 4.1.2 compiler with streaming SIMD extensions 3 (SSE 3), also known as Prescott New Instructions. (2) The dual AMD Opteron 2335 (Barcelona) quadcore platform dual AMD Opteron x86_64 2350 quadcore processors, running at 2.0 GHz. The system had a 16-GB DDR2 memory shared by all the cores, and the operating system was CentOS Version 5 Linux. We also used the GCC 4.1.2 compiler on this platform. (3) The quad AMD Opteron 8358 (Barcelona) quadcore platform had four AMD Opteron x86_64 8358 quadcore processors, running at 1.2 GHz. The system had 64 GB shared memory and the operating system was Red Hat Enterprise Linux Server Release 5.3 (Tikanga). We also used GCC 4.1.2 compiler on this platform.

8.4.5.1   Baseline schedulers

To evaluate the proposed method, we implemented six baseline methods and compared them along with the proposed one using the same input-task dependency graphs on various multicore platforms.

(1) Centralized scheduling with shared core (**Cent shared**): This scheduling method consists of a *scheduler thread* and several *execution threads*. Each execution thread was bound to a core, while the scheduler thread can be executed on any of these cores. In this scheduling method, the input DAG was *local* to the scheduler thread. Each execution thread had a ready task list *shared* with the scheduler thread. In addition, there was a Completed Task ID buffer *shared* by all the threads. The scheduler thread was in charge of all the activities related to scheduling, including updating task dependency degrees, fetching ready-to-execute tasks, and evenly distributing tasks to the ready task lists. Each execution thread fetched tasks from its ready task list for execution. The IDs of completed tasks were inserted into the Completed Task ID buffer, so that the scheduler thread could fetch new ready-to-execute tasks from the successors of tasks in the Completed Task ID buffer. After a given number (i.e., δ_M in Algorithm 8.1) of tasks in the ready task list were processed, the execution thread invoked the scheduler thread and then went to sleep. When a task was allocated to the ready task list of a sleeping thread, the scheduler invoked the corresponding execution thread. Spinlocks were used for the ready task lists and Completed Task ID buffer.

(2) Centralized scheduling with dedicated core (`Cent ded`): This scheduling method was adapted from the centralized scheduling with shared core. The only difference is that a core was dedicated to the scheduler thread, i.e., each thread was bound to a separate core. Similar to `Cent shared`, the input DAG was *local* to the scheduler. Each execution thread had a ready task list *shared* with the scheduler thread. There was a Completed Task ID buffer *shared* by all the threads. The scheduler thread was also in charge of all the activities related to scheduling, and the execution threads executed assigned tasks only. Spinlocks were used for the ready task lists and Completed Task ID buffer.

(3) Distributed scheduling with shared ready task list (`Dist shared`): In this method, we distributed the scheduling activities across the threads. This method had a *shared* global task list and a *shared* ready task list. Each thread had a *local* Completed Task ID buffer. The schedulers integrated into each thread fetched ready-to-execute tasks from the global task list and inserted the tasks into the shared ready task list. If the ready task list was not empty, each thread fetched tasks from the ready task list for execution. Each thread inserted the IDs of completed tasks into its Completed Task ID buffer. Then, the scheduler in each thread updated the dependency degree of the successors of tasks in the Completed Task ID buffer and fetched the tasks with dependency degree equal to 0 for allocation. Pthreads spinlocks were used for the global task list and the ready task list.

(4) Collaborative scheduling with *lock-based* local task list (`ColSch lock`): This was the collaborative scheduling (Section 8.4.1) without any optimization discussed in Section 8.4.3. Rather than using the lock-free data structures, we used mutex locks to avoid concurrent write to the local task lists and weight counters. Spinlocks were used to protect task dependency degrees in the shared global task list.

(5) Collaborative scheduling with *lock-free* local task list (`ColSch lockfree`): This was *not* a baseline, but the proposed method (Section 8.4.3) with lock-free weight counters and local task lists (Section 8.4.3). We used spinlocks to protect task-dependency degrees in the shared global task list.

(6) DAG Scheduling using the Cilk (`Cilk`): This baseline scheduler performed work-stealing-based scheduling using the Cilk runtime system. Unlike the above scheduling methods where we bound a thread to a core of a multi-core processor and allocated tasks to the threads, we dynamically created a thread for each ready-to-execute task and then let the Cilk runtime system schedule the threads onto cores. Although Cilk can generate a DAG dynamically, we used a given task-dependency graph stored in a *shared* global list for the sake of fair comparison. Once a task completed, the corresponding thread reduced the dependency degree of the successors of the task and created new threads for the successors with a resulting dependency degree equal to 0. We used spinlocks for the dependency degrees to prevent concurrent write.

(7) Our implementation of the *work-stealing*-based scheduler (`Stealing`): Although the above baseline `Cilk` is also a work-stealing scheduler, it used the

Cilk runtime system to schedule the threads, each corresponding to a task. On the one hand, the Cilk runtime system has various additional optimizations; on the other hand, scheduling the threads onto cores incurs overhead due to context switching. Therefore, for the sake of fair comparison, we implemented the `Stealing` baseline; we distributed the scheduling activities across the threads, each having a *shared* ready task list. The global task list was *shared* by all the threads. If the ready task list of a thread was not empty, the thread fetched a task from it at the top for execution and, upon completion, updated the dependency degree of the successors of the task. Tasks with dependency degree equal to 0 were placed into the top of its ready task list by the thread. When a thread ran out of tasks to execute, it randomly chose a ready list to steal a task from its bottom unless all tasks were completed. The data for randomization were generated offline to reduce possible overhead due to a random number generator. Pthreads spinlocks were used for the ready task lists and global task list.

8.4.5.2   Data sets and task types

We used both synthetic and real data sets to evaluate the proposed scheduling methods. We built a task-dependency graph for exact inference in a Bayesian network used by a real application called QuickMedical Reference decision theoretic version (QMR-DT), a microcomputer-based decision support tool for diagnosis in internal medicine [22]. There were 1,000 nodes in this network of two layers, one representing diseases and the other symptoms. Each disease has one or more edges pointing to the corresponding symptoms. All random variables (nodes) were binary. The resulting task-dependency graph for exact inference had 228 tasks, each corresponding to a series of computations called node-level primitives. These primitives consist of addition, multiplication, and division operations among single-precision floating-point data in one or two tables [23]. Given the number of random variables involved in each task and the number of states of the random variables, denoted by W_c and r, respectively, the table size is $4r^{W_c}$ bytes. Each table entry was a 4-byte single-precision floating-point data point. In this data set, r was 2 and the average value for W_c was 10. The average number of successors d for each task was also 2. The task weights were estimated using W_c, r and d (see [23]).

We used extensive synthetic data sets to investigate the impact of various parameters of task-dependency graphs on the scheduling performance. Our graph generator synthesized DAGs to simulate the structure of programs: Each node represented a code segment. The edges showed the data dependency between the code segments. We assumed that the loops in a program must either be inside a code segment or unrolled. Let N denote the number of nodes in the DAG and d the average node degree. Let $\alpha_0, \alpha_1, \cdots, \alpha_{N-1}$ denote the nodes. We started with an empty set of edges. When node α_i was visited, the generator spawned $max(0, (d - d_{in} + \delta))$ outgoing edges for α_i, where d_{in} is the number of incoming edges, and $\delta \in (\lfloor -d/2 \rfloor, \lceil d/2 \rceil)$ is a random integer. The outgoing edges were connected to nodes selected randomly from α_{i+1} to α_{N-1}: the probability that α_j, $i < j < N$, was selected is $P(\alpha_i \to \alpha_j) = e^{-1/(j-i)}$. That is, for a given node α_i, the nodes close to α_i had higher probability of selection. Finally, all the nodes with no parent, except for α_0, were connected to α_0. We generated DAGs with 10,000 nodes. The average degree of the nodes was 8. The threshold for the Task ID buffer δ_M was 5.

We experimented with three types of tasks for *each node* of the generated DAGs in our experiments: (1) *Dummy task*: Dummy task was actually a timer, which allowed us to assign the execution time of a task. Dummy tasks also helped us analyze the scheduling overhead since we could easily calculate the task execution time. (2) *Computation intensive task*: This simulated a number of real tasks. We performed matrix multiplication $Y = X^T X$, where X was a 256×256 matrix. (3) *Memory-access-intensive task*: This simulated another common type of real task, which performed irregular memory access. In our experiments, we updated each element $x[i]$ of an array with 128k entries by performing $x[i] = x[y[i]]$, where y was an index array for x.

8.4.5.3 Experimental results

We first compared the proposed scheduling method (i.e., `ColSch lockfree`) with the two work-stealing-based methods `Stealing` and `Cilk` using the task-dependency graph for exact inference in the QMR-DT Bayesian network on the platform with dual AMD Opteron 2335 quadcore processors. In each comparison, the input task-dependency graphs were *identical* for all the methods. In Figure 8.4(a), we show the execution time for scheduling exact inference using the task-dependency graph. The corresponding speedup is shown in Figure 8.4(b). According to Figure 8.4(a), when the number of threads is less than four, almost no difference in execution time can be observed. However, as the number of threads increases, we can observe that the proposed method leads to less execution time than the two baseline methods. The difference is clear in Figure 8.4(b), where we converted the execution time into speedup. A reason for the difference in the speedups is that the proposed scheduling method balanced the workload across the threads according to the task weights. However, neither of the baseline methods considered the task weights. Instead, they stole tasks from other threads to feed the underutilized threads. Since the DAG had 228 nodes and the number of threads was 8, the number of tasks per thread was 28.5 on the average. Thus, the number of available tasks for stealing at a particular time was even less. In this scenario, there is no guarantee for work stealing to achieve load balance across the threads. Note that, although the results of `Stealing` and `Cilk` were similar in Figure 8.4(b), the performance of `Stealing` was slightly better than `Cilk`.

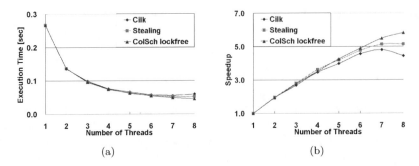

FIGURE 8.4: Comparison of the proposed scheduler with work-stealing-based schedulers.

One reason for the performance difference was the overhead due to thread creation and the context switch. For `Cilk`, we dynamically created threads for the tasks. This incurs thread-creation overheads. In addition, we let the runtime system swap the threads in the cores, which resulted in context-switch overhead. However, for the other methods, we bound the threads to separate cores and allocated tasks to the threads in user space, where no thread creation or swap in/out occurred.

We conducted the following four experiments using synthetic datasets on multicore platforms to investigate the impact of various parameters of input task-dependency graphs on scheduling performance. The first experiment compared the proposed scheduling method and various baseline methods. We used our graph generator to construct a DAG with 10,000 nodes, where each node was assigned a dummy task taking 50 microseconds. We executed the DAG using the five task-sharing scheduling methods on a dual Intel Xeon 5335 (Clovertown) quadcore platform and a quad AMD Opteron 8358 (Barcelona) quadcore platform, where up to 8 and 16 concurrent threads were supported, respectively. Given various available threads, we measured the overall execution time on both platforms and converted it into speedups in Figures 8.5(a) and 8.6(a), where the speedup is defined as the serial execution time over parallel execution time.

The second experiment investigated the overhead due to the proposed lock-free collaborative scheduling method with respect to the number of threads and task size on various platforms. As in the above experiment, we used a DAG with 10,000 nodes and dummy tasks. However, we varied the time delay for each dummy task. We show the ideal speedup and the experimental results with dummy tasks taking 50, 100, 500, and 1,000 microseconds in Figures 8.5(b) and 8.6(b).

The third experiment showed the impact of the error in the estimated task-execution time. We again used a DAG with 10,000 dummy task nodes. However, instead of using the real execution time as the task weight, we used

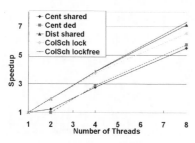

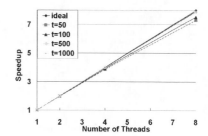

(a) Comparison with baseline scheduling methods.

(b) Impact of the size of tasks on the speedup achieved by `ColSch lockfree`.

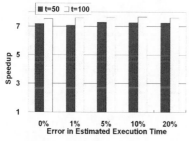

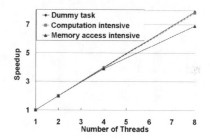

(c) Impact of error in the estimated task-execution time on the speedup achieved by `ColSch lockfree`.

(d) Performance of `ColSch lockfree` with respect to task types.

FIGURE 8.5: Experimental results on dual Intel Xeon 5335 (Clovertown) quadcore platform.

estimated execution time with errors. Let t denote the real task-execution time and t' the estimated execution time. The real task-execution time t was randomly selected from range $((1 - r/100)t', (1 + r/100)t')$, where r is the absolute percentage error in estimated task-execution time. We conducted experiments with $r = 0, 1, 5, 10, 20$ and show the results Figures 8.5(c) and 8.6(c). Note that the definition of speedup in the above two figures is the serial execution time over 8-thread parallel execution time.

The last experiment examined the impact of task types. We used the lock-free collaborative scheduler to schedule a DAG with 10,000 tasks. In the experiments, we used dummy tasks, computation-intensive tasks, and memory-access-intensive tasks, discussed in Section 8.4.5.2. The results are shown in Figures 8.5(d) and 8.6(d).

According to Figures 8.5 and 8.6, the proposed method showed superior performance compared with baseline methods on various platforms as the number of cores increased. For example, in Figure 8.6(a), when 16 cores were used, the speedup achieved by our method (15.12×) was significantly higher than others methods (up to 8.77×). This was because more cores lead

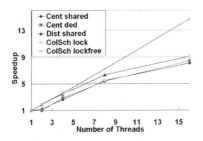

(a) Comparison with baseline scheduling methods.

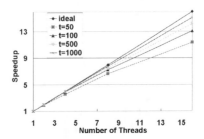

(b) Impact of the size of tasks on the speedup achieved by `ColSch lockfree`.

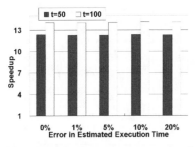

(c) Impact of error in the estimated task-execution time on the speedup achieved by `ColSch lockfree`.

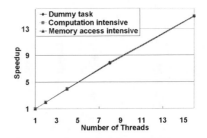

(d) Performance of `ColSch lockfree` with respect to task types.

FIGURE 8.6: Experimental results on quad AMD Opteron 8358 (Barcelona) quadcore platform.

to higher conflict in accessing the task lists. The proposed method allowed such concurrent access, but the baseline methods serialized the accesses. The baseline `Cent shared` showed relatively low performance due to the overhead of frequent context switch between the scheduler thread and the execution thread. `Cent ded` also showed limited speedup, since the computing capability of the core dedicated to the scheduler thread was not used efficiently. The proposed method exhibited good scalability for various task weights. The speedups were close to the ideal speedup for large task weights. The results showed that our technique was tolerant to the error in estimation of task weights. This is because the errors in estimation of task weights can counteract each other. In addition, we observed that the scheduling overhead was less than 1% for the proposed method in all the experiments. Finally, the proposed method achieved almost linear speedup for tasks of various types, including computation-intensive tasks and memory-access-intensive tasks, which implied that the proposed scheduler can work well in real-life scenarios.

8.5   Hierarchical Scheduling with Dynamic Thread Grouping

In order to improve scheduling performance on many-core processors, we propose a hierarchical scheduling method with dynamic thread grouping, which schedules DAG-structured computations at three different levels. At the top level, a supermanager separates threads into groups, each consisting of a manager thread and several worker threads. The supermanager dynamically merges and partitions the groups to adapt the scheduler to the input task-dependency graphs. Through group merging and partitioning, the proposed scheduler can dynamically adjust to become a centralized scheduler, a distributed scheduler, or somewhere in between, depending on the input graph. At the group level, managers collaboratively schedule tasks for their workers. At the within-group level, workers perform self-scheduling within their respective groups and execute tasks.

8.5.1   Organization

We illustrate the components of the hierarchical scheduler in Figure 8.7. The boxes with rounded corners represent thread groups. Each group consists of a *manager* thread and several *worker* threads. The manager in $Group_0$ is also the *supermanager*. The components inside a box are private to the group, while the components out of the boxes are shared by all groups.

The representation of the *global list* (GL) is discussed in Section 8.4.1 (see Figure 8.1). The *global ready list* (GRL) in Figure 8.7 stores the IDs of tasks with dependency degree equal to 0. These tasks are ready to be executed. During the scheduling process, a task is put into this list by a manager thread once the dependency degree of the task becomes 0.

The *local ready list* (LRL) in each group stores the IDs of tasks allocated to the group by the manager of the group. The workers in the group fetch tasks from LRL for execution. Each LRL is associated with a *workload indicator* (WI) to record the overall workload of the tasks currently in the LRL. Once a task is inserted into (or fetched from) the LRL, the indicator is updated.

The *local completed task list* (LCL) in each group stores the IDs of tasks completed by a worker thread in the group. The list is read by the manager thread in the group for decreasing the dependency degree of the successors of the tasks in the list.

The arrows in Figure 8.7 illustrate how each thread accesses a component (read or write). As we can see, GL and GRL are shared by all the managers for both read and write. For each group, the LRL is write-only for the manager and read-only for the workers, while LCL is write-only for the workers and read-only for the manager. WI is local to the manager in the respective group only.

224 Multicore Computing: Algorithms, Architectures, and Applications

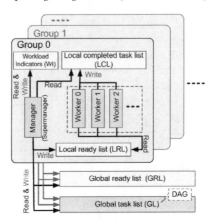

FIGURE 8.7: Components of the hierarchical scheduler.

8.5.2 Dynamic Thread Grouping

The scheduler organization shown in Figure 8.7 supports dynamic thread grouping, which means that the number of threads in a group can be adjusted at runtime. We adjust groups by either merging two groups or partitioning a group. The proposed organization ensures efficient group merging and partitioning.

Figure 8.8(a) illustrates the merging of $Group_i$ and $Group_j$, $i < j$. The two groups are merged by converting all threads of $Group_j$ into the workers of $Group_i$ and merging WIs, LCLs, and LRLs accordingly. Converting threads of $Group_j$ into the workers of $Group_i$ is straightforward: $Manager_j$ stops allocating tasks to $Group_j$ but performs self-scheduling as a worker thread. Then, all the threads in $Group_j$ access tasks from the merged LRL and LCL. To combine WI_i and WI_j, we add the value of WI_j to WI_i. Although WI_j is not used after merging, we still keep it updated for the sake of possible future group partitioning. Merging the lists, i.e., LCLs and LRLs is efficient. Note that both LCL and LRL are circular lists, each having a head and a tail pointer to indicate the first and last tasks stored in the list, respectively. Figure 8.8(b) illustrates the approach to merging two circular lists. We need to update two links only, i.e., the bold arrows shown in Figure 8.8(b). None of the tasks stored in the lists are moved or duplicated. The head and tail of the merged list are $Head_i$ and $Tail_j$, respectively. Note that two merged groups can be further merged into a larger group.

We summarize the procedure in Algorithm 8.2. Since the queues and weight indicators are shared by several threads, locks must be used to avoid concurrent write. For example, we lock LRL_i and LRL_j immediately before Line 1 and unlock them after Line 3. Algorithm 8.3 does not explicitly assign the threads in $Group_i$ and $Group_j$ to $Group_k$ since this algorithm is executed

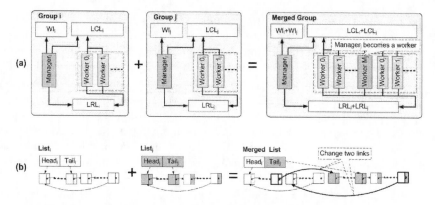

FIGURE 8.8: (a) Merge Group$_i$ and Group$_j$. (b) Merge circular lists $List_i$ and $List_j$. The head (tail) points to the first (last) tasks stored in the list. The blank elements have no task stored yet.

only by the supermanager. Each thread dynamically updates its group information and decides if it should be a manager or worker (see Algorithm 8.3).

Algorithm 8.2 Group merge

Input: Group$_i$ and Group$_j$.

Output: Group$_k$ = Group$_i$ + Group$_j$

    {Merge LRL_i and LRL_j}

1: Let $LRL_j.Head.Predecessor$ points to $LRL_i.Tail.Successor$
2: Let $LRL_i.Tail.Successor$ points to $LRL_j.Head$
3: $LRL_k.Head = LRL_i.Head$, $LRL_k.Tail = LRL_j.Tail$

    {Merge LCL_i and LCL_j}

4: Let $LCL_j.Head.Predecessor$ points to $LCL_i.Tail.Successor$
5: Let $LCL_i.Tail.Successor$ points to $LCL_j.Head$
6: $LCL_k.Head = LCL_i.Head$, $LCL_k.Tail = LCL_j.Tail$

    {Merge WI_i and WI_j}

7: $WI_k = WI_i + WI_j$

Group$_i$ and Group$_j$ can be restored from the merged group by partitioning. As a reverse process of group merging, group partitioning is also straightforward and efficient. Due to space limitations, we do not elaborate here. Group merging and partitioning can be used for groups with an arbitrary number of threads. We assume the number of threads per group is a power of two hereinafter for the sake of simplicity.

8.5.3 Hierarchical Scheduling

Using the proposed data organization, we schedule a given DAG-structured computation at three levels. The top level is called the *meta-level*, where we have a supermanager to control group merging/partitioning. At this level, we are not scheduling tasks directly, but reconfiguring the scheduler according to the characteristics of the input tasks. Such a process is called *meta-scheduling*. The supermanager is hosted along with the manager of $Group_0$ by $Thread_0$. Note that $Manager_0$ can never become a worker as discussed in Section 8.5.2.

The mediate level is called the *group level*, where the manager in each group collaborates with each other and allocates tasks for the workers in the group. The purpose of collaborating between managers is to improve the load balance across the groups. Specifically, the managers ensure that the workload in the local ready lists is roughly equal for all groups. A manager is hosted by the first thread in a group.

The bottom level is called the *within-group level*, where the workers in each group perform self-scheduling. That is, once a worker finishes a task execution and updates LCL in its group, it fetches a new task, if any, from LRL immediately. Self-scheduling keeps all workers busy unless the LRL is empty. Each worker is hosted by a separate thread.

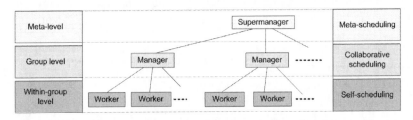

FIGURE 8.9: The hierarchical relationship between the supermanager, managers and workers, and the corresponding scheduling schemes.

The hierarchical scheduler behaves between centralized and distributed schedulers so that it can adapt to the input task graph. Note that each group consists of a manager thread and several worker threads. When all the groups are merged into a single group, the proposed method becomes a centralized scheduler; when multiple groups exist, the proposed method behaves as a distributed scheduler.

8.5.4　Scheduling Algorithm and Analysis

We propose a sample implementation of the hierarchical scheduler presented in Section 8.5.3. Based on the organization shown in Section 8.5.1, we use the following notations to describe the implementation. Assume there are P threads, each bound to a core. The threads are divided into groups consisting of a manager and several workers. GL and GRL denote the global task list and global ready list, respectively. LRL_i and LCL_i denote the local ready list and local completed task list of $Group_i$, $0 \le i < P$. d_T and w_T represent the dependency degree and the weight of task T, respectively. WI_i is the workload indicator of $Thread_i$. Parameters δ_M, δ_+, and δ_- are given thresholds. The boxes show the statements that access variables shared by all groups.

Algorithm 8.3 illustrates the framework of the hierarchical scheduler. In Lines 1–3, thread groups are initialized, each with a manager and a worker along with a set of ready-to-execute tasks stored in LRL_j, where the overall task weight is recorded in WI_j. A boolean flag f_{exit} in Line 3 notifies us if the threads can exit the scheduling iteration (Lines 5–15). *rank* controls the size of groups: Increasing *rank* leads to the merging of two adjacent groups, while decreasing *rank* leads to partitioning of current groups. $rank = 1$ corresponds to the minimum group size, i.e., two threads per group. Thus, we have $1 \leq rank \leq \log P$. The group size Q is therefore given by

$$Q = \frac{P}{2^{\log P - rank}} = 2^{rank} \qquad (8.1)$$

Line 4 in Algorithm 8.3 starts all the threads in parallel. The threads perform various scheduling schemes according to their thread IDs. The first thread in each group is a manager (Line 8). In addition, the first thread in Group$_0$, i.e., Thread 0 performs as the supermanager (Line 10). The rest of the threads are workers (Line 13). Given thread ID i, the corresponding group is $\lfloor i/Q \rfloor$.

Algorithm 8.4 shows the meta-scheduling method for the supermanager. The algorithm consists of two parts: updating *rank* (Lines 1–2) and regrouping (Lines 3–11). We use a heuristic to update *rank*. Note that WI_j is the computational workload for Group$_j$. A large WI_j requires more workers for task execution. $|LCL_j|$ is the number of completed tasks, and d is the average number of successive tasks. For each completed task, the manager reduces the dependency degree of the successive tasks and moves ready-to-execute tasks to LRL_j. Thus, $(|LCL_j| \cdot d)$ represents the workload for the scheduler. A larger $(|LCL_j| \cdot d)$ requires more managers for task scheduling. In Line 1, the ratio r tells us if we need more managers or more workers. If more workers are needed, we increase *rank* in Line 2. In this case, groups are merged to provide more workers per manager. Otherwise, *rank* decreases. Line 2 also ensures that *rank* is valid by checking the boundary values. d, δ_+, and δ_- are given as inputs. The regrouping depends on the value of *rank*. If *rank* increases, two groups merge (Line 5); if *rank* decreases, the merged group is partitioned (Line 9). The two operators Merge($\cdot$) and Partition($\cdot$) are discussed in Section 8.5.2. Line 12 flips f_{exit} if no task remains in GL. This notifies all of the threads to terminate (Line 5 in Algorithm 8.4).

Algorithm 8.5 shows an iteration of the group-level scheduling for managers. Each iteration consists of three parts: updating WI_i (Lines 1–2 and 15), maintaining precedence relationship (Lines 3–8), and allocating tasks (Lines 9–14). Lines 3–8 check the successors of all tasks in LCL_i in batch mode to reduce synchronization overhead. Let $m = 2^{rank} - 1$ denote the number of workers per group. In the batch task-allocation part (Lines 9–14), we first fetch m tasks from GRL. Line 12 is an adaptive step of this algorithm. If the overall workload of the m tasks is too light ($\sum_{T \in S'} w_T < \Delta W$) or the current tasks in LRL_i are not enough to keep the workers busy ($WI_i < \delta_M$), more tasks are fetched for the next iteration. This dynamically adjusts the workload

---

Algorithm 8.3 A Sample Implementation of Hierarchical Scheduler

---

Input: P threads; Task-dependency graph stored in GL; Thresholds δ_M, δ_+
        and δ_-.

Output: Assign each task to a worker thread

    {Initialization}
 1: Group$_j$={*Manager*: Thread$_{2j}$, *Worker*: Thread$_{2j+1}$}, $j = 0, 1, \cdots, P/2-1$
 2: Evenly distribute tasks $\{T_i | T_i \in GL \text{ and } d_i = 0\}$ across LRL_j, $WI_j =$
        $\sum_{T \in LRL_j} w_T$, $\forall j = 0, 1, \cdots, P/2 - 1$
 3: f_{exit} =**false**, $rank = 1$

    {Scheduling}
 4: **for** Thread $i = 0, 1, \cdots, P - 1$ **pardo**
 5:     **while** f_{exit} =**false do**
 6:         $Q = 2^{rank}$
 7:         **if** $i\%Q = 0$ **then**

            {Manager thread}
 8:             Group-level scheduling at Group$_{\lfloor i/Q \rfloor}$ (*Algorithm* 8.5)
 9:             **if** $i = 0$ **then**

                {Supermanager thread}
10:                 Meta-level scheduling (*Algorithm* 8.4)
11:             **end if**
12:         **else**

                {Worker thread}
13:             Within-group-level scheduling at Group $_{\lfloor i/Q \rfloor}$ (*Algorithm* 8.6)
14:         **end if**
15:     **end while**
16: **end for**

---

distribution and prevents possible starvation for any groups. In Line 10, the manager inspects a set of tasks and selects m tasks with relatively more successors. This is a widely used heuristic for scheduling [2]. Several statements in Algorithm 8.5 are put into boxes, where the managers access shared components across the groups. Synchronization costs of these statements vary as the number of groups changes.

The workers schedule tasks assigned by their manager (Algorithm 8.6). This algorithm is a straightforward self-scheduling scheme, where each idle worker fetches a task from LRL_i and then puts the tasks into LCL_i after execution. Although LRL_i and LCL_i are shared by the manager and worker threads in the same group, no worker accesses any variables shared between groups.

Algorithm 8.4 Meta-Level Scheduling for Supermanager

{Update rank}
1: $r = \sum_{j=0}^{P/Q}(WI_j/(|LCL_j| \cdot d))$, $rank_{old} = rank$
2: $rank = \begin{cases} \min(rank+1, \log P), & r > \delta_+ \\ \max(rank-1, 1), & r < \delta_- \end{cases}$

{regrouping}
3: **if** $rank_{old} < rank$ **then**
   {Combine Groups}
4:    **for** $j = 0$ to $P/(2 \cdot Q) - 1$ **do**
5:       $\text{Group}_j = \text{Merge}(\text{Group}_{2j}, \text{Group}_{2j+1})$
6:    **end for**
7: **else if** $rank_{old} > rank$ **then**
   {Partition Group}
8:    **for** $j = P/Q - 1$ downto 0 **do**
9:       $(\text{Group}_{2j}, \text{Group}_{2j+1}) = \text{Partition}(\text{Group}_j)$
10:    **end for**
11: **end if**
12: **if** $GL = \emptyset$ **then** $f_{exit} =$**true**

Algorithm 8.5 Group-Level Scheduling for the Manager of Group$_i$

{Update workload indicator}
1: $\Delta W = \sum_{\tilde{T} \in LCL_i} w_{\tilde{T}}$
2: $WI_i = WI_i - \Delta W$

{Update precedence relations}
3: **for all** $T \in \{$successors of $\tilde{T}, \forall \tilde{T} \in LCL_i\}$ **do**
4:    $\boxed{d_T = d_T - 1}$
5:    **if** $d_T = 0$ **then**
6:       $\boxed{GRL = GRL \cup \{T\}; GL = GL \setminus \{T\}}$
7:    **end if**
8: **end for**

{Batch task allocation}
9: **if** LRL_i is not full **then**
10:    $\boxed{S' \Leftarrow \text{fetch } m \text{ tasks from } GRL, \text{ if any}}$
11:    **if** $\sum_{T \in S'} w_T < \Delta W$ or $WI_i < \delta_M$ **then**
12:       $\boxed{\text{Fetch more tasks from } GRL \text{ to } S' \text{ so that } \sum_{T \in S'} w_T \approx \Delta W + \delta_M}$
13:    **end if**
14:    $LRL_i = LRL_i \cup \{S'\}$
15:    $WI_i = WI_i + \sum_{T \in S'} w_T$
16: **end if**

Algorithm 8.6 Within-Group-Level Scheduling for a Worker of Group$_i$

{Perform task execution}
1: Fetch T from LRL_i
2: **if** $T \neq \emptyset$ **then**
3:    Execute task T
4:    $LCL_i = LCL_i \cup \{T\}$
5: **end if**

8.5.5   Experiments

The Sun UltraSPARC T2 (Niagara 2) platform was a Sunfire T2000 server with a Sun UltraSPARC T2 multithreading processor [8]. UltraSPARC T2 has eight hardware multithreading cores, each running at 1.4 GHz. In addition, each core supports up to eight hardware threads with two shared pipelines. Thus, there are 64 hardware threads. Each core has its own L1 cache shared by the threads within a core. The L2 cache size is 4 MB, shared by all hardware threads. The platform had 32 GB DDR2 memory shared by all the cores. The operating system was Sun Solaris 11, and we used Sun Studio CC with Level-4 optimization (-xO4) to compile the code.

8.5.5.1   Baseline schedulers

To compare the performance of the proposed method, we performed DAG-structured computations using Charm++ [24], Cilk [13], and OpenMP [19]. In addition, we implemented three typical schedulers called `Cent ded`, `Dist shared`, and `Steal`, respectively. We evaluated the baseline methods along with the proposed scheduler using the same input task-dependency graphs.

(a) Scheduling DAG-structured computations using Charm++ (`Charm++`): The Charm++ runtime system employs a phase-based dynamic load-balancing scheme facilitated by virtualization, where the computation is monitored for load imbalance, and computation objects (tasks) are migrated between phases by message passing to restore balance. Given a task-dependency graph, each task is packaged as an object called *chore*. Initially, all tasks with a dependency degree equal to 0 are submitted to the runtime system. When a task completes, it reduces the dependency degree of the successors. Any successors with reduced dependency degree equal to 0 are submitted to the runtime system for scheduling.

(b) Scheduling DAG-structured computations using Cilk (`Cilk`): This baseline scheduler performs work-stealing-based scheduling using the Cilk runtime system. Unlike the proposed scheduling methods where we bound a thread to a core of a multicore processor and allocated tasks to the threads, we dynamically created a thread for each ready-to-execute task and then let the Cilk runtime system schedule the threads onto cores. Although Cilk can generate a DAG dynamically, we used a given task-dependency graph stored in a *shared* global list for the sake of fair comparison. Once a task completed, the corresponding thread reduced the dependency degree of the successors of the task and created new threads for the successors with dependency degree equal to 0. We used spinlocks for the dependency degrees to prevent concurrent write.

(c) Scheduling DAG-structured computation using OpenMP (`OpenMP`): This baseline initially inserted all tasks with dependency degree equal to 0 into a ready queue. Then, using the OpenMP pragma directives, we created threads to execute these tasks in parallel. While executing the tasks in the ready queue, we inserted new ready-to-execute tasks into another ready queue for parallel execution in the next iteration. Note that the number of tasks in the ready queue can be much greater than the number of cores. We let the OpenMP runtime system dynamically schedule tasks to underutilized cores.

(d) Centralized scheduling with dedicated core (`Cent ded`), (e) Distributed scheduling with shared ready task list (`Dist shared`), and (f) Task-stealing-based scheduling with distributed ready task list (`Steal`) are discussed in Items (2), (3), and (7) in Section 8.4.5.1, respectively.

8.5.5.2   Data sets and data layout

We experimented with both synthetic and real data sets to evaluate the performance of the proposed scheduler. For the synthetic data sets, we varied the task-dependency graphs to evaluate our scheduling method using task-dependency graphs with various graph topologies, sizes, task workload, task types, and accuracies in estimating task weights. For the real data sets, we used task-dependency graphs for blocked matrix multiplication (BMM), LU, and Cholesky decomposition. In addition, we also used the task-dependency graph for exact inference, a classic problem in artificial intelligence, where each task consists of data-intensive computations between a set of probabilistic distribution tables (also known as *potential tables*) involving both regular and irregular data accesses [25].

We used the following data layout in the experiments: the task-dependency graph was stored as an array in the memory, where each element represents a task with a task ID, weight, number of successors, a pointer to the successor array, and a pointer to the task metadata. Thus, each element took 32 bytes, regardless of what the task consisted of. The task metadata was the data used for task execution. For LU decomposition, the task metadata is a matrix block; for exact inference, it is a set of potential tables. The lists used by the scheduler, such as GRL, LRLs, and LCLs, were circular lists, each having a head and a tail pointer. In case any list was full during scheduling, new elements were inserted on the fly.

8.5.5.3   Experimental results

We compared the performance of the proposed scheduling method with three state-of-the-art parallel programming systems, i.e., Charm++[24],

Cilk [13], and OpenMP [19]. We used a task-dependency graph for which the structure was a random DAG with 10,000 tasks, and there was an average of eight successors for each task. Each task was a dense operation, e.g., multiplication of two 30×30 matrices. For each scheduling method, we varied the number of available threads so that we could observe the achieved scalability. The results are shown in Figure 8.10. Similar results were observed for other tasks. Given the number of available threads, we repeated the experiments five times. The results were consistent; the standard deviations of the results were almost within 5% of the execution time. In Figure 8.10(a), all the methods exhibited scalability, though Charm++ showed a relatively large overhead. A reason for the significant overhead of Charm++ compared with other methods is that the Charm++ runtime system employs a message-passing-based mechanism to migrate tasks for load balancing (see Section 8.5.5.1). This increased the amount of data transferring on the system bus. Note that the proposed method required at least two threads to form a group. In Figure 8.10(b), where more threads were used, our proposed method still showed good scalability; while the performance of the OpenMP and Charm++ degraded significantly. As the number of threads increased, the Charm+ required frequent message-passing-based task migration to balance the workload. This stessed the system bus and caused the performance degradation. The performance of OpenMP degraded as the number of threads increased because it can only schedule the tasks in the ready queue (see Section 8.5.5.1), which limits the parallelism. Cilk showed scalability close to the proposed method, but the execution time was higher.

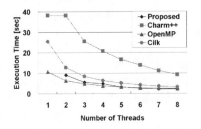

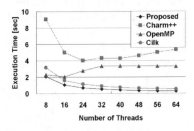

(a) Scalability with respect to 1–8 threads.   (b) Scalability with respect to 8–64 threads.

FIGURE 8.10: Comparison of average execution time with existing parallel programming systems.

We compared the proposed scheduling method with three typical schedulers, a centralized scheduler, a distributed scheduler, and a task-stealing-based scheduler addressed in Section 8.5.5.1. We used the same data set as in the previous experiment, but the matrix sizes were 50×50 (*large*) and 10×10 (*small*) for Figures 8.11(a) and (b), respectively. We normalized the throughput of each experiment for comparison. We divided the throughput of each experiment by the throughput of the proposed method using eight threads.

The results exhibited *inconsistencies* for the two baseline methods: `Cent ded` achieved much better performance than `Dist shared` with respect to large tasks but significantly poorer performance with respect to small tasks. Such inconsistencies imply that the impact of the input task-dependency graphs on scheduling performance can be significant. An explanation for this observation is that the large tasks require more resources for task execution, but `Dist shared` dedicates many threads to scheduling, which limits the resources for task execution. In addition, many schedulers frequently accessing shared data leads to significant overheads due to coordination. Thus, the throughput decreased for `Dist shared` as the number of threads increased. When scheduling small tasks, the workers completed the assigned tasks quickly, but the single scheduler of `Cent ded` could not process the completed tasks and allocate new tasks to all the workers in time. Therefore, `Dist shared` achieved higher throughput than `Cent ded` in this case. When scheduling large tasks, the proposed method dynamically merged all the groups and therefore became the same as `Cent ded` (Figure 8.11(a)). When scheduling small tasks, the proposed scheduler became a distributed scheduler by keeping each core (eight threads) as a group. Compared with `Dist shared`, eight threads per group led to the best throughput (Figure 8.11(b)). `Steal` exhibited increasing throughput with respect to the number of threads for large tasks. However, the performance tapered off when more than 48 threads were used. One reason for this observation is that, as the number of thread increases, the chance of stealing tasks also increases. Since a thread must access shared variables when stealing tasks, the coordination overhead increases accordingly. For small tasks, `Steal` showed limited performance compared with the proposed method. As the number of threads increased, the throughput was adversely affected. The proposed method dynamically changed the group size and merged all the groups for the large tasks. Thus, the proposed method became `Cent ded` except for the overhead of grouping. The proposed scheduler kept each core (eight threads) as a group when scheduling the small tasks. Thus, the proposed method achieved almost the same performance as `Cent ded` in Figure 8.11(a) and the best performance in Figure 8.11(b).

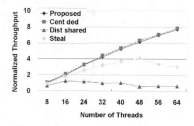 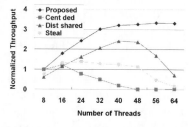

(a) Performance with respect to large tasks (50×50 matrix multiplication for each task).

(b) Performance with respect to small tasks (10×10 matrix multiplication for each task).

FIGURE 8.11: Comparison with baseline scheduling methods using task graphs of various task sizes.

We experimentally show the importance of adapting the group size to the task-dependency graphs in Figure 8.12. In this experiment, we modified the proposed scheduler by fixing the group size. For each fixed group size, we used the same data set in the previous experiment and measured the performance as the number of threads increases. According to Figure 8.12, larger group size led to better performance for large tasks, while for the small tasks, the best performance was achieved when the group size was four or eight. Since the optimized group size varied according to the input task-dependency graphs, it is necessary to adapt the group size to the input task-dependency graph.

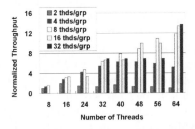

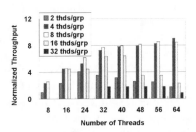

(a) Performance with respect to large tasks (50 × 50 matrix multiplication for each task).

(b) Performance with respect to small tasks (10 × 10 matrix multiplication for each task).

FIGURE 8.12: Performance achieved by the proposed method without dynamically adjusting the scheduler group size (number of threads per group, thds/grp) with respect to task graphs of various task sizes.

In Figure 8.13, we illustrate the impact of various properties of task-dependency graphs on the performance of the proposed scheduler. We studied the impact of the topology of the graph structure, the number of tasks in the graph, the number of successors, and the size of the tasks. We modified these parameters of the data set used in the previous experiments. The topologies used in Figure 8.13(a) included a random graph (Rand), an eight-dimensional grid graph (8D-grid), and the task graph of blocked matrix multiplication (BMM). Note that we only used the topology of the task-dependency graph for BMM in this experiment. Each task in the graph was replaced by a matrix multiplication. According to the results, for most of the scenarios, the proposed scheduler achieved almost linear speedup. Note that the speedup for a 10 × 10 task size was relatively lower than others. This was because synchronization in scheduling was relatively large for the task-dependency graph with small task sizes. Note that we used the speedup as the metric in Figure 8.13. By speedup, we mean the serial execution time over the parallel execution time, when all the parameters of the task-dependency graph are given.

In Figure 8.13(e), we investigated the impact of task types on scheduling performance. The computation-intensive tasks (label *Computation*) were

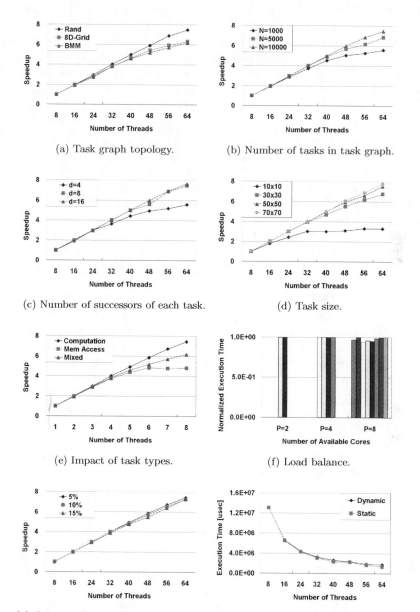

(a) Task graph topology.

(b) Number of tasks in task graph.

(c) Number of successors of each task.

(d) Task size.

(e) Impact of task types.

(f) Load balance.

(g) Impact of error in estimated task weight.

(h) Overhead of the proposed method.

FIGURE 8.13: Impact of various properties of task-dependency graphs on speedup achieved by the proposed method.

matrix multiplications, for which the complexity was $O(N^3)$, assuming the matrix size was $N \times N$. In our experiments, we had $N = 50$. The memory-access-intensive tasks (Mem Access) summed an array of N^2 elements using $O(N^2)$ time. For the last task type (Mixed), we let all the tasks with an even ID perform matrix multiplication and the rest sum an array. We achieved speedup with respect to all task types. The speedup for memory-access-intensive tasks was relatively lower due to the latency of memory access.

Figure 8.13(f) reflects the efficiency of the proposed scheduler. We measured the execution time of each thread to check if the workload was evenly distributed and normalized the execution time of each thread for the sake of comparison. The underlying graph was a random graph. We also limited the number of available cores in this experiment to observe the load balance in various scenarios. Each core had eight threads. As the number of cores increased, there was a minor imbalance across the threads. However, the percentage of the imbalanced work was very small compared with the entire execution time.

For real applications, it is generally difficult to estimate the task weights accurately. To study the impact of the error in estimated task weight, we intentionally added noise to the estimated task weight in our experiments. We included noise that added 5%, 10%, and 15% of the real task-execution time. The noise was drawn from uniform distribution using the POSIX math library. According to the results in Figure 8.13(f), the impact was not significant.

In Figure 8.13(h), we investigated the overhead of the proposed scheduler. Using the same data set used in the previous experiment, we first performed hierarchical scheduling and recorded to which thread a task was allocated. According to such allocation information, we performed static scheduling to eliminate the overhead due to the proposed dynamic scheduler. We illustrate the execution time in Figure 8.13(h). Unlike the previous experiments, we show the results with respect to execution time to compare both the scalability and the scheduling overhead for a given number of threads. The overhead due to dynamic scheduling was very small.

8.6 Conclusion

In this chapter, we discussed two scheduling methods for DAG-structured computations on general-purpose multicore processors and many-core processors. The first method is called collaborative scheduling, and the second is called hierarchical scheduling. Both methods consist of a set of plug-and-play components. Various heuristics could be used for optimization within each component. For example, randomization techniques could be used for the Allocate module [2].

Collaborative scheduling is designed for general-purpose multicore processors. In this method, we developed lock-free local task lists and weight counters that are different from traditional lock-free data structures. The lock-free mechanism reduces scheduling overhead due to synchronization and contention across the cores. The hierarchical scheduling is designed for homogeneous many-core processors, where we divided the threads into groups, each having a manager to perform scheduling at the group level and several workers to perform self-scheduling for the tasks assigned by the manager. A supermanager was used to dynamically adjust the group size so that the scheduler could adapt to the input task-dependency graph.

In the future, we plan to reduce the number of variables used by the lock-free data structure and explore the heuristics for each component in the proposed scheduler. For example, for the task-fetch module, we plan to interleave computationally intensive tasks with memory-access-intensive tasks for the threads assigned to the same core to improve the overall performance. We also plan to study data layout for high throughput processors to efficiently use the data cache of the UltraSPARC processors since the L2 cache is no more than 4 MB, shared by up to 64 hardware threads.

Bibliography

[1] I. Ahmad, S. Ranka, and S.U. Khan. Using game theory for scheduling tasks on multi-core processors for simultaneous optimization of performance and energy. *IEEE International Symposium on Parallel and Distributed Processing (IPDPS)*, pages 1–6, April 2008.

[2] Yu-Kwong Kwok and Ishfaq Ahmad. Static scheduling algorithms for allocating directed task graphs to multiprocessors. *ACM Comput. Surv.*, 31(4):406–471, 1999.

[3] Huey-Ling Chen and Chung-Ta King. Eager scheduling with lazy retry in multiprocessors. *Future Gener. Comput. Syst.*, 17(3):215–226, 2000.

[4] M. De Vuyst, R. Kumar, and D.M. Tullsen. Exploiting unbalanced thread scheduling for energy and performance on a CMP of SMT processors. In *IEEE International Symposium on Parallel and Distributed Processing (IPDPS)*, pages 1–6, 2006.

[5] Xinan Tang and Guang R. Gao. Automatically partitioning threads for multithreaded architectures. *J. Parallel Distrib. Comput.*, 58(2):159–189, 1999.

[6] S.R. Alarm, R.F. Barrett, J.A. Kuehn, P.C. Roth, and J.S. Vetter. Characterization of scientific workloads on systems with multicore processors. In *IEEE International Symposium on Workload Characterization*, pages 225–236, 2006.

[7] Guojing Cong and David A. Bader. Designing irregular parallel algorithms with mutual exclusion and lock-free protocols. *Journal of Parallel and Distributed Computing*, 66:854–866, 2006.

[8] Denis Sheahan. Developing and tuning applications on UltraSPARC T1 chip multithreading systems. Technical report, 2007.

[9] I. Ahmad, Yu-Kwong Kwok, and Min-You Wu. Analysis, evaluation, and comparison of algorithms for scheduling task graphs on parallel processors. In *Proceedings of the 1996 International Symposium on Parallel Architectures, Algorithms and Networks*, page 207, 1996.

[10] Michael R. Garey and David S. Johnson. *Computers and Intractability; A Guide to the Theory of NP-Completeness*. W. H. Freeman & Co., New York, NY, 1990.

[11] Christos Papadimitriou and Mihalis Yannakakis. Towards an architecture-independent analysis of parallel algorithms. In *Proceedings of the Twentieth Annual ACM Symposium on Theory of Computing*, pages 510–513, 1988.

[12] M. S. Squillante and R. D. Nelson. Analysis of task migration in shared memory multiprocessor scheduling. In *ACM Conference on the Measurement and Modeling of Computer Systems*, pages 143–155, 1991.

[13] R. D. Blumofe, C. F. Joerg, B. C. Kuszmaul, C. E. Leiserson, K. H Randall, and Y. Zhou. Cilk: An efficient multithreaded runtime system. Technical report, Cambridge, 1996.

[14] E. G. Coffman. *Computer and Job-Shop Scheduling Theory*. John Wiley and Sons, New York, NY, 1976.

[15] Olivier Beaumont, Larry Carter, Jeanne Ferrante, Arnaud Legr, Loris Marchal, and Yves Robert. Centralized versus distributed schedulers for multiple bag-of-task applications. In *International Parallel and Distributed Processing Symposium (IPDPS)*, pages 1–10, 2006.

[16] Fengguang Song, Asim YarKhan, and Jack Dongarra. Dynamic task scheduling for linear algebra algorithms on distributed-memory multicore systems. In *International Conference for Hight Performance Computing, Networking Storage and Analysis*, 2009.

[17] Henan Zhao and R. Sakellariou. Scheduling multiple DAGs onto heterogeneous systems. In *IEEE International Symposium on Parallel and Distributed Processing (IPDPS)*, pages 1–12, 2006.

[18] Intel Threading Building Blocks. http://www.threadingbuldingblocks.org/, 2013.

[19] OpenMP Application Programming Interface. http://www.openmp.org/, 2013.

[20] James H. Anderson and Srikanth Ramamurthy. Lock-free and practical doubly linked list-based deques using single-word compare-and-swap. In *Proceedings of the 8th International Conference on Principles of Distributed Systems*, pages 240–255, 2005.

[21] Jason Liu, David M. Nicol, and King Tan. Lock-free scheduling of logical processes in parallel simulation. In *Proceedings of the 2000 Parallel and Distributed Simulation Conference*, pages 22–31, 2001.

[22] Tommi S. Jaakkola and Michael I. Jordan. Variational probabilistic inference and the QMR-DT network. *Journal of Artificial Intelligence Research*, 10(1):291–322, 1999.

[23] Yinglong Xia and Viktor K. Prasanna. Scalable node-level computation kernels for parallel exact inference. *IEEE Trans. Comput.*, 59(1):103–115, 2010.

[24] Charm++ programming system. http://charm.cs.uiuc.edu/research/charm/, 2013.

[25] Yinglong Xia, Xiaojun Feng, and Viktor K. Prasanna. Parallel evidence propagation on multicore processors. In *The 10th International Conference on Parallel Computing Technologies*, pages 377–391, 2009.

Chapter 9

Evaluating Multicore Processors and Accelerators for Dense Numerical Computations

Seunghwa Kang

Georgia Institute of Technology

Nitin Arora

Georgia Institute of Technology

Aashay Shringarpure

Georgia Institute of Technology

Richard W. Vuduc

Georgia Institute of Technology

David A. Bader

Georgia Institute of Technology

In this chapter, we empirically evaluate fundamental design trade-offs among current multicore processors and accelerator technologies and their impact on dense numerical computations. The main objectives of this work are to understand the differences in the implementation techniques required to achieve good performance on a variety of current multicore and accelerator platforms and to aid application designers in better mapping their software to the most suitable architecture. We also aim to influence future computing system design. We present interarchitectural comparisons of dense numerical kernels from computational statistics and direct n-body problems using a spectrum of multicore and accelerator platforms, including those based on the Intel Harpertown and Nehalem architectures, the AMD Barcelona architecture, the Sony-Toshiba-IBM Cell Broadband Engine, and the second-generation PowerXCell/8i and the NVIDIA Tesla C870 and C1060. We illustrate the software implementation process on each platform; measure and analyze the performance, coding complexity, and energy efficiency of each implementation; and discuss the impact of different architectural design choices on each implementation.

9.1 Introduction

Driven by the limits of single-core performance due to power constraints, multicore processors and accelerator architectures are becoming ubiquitous in every computing area. These systems are attractive to application developers because of their impressive peak computational potential and (in several cases) their energy-efficient processing capabilities. However, the architectures themselves are diverse and reflect a wide variety of design trade-offs. Consequently, we might expect the performance of a given application to be an even more sensitive function of the architecture than in previous-generation single-core general-purpose processors, which in turn is expected to affect software development costs significantly.

The research literature on software optimization for these multicore systems is growing rapidly, particularly for platforms based on the Sony-Toshiba-IBM (STI) Cell Broadband Engine (Cell/B.E.) [1, 2, 3] and for GPUs [4, 5]. Ryoo et al. [6, 7] have published extensively on generalizing optimization principles for the NVIDIA GPUs based on the CUDA framework [8]. In addition to the optimization research for a single platform, several interarchitecture comparisons are published as well. Williams et al. [9] compare the performance of emerging multicore platforms, including the AMD dual-core Opteron processor, the Intel quad-core Harpertown processor, the Sun Niagara processor, and the Cell/B.E. processor for sparse matrix-vector multiplication. Also, there are a few studies that compare the performance of CPUs, GPUs, FPGAs, the Cell/B.E. processor, and the Cray MTA-2 [10, 11, 1].

This chapter is based on the two papers [12, 13] written by the authors and aims to evaluate the impact of fundamental architectural design trade-offs (see Section 9.2) on a set of dense numerical kernels and to understand the differences in the implementation techniques required to achieve good performance on a variety of modern multicore CPU and accelerator desktop platforms. These platforms include 8-core (dual-socket quad-core) Intel Harpertown (one thread per core) and Nehalem (two threads per core) processor-based systems (8 and 16 effective threads, respectively); a 16-core (quad-socket quad-core) AMD Barcelona-based system; IBM dual-socket QS20 (based on the STI Cell/B.E.) and QS22 (based on the IBM PowerXCell/8i) blades; and NVIDIA Tesla C870 and C1060 graphics processing unit (GPU) systems.

We evaluate these systems experimentally using three kernels from computational statistics first (see Sections 9.3, 9.4, and 9.5). We create two kernels by extracting the most computationally intensive part of the covariance/correlation computation from the R statistics package [14]. These kernels are based on Pearson's method and Kendall's method [15], which we hereafter refer to as Kernel1 and Kernel2, respectively. The third kernel (Kernel3) is created by modifying Kernel2 to highlight each system's capability in processing highly floating-point-intensive computations. As the test systems have varying numbers of chips, die size, clock frequencies, prices, and power consumptions, we focus on architectural design trade-offs and their impacts on different kernels rather than identifying the best-performing processor.

We discuss our implementation and software-optimization process to highlight the challenges and complexities of software development for each architecture. However, we consider the main contribution of this work to be our interarchitectural analysis, not the optimization work. Our experimental results highlight the performance of each system in executing different mix of instructions for compute-bound and data-intensive cases. We consider both single-precision and double-precision performance for floating-point operations and attempt to characterize the resulting performance in terms of each system's design choices.

We also study direct n-body computations primarily because of our interest in this application domain (see Sections 9.7 and 9.8). The structure of the computation is a fundamental building block in larger applications [16, 17] as well as approximate hierarchical tree-based algorithms for larger systems, e.g., Barnes-Hut or fast multipole method (FMM) [18, 19, 20]. Moreover, lessons learned in implementing these kernels for physics applications readily extend to new application domains in statistical data analysis, search, and mining [21].

For the n-body simulation, we initially believed it would be simple to achieve near-peak performance on these platforms. The key computational bottleneck is the $O(n^2)$ evaluation of pairwise interaction forces among a system of n particles. This kernel is highly regular and floating-point (flop) intensive and therefore well-suited to accelerators like GPUs or PowerXCell/8i. However, to our surprise, each implementation required more significant tuning effort than expected. We describe these implementations and some "lessons learned" in this chapter. As far as we know, ours is among the first comprehensive *cross-platform* multicore studies for computations in this domain and is, in particular, unique in its contrasting of GPU and Cell-based accelerated systems.

Given that we consider only the $O(n^2)$ direct evaluation, our study for direct n-body computations will be limited in several ways. First, we achieve and highlight the best performance at large values of n. Small values of n, which are of great interest in, for instance, hierarchical tree-based approximation n-body algorithms, may require a different emphasis on low-level tuning techniques. Secondly, the computation is compute-intensive with largely regular access patterns, so we do not stress the memory system. Still, we believe this study can make a useful contribution to other computations more generally. In the numerical domain, hardware reciprocal square root is typically fast for single precision but not double, making our performance far from peak even though the computations are relatively compute-bound. Furthermore, for our accelerator architectures, we still need to carefully manage the local store to make effective use of available memory-system resources.

9.2   Interarchitectural Design Trade-offs

This section describes seven multicore systems of interest in this study, which are summarized in Table 9.1. In particular, each subsection considers a particular design dimension and qualitatively summarizes the differences among the architectures. For additional processor details, we refer interested readers elsewhere [22, 23, 24].

Among "conventional" general-purpose multicore microprocessors, we consider the Intel quad-core E5420 Harpertown processor, the Intel quad-core X5550 Nehalem processor, and AMD's quad-core 8350 Barcelona. The Intel Harpertown, Intel Nehalem, and AMD Barcelona processors share a similar microarchitecture but take distinct approaches to cache hierarchy and memory subsystem design. The Intel Harpertown has a 12-MB L2 cache memory, and two cores among four cores in a chip share a 6-MB L2 cache. The Intel Nehalem has a dedicated 256-KB L2 cache per core and an 8-MB L3 cache shared by all four cores in a chip. The AMD Barcelona has a 512-KB L2 cache per core in addition to a 2-MB shared L3 cache. Also, the Intel Nehalem and the AMD Barcelona support NUMA (Non-Uniform Memory Access) architecture while the Intel Harpertown is based on UMA (Uniform Memory Access) architecture. The Nehalem architecture supports 2-way simultaneous multithreading as well.

Among the multicore accelerator systems, we consider two generations of the STI Cell/B.E. architectures and NVIDIA Tesla C870 and C1060 GPUs. The Cell/B.E. processor is a heterogeneous multicore processor with one conventional PowerPC core ("PPE") and eight specialized single-instruction multiple-data (SIMD) accelerators ("SPEs"). Each SPE has a 256-KB local store scratchpad memory. The Tesla C870's peak performance is 512 Gflop/s in single precision. The C870 does not support native double-precision arithmetic. The C870 has 16 Streaming Multiprocessors ("SMs") and each SM has eight streaming processors ("SPs") and two special function units ("SFUs") for transcendental functions and attribute interpolation. Each SM has a 16-KB shared memory as well. The Tesla C1060 delivers 933 Gflop/s in single-precision with native 78 Gflop/s double-precision support. The C1060, based on Tesla architecture [24], has 30 SMs.

9.2.1   Requirements for Parallelism

The Intel Harpertown and Nehalem, the AMD Barcelona, and the first- and second-generation Cell/B.E. processors have four to nine cores per socket (two threads per core for the Nehalem and the Cell/B.E.'s PPE), and each core supports SIMD instructions for acceleration. To exploit the parallelism in these processors, these chips require coarse-grain parallelism to exploit their multiple cores in addition to SIMD parallelism.

Criteria	NVIDIA Tesla C870	NVIDIA Tesla C1060	IBM Cell/B.E. QS20 blade	IBM PowerXCell8i QS22 blade	2P Intel Xeon E5420 quad-core "Harpertown"	2P Intel Xeon X5550 quad-core "Nehalem"	4P AMD Opteron 8350 quad-core "Barcelona"
Clock (GHz)	1.35	1.296	3.2	3.2	2.5	2.66	2.0
SP Gflop/s	512	993	409.6	409.6	160	170	256
DP Gflop/s	N/A	78	29.2	204.8	80	85	128
Peak Power (W)	170	200	315	250	2 × 80	2 × 95	4 × 75
Number of sockets	(board) 1	(board) 1	(blade) 2	(blade) 2	(chip) 2	(chip) 2	(chip) 4
Number of cores per socket	16 (SM)	30 (SM)	1 (PPE) + 8 (SPE)	1 (PPE) + 8 (SPE)	4	4	4
Number of threads per core	768	1024	2 (PPE) or 1 (SPE)	2 (PPE) or 1 (SPE)	1	2	1
On-chip memory (per core)	16 KB (shared memory)	16 KB (shared memory)	256 KB (local store)	256 KB (local store)	128 KB (L1)	256 KB (L2)	512 KB (L2)
On-chip memory (shared)	N/A	N/A	N/A	N/A	6 + 6 MB (L2)	8 MB (L3)	2 MB (L3)
DRAM type	GDDR3	GDDR3	Rambus XDR	DDR2	DDR2	DDR3	DDR2
Shared DRAM access	N/A	N/A	NUMA	NUMA	UMA	NUMA	NUMA
Latency hiding	multithreading	multithreading	multilevel (e.g., double) buffering	multilevel (e.g., double) buffering	prefetching	prefetching + multithreading	prefetching
Peak bandwidth (GB/s)	76.8	102	51.2	51.2	21.4	51.2	42.8

TABLE 9.1: Summary of the evaluation platforms.

In contrast, the Tesla C870 and C1060 GPUs have 128 SPs and 32 SFUs and 240 SPs and 60 SFUs for single precision, respectively. In addition, discussed in detail in Section 9.2.4, the Tesla architecture adopts massive hardware multithreading to hide the DRAM access latency. The Tesla C870 and C1060 require at least several thousand–way parallelism to exploit its architectural features. Also, the Tesla C870 and C1060 require SIMT (Single Instruction Multiple Threads) parallelism, and every thread in a single warp (a group of 32 threads) needs to agree on the execution path to maximize chip utilization.

9.2.2 Computation Units

Each core of the Intel Harpertown, Intel Nehalem, or AMD Barcelona processor can retire up to two SIMD floating-point instructions (one SIMD add and one SIMD multiply) [25, 26]; thus, each core can deliver four double-precision floating-point operations per cycle. Also, the Intel and AMD cores can execute integer instructions in parallel with floating-point instructions. Each SPE in the Cell/B.E. processor has even and odd pipelines. The even pipeline can execute floating-point instructions, fixed-point arithmetics, and logical and word-granularity shift and rotate instructions. The odd pipeline executes load/store instructions and fixed-point byte-granularity shift, rotate mask, and shuffle instructions. Each SPE can retire one SIMD FMA (fused-multiply-and-add) instructions per cycle. Each SP in the Tesla C870 and C1060 executes one scalar instruction per cycle, including a single-precision FMA instruction. Each SFU can also execute four single-precision multiply instructions per cycle. Each SM in the C1060 has one SP for double-precision support, and one SP can retire one double-precision FMA instruction per cycle.

The sustainable flop-rate is highly affected by the mix of different instruction types in an execution stream and the structure of the computation unit. The Intel and AMD cores have separate multiply and add units instead of an FMA unit and also can run integer instructions in parallel; thus, they can more flexibly execute a different combination of integer and floating-point instructions. Still, to fully utilize both floating-point multiply unit and add unit, this chip requires a 1:1 ratio of multiply and add instructions. One SPE of the Cell/B.E. can issue only one floating-point instruction in a single cycle (whether it is FMA or not), so if multiply cannot be fused with add, the achievable peak flop-rate becomes halved. Still, the Cell/B.E. can run several types of fixed-point instructions in parallel with floating-point operations. Fully exploiting the SMs in the Tesla C870 and C1060 requires FMA instructions and additional multiply instructions for the SFUs. Also, one SP can execute only one instruction per cycle.

9.2.3   Start-up Overhead

Kernel launching is faster on the Intel and AMD processors than on the Cell/B.E. or the Tesla C870 and C1060. Thus, for a small amount of computation, these general-purpose processors outperform the accelerators, while the accelerator architectures often exhibit impressive performance for larger data sets [10].

The test systems also incur different levels of data off-loading overhead. In the Harpertown, Nehalem, Barcelona, and Cell/B.E. processors, the offloading overhead is largely determined by the offchip memory bandwidth. In contrast, GPU systems incur additional host memory to the onboard device memory data transfer via a slower PCI Express bus. While GPUs can partially hide the off-loading overhead with asynchronous data transfer (i.e., double-buffering), this mechanism currently works only for page-locked memory and incurs additional programming overhead [27]. To amortize the offloading overhead, GPUs require higher computational intensity than other processors [28, 29, 10]. However, the Tesla GPUs' onboard memory is much larger (1.5 GB for C870 and 4 GB for C1060) than the Harpertown, Nehalem, or Barcelona's cache memory (12, 8, or 2 MB) or the Cell/B.E.'s local store (256 KB per SPE $\times$ 8 SPEs). Accordingly, the Tesla GPU can fit larger data into its onboard memory to minimize the data transfer over the PCI Express bus.

9.2.4   Memory Latency Hiding

The Intel Harpertown and Nehalem and AMD Barcelona processors hide memory latency via cache memory and prefetching mechanisms. The Cell/B.E. overlaps computation with communication via double or triple buffering. Double buffering efficiently hides the latency but requires explicit software intervention.

The Tesla C870 and C1060 tolerate several-hundred-cycle DRAM access latency via massive hardware multithreading. The register usage of SMs on the Tesla GPU limits the maximum number of active threads. Very high register usage decreases the occupancy (utilization) of each multiprocessor, as it limits the number of threads that can be active simultaneously on the multiprocessor. This may affect the performance of the kernel, especially if the algorithm is not compute-bound. Optimizations like loop unrolling can also affect the performance by increasing or decreasing register usage. To effectively hide the DRAM access latency, programmers need to consider register usage on the Tesla GPU. The Tesla GPUs also have per SM shared memory (16 KB) in addition to constant cache and texture cache. Yet, as each SM can run hundreds of threads in parallel, these on-chip memories have minimum performance impact if there is only a low degree of data sharing among different threads.

9.2.5   Control over On-Chip Memory

For the cache-based multicore processors, cache memory is managed by hardware using the LRU (Least Recently Used) policy (or its variants), and programmers have essentially no control over cache partitioning. By contrast, programmers can explicitly manage on-chip (local store) memory on the Cell/B.E. The Tesla C870 and C1060, along with the NVIDIA CUDA framework, also allow programmers to control the placement of data arrays to the chip's different types of memories.

9.2.6   Main-Memory Access Mechanisms and Bandwidth Utilization

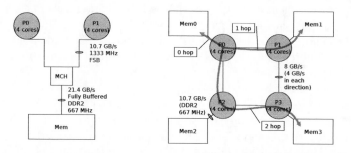

FIGURE 9.1: The 2P Intel Harpertown system with UMA architecture (left) and the 4P AMD Barcelona system with NUMA architecture (right).

The Intel Nehalem, AMD Barcelona, and Cell/B.E. processors use a NUMA (Non-Uniform Memory Access) architecture, while the Intel Harpertown processor adopts a UMA (Uniform Memory Access) architecture. UMA is conceptually simpler, but NUMA has scalability advantages if cores running on different chips access distinct data arrays. In particular, by locating data to the chip's local main memory, we can minimize the contention and interference in the main-memory interface. For instance, if a computation accesses read-only data multiple times, we can replicate the data to each processor's local DRAM to maximize the bandwidth for accessing the data. The Tesla C870 or C1060's device memory does not support shared memory access over two or more GPUs.

The systems also differ in their deliverable memory bandwidth [30]. In terms of peak aggregate bandwidth, the 2P Harpertown system can deliver 21.4 GB/s, the 2P Nehalem system supports 51.2 GB/s, the 4P Barcelona system supports 42.8 GB/s, and the QS20 and QS22 blades support 51.2 GB/s for main memory access. The Tesla C870 and C1060 support 76.8 GB/s and 102 GB/s peak bandwidth to their onboard device memories, respectively. However, there is often a significant gap between the peak bandwidth and the sustainable bandwidth [31, 32]. The gap is even larger for the multicore processors due to the interference among multiple threads performing data accesses [33, 34, 35].

The systems adopt different memory-controller architectures. In general, most memory controllers are designed to deliver the highest bandwidth when accessing a large contiguous chunk of data, in particular by exploiting maximum locality of a row buffer and bank-level parallelism. Switching between DRAM read and write should also be minimized to achieve the highest bandwidth utilization. However, even for simple computations in which each thread is reading a linear array with stride one, memory requests coming from multiple cores can be intermixed, thereby destroying the locality and parallelism of a DRAM chip. For the Intel Harpertown and Nehalem and AMD Barcelona processors, the granularity of memory access is the lowest-level-cache line size (64 byte). For data-intensive applications, memory access requests from multiple cores with a size of 64 bytes can be heavily interleaved. The situation is even worse for the Intel Harpertown processor with its UMA configuration, as the memory controller hub must mix memory access requests coming out from two different chips. For NVIDIA GPUs, the access granularity is 32, 64, or 128 bytes [27].

By contrast, the Cell/B.E. adopts a different memory subsystem. First, each SPE generates DMA requests with a significantly larger size of up to 16 KB. Even for the data-intensive applications, each SPE issues DMA requests in an intermittent fashion. This minimizes the inter-core interference in DRAM accesses. Therefore, programmers can maximize the bandwidth utilization by increasing the size of DMA accesses [2, 3]. Thus, the Cell/B.E. architecture fundamentally lends itself to higher bandwidth utilization than other systems in spite of the significant effort toward increasing bandwidth utilization in general-purpose multicore processors [33, 36, 31]. The AMD Barcelona processor claims to adopt optimized scheduling algorithms, especially for interleaved DRAM access streams as well [25]. Williams et al. also demonstrate the first-generation Cell/B.E.'s high bandwidth utilization [9], though an open question is what the ultimate impact of adopting different DRAM technologies will be (i.e., first- and second-generation Cell/B.E. architectures adopt different DRAM technologies, namely, XDR and DDR2, respectively). Finally, the Cell/B.E. has an additional advantage in optimizing memory controller, as this processor targets streaming applications that are highly latency tolerant. The Cell/B.E.'s memory controller can focus on bandwidth utilization, while general-purpose multicore processors attempt to address the significantly more difficult problem of balancing bandwidth utilization with fairness and latency issues [32, 33, 34].

9.2.7   Ideal Software Implementations

To optimize the code for the Intel Harpertown and Nehalem, AMD Barcelona, and Cell/B.E. processors, one first needs to identify coarse-grain parallelism and partition data to exploit all the cores. One then needs to consider data layout for higher data transfer and vectorization efficiency and vectorization to exploit SIMD units. The Harpertown, Nehalem, and Barcelona processors are significantly less sensitive to data alignment than the Cell/B.E. since they support multiple additional instructions for unaligned data accesses; however, data layout still affects the performance in a nonnegligible amount. At a high level, optimizing for the Cell/B.E. does not differ much from the Harpertown, Nehalem, and Barcelona processors, but the actual implementation is significantly more complex as programmers need to explicitly program for data transfer within the local store size limit of 256 KB. In addition, the gap between the performance of baseline and optimized code is significantly higher for the Cell/B.E., and this often requires manual optimization.

The optimization process for the Tesla C870 and C1060 is largely different from the above four processors. For the Tesla C870 and C1060, easily identifiable coarse-grain parallelism does not suffice to fully exploit the chip. Thus, the optimization should focus on extracting additional levels of parallelism. To benefit from the high-bandwidth and low-latency on-chip memories, programmers need to modify an algorithm to maximize data sharing among multiple cores. Data coalescing and broadcasting mechanisms are also crucial for the high performance, and this also needs to be considered in algorithm design. For the Tesla C870 and C1060 or other CUDA enabled GPUs, the key challenge arises from high-level algorithm design, and the actual implementation is less complex in terms of code size.

For the NUMA-based systems, one can gain significant speedup for bandwidth-intensive algorithms by controlling thread binding and data allocation. The optimization result for the Cell/B.E. often is more predictable than the x86-based architectures or the Tesla C870 and C1060 owing to its simpler architecture. For the x86-based architectures, the multilevel memory hierarchy with different latency, size, and associativity in each level and complex and adaptive prefetching mechanisms across the memory hierarchy significantly complicate the performance analysis. The Tesla GPU optimization is complicated by its large search space as well, which is nonlinear in nature [7].

9.3 Descriptions and Qualitative Analysis of Computational Statistics Kernels

For our evaluation, we consider two versions of covariance computation based on Pearson's method and Kendall's method, as implemented in the open-source R statistics package [14]. We also create the third Kernel by modifying the second kernel based on Kendall's method. Given two test data sets, represented by an $n_X \times n$ matrix X, an $n_Y \times n$ matrix Y, and precomputed mean vectors $\bar{x}$ and $\bar{y}$ of length n_X and n_Y, respectively, the basic covariance computation (based on Pearson's method) produces an $n_X \times n_Y$ matrix C such that

$$C_{ij} \leftarrow \frac{1}{n-1} \sum_{k=1}^{n} (X_{ik} - \bar{x}_i) \cdot (Y_{jk} - \bar{y}_j) \tag{9.1}$$

In this section, we describe three kernels we considered, and explain their high-level characteristics.

9.3.1   Conventional Sequential Code

We show the C implementation of the basic covariance kernel based on Pearson's method in Listing 9.1. We refer to this code as Kernel1. Kernel2 computes covariance using Kendall's method, and Kernel3 is artificially created by modifying Kernel2. Kernel2 and Kernel3 are shown in Listing 9.2. Kernel1 and Kernel2 codes are adopted from the R project source code [14].

In Kernel1, we can first subtract the mean vector from each matrix operand to remove the redundant subtracts. Then, transposing matrix Y converts this algorithm to a dense matrix multiplication problem, which is extensively studied, and there is also a highly optimized BLAS library for the problem. Kernel2 and Kernel3 have more complex data access patterns but still can be optimized based on a cache-blocking approach. Initially, *we intentionally ignore these particular optimization opportunities for the following two reasons.*

```
// p_x:   a pointer for X
// p_y:   a pointer for Y
// p_xm:  a pointer for $\bar{x}$
// p_ym:  a pointer for $\bar{y}$
// p_ans: a pointer for C
for(i = 0 ; i < n_X ; i++) {
  p_xx = &p_x[i * n];
  xxm = p_xm[i];
  for(j = 0 ; j < n_Y ; j++) {
    p_yy = &p_y[j * n];
    yym = p_ym[j];
    sum = 0.0;
    for(k = 0 ; k < n ; k++) {
      sum += (p_xx[k] - xxm)
        * (p_yy[k] - yym);
    }
    p_ans[i * n_Y + j] = sum / (n - 1);
  }
}
```

LISTING 9.1: C code for Kernel1.

```
// p_x:    a pointer for X
// p_y:    a pointer for Y
// p_ans: a pointer for C
for(i = 0 ; i < n_X ; i++) {
  p_xx = &p_x[i * n];
  for(j = 0 ; j < n_Y ; j++) {
    p_yy = &p_y[j * n];
    sum = 0.0;
    for(k = 0 ; k < n ; k++) {
      for(n1 = 0 ; n1 < n ; n1++) {
#if SIGN// Kernel2
        sum += sign((p_xx[k] - p_xx[n1])
          * (p_yy[k] - p_yy[n1]));
#else// Kernel3
        sum += (p_xx[k] - p_xx[n1])
          * (p_yy[k] - p_yy[n1]));
#endif
      }
    }
    p_ans[i * n_Y + j] = sum;
  }
}
```

LISTING 9.2: C code for Kernel2 and Kernel3.

First, we wish to stress the memory systems experimentally, and secondly, we wish to show the more typical and intuitive optimization process that will be common in practice. Then, if memory bandwidth turns out to be a performance bottleneck, we implement the blocking approach. Our focus is on highlighting the impact of architectural design trade-offs on performance and programmability. In particular, we do not intend to conclude which system is the best for computing covariance, nor do we claim to have implemented the best possible covariance code.

9.3.2 Basic Algorithmic Analysis

The memory footprint of all three kernels is $O((n_X + n_Y) \times n)$, and the size of two input matrices are typically much larger than mean vectors or the output matrix. The computational complexity is $O(n_X \times n_Y \times n)$ for Kernel1 and $O(n_X \times n_Y \times n^2)$ for Kernel2 and Kernel3. While these kernels are compute-intensive in their asymptotic notations, if the entire memory footprint does not fit into the on-chip memory of the test systems, then these kernels can be bandwidth-bound. All three kernels have obvious $n_X \times n_Y$–way parallelism as every pair of rows from matrix X and Y can be computed independently. Also, if we ignore the floating-point associativity issues, we can also trivially parallelize the innermost loop of Kernel1 and the second innermost loop of Kernel2 and Kernel3. For Kernel1, if we execute the code in a sequential way, there is higher temporal locality in the row data of matrix X than the row data of matrix Y. For Kernel2 and Kernel3, if we can place two rows from matrix X and Y on on-chip memory, we can perform $O(n^2)$ computation over $O(n)$ data without off-chip memory access. The total number of flops executed by Kernel1 is $(n_X + n_Y) \times n + n_X \times n_Y \times n \times 2$, and $n_X \times n_Y \times n^2 \times 4$ for Kernel2 and Kernel3.

9.4   Baseline Architecture-Specific Implementations for the Computational Statistics Kernels

For subsequent evaluations, we create a basic parallel implementation for each architecture, described in this section. These implementations include baseline parallelization and tuning, meaning they include some degree of platform-specific tuning but are not extensively tuned. Again, as Section 9.3.1 states, our focus is on system evaluation and not on kernel optimization.

9.4.1   Intel Harpertown (2P) and AMD Barcelona (4P) Multicore Implementations

We can easily parallelize the outermost loop of all three kernels with OpenMP or pthreads for our 8- and 16-core systems, assuming a sufficiently large n_X (n_Y). For Kernel1, we apply autovectorization with two directives, `#pragma unroll(16)` and `#pragma vector aligned`, achieving comparable performance to an intrinsics-based vectorization approach.

For Kernel2, the `sign()` function involves branches, lowering the performance significantly. We replace the branch with an SSE compare (e.g., `_mm_cmpgt_pd()` and `_mm_cmplt_pd()`) and bitwise operations (e.g., `_mm_and_pd()` and `_mm_or_pd()`). The Intel icc compiler fails to perform this replacement automatically, and so we hand-code this translation to use SIMD intrinsics. The Kernel3 code can be trivially vectorized in the same way.

For the 4P Barcelona system, which is NUMA-based, we replicate the input matrices to all four chips' local DRAM and pin the threads to each core. The replication cost can be amortized with the multiple reads, and this optimization maximizes the available bandwidth while minimizing the interference.

Even though our test kernels are asymptotically compute-intensive, if the input matrices do not fit into the on-chip cache memory, these algorithms can be bandwidth-bound. A blocking approach can reduce the amount of off-chip data transfer at the cost of increased implementation complexity. Also, for the Harpertown and Barcelona processors, selecting the optimal block size requires an exhaustive search over parameter space, as it is a complex function of the multiple levels of cache hierarchy and their size and associativity. This exhaustive search is beyond the scope of our work, and we find the block size based on heuristics.

9.4.2 STI Cell/B.E. (2P) Implementation

The Cell/B.E. implementation resembles the 4P AMD Barcelona implementation, though it provides an additional opportunity for fine tuning owing to the higher level of control over on-chip memory supported by the architecture. In particular, observe that a row of data of matrix X has, assuming the given loop order, a higher temporal locality than a row of data of matrix Y. Thus, we can assign a larger buffer for matrix X than Y. Furthermore, to reduce the bandwidth requirement even when a single row does not fit into the local store, we allocate additional small buffers for streaming. In this case, we read data from the larger buffer for matrix X and Y for accessing the initial part of the row (which fits into the local store), and then we switch to the streaming mode with the smaller buffers for the remaining.

However, fine-grained control over on-chip memory significantly increases the coding complexity, especially when the on-chip memory requirement varies as a function of a input data size. A blocking approach, even though it adds additional complexity in high level, fixes the on-chip memory requirement regardless of a input data size. Accordingly, a blocking approach can reduce the coding complexity for the Cell/B.E. in addition to the improved performance. For the Cell/B.E., the impact of different block size is easier to understand owing to its simple memory subsystem. Larger block height reduces the amount of traffic whereas larger block width increases the iteration count of the innermost loop to improve the compute efficiency. We can also simply pick the largest block size that fits into the local store instead of considering different cache sizes in the memory hierarchy.

9.4.3 NVIDIA Tesla C1060 Implementation

For the NVIDIA Tesla C1060, $n_X \times n_Y$–way parallelism may not be sufficient for practical data set sizes. Even when $n_X \times n_Y$ is very large, having every thread processes a distinct pair of rows can lead to poor bandwidth utilization (no coalescing in data transfer) or low on-chip cache utilization (no data sharing). For Kernel1, we partition the innermost loop with chunks of size 16 elements (a half-warp, as high memory bandwidth utilization is achieved when the memory accesses from a half-warp can be coalesced [27]). Each thread in a half-warp processes one element out of 16 elements in a chunk to maximize the coalescing. For Kernel2 and Kernel3, we partition the second innermost loop identical to the case of Kernel1. In this case, every thread in a half-warp traverses the same row of data in a synchronized way (in the innermost loop of Listing 9.2, array index k remains constant, and only array index n1 changes. In our optimized code, every thread in the same half-warp accesses p_xx and p_yy with the same n1 but different k), and we can use the on-chip shared memory to exploit this fact. As each thread accesses the same data element, we can use the shared memory's broadcasting mechanism as well.

One critical issue is the `sign()` function, which involves branch instructions. The NVIDIA CUDA compiler replaces branch instructions with predicates when the number of instructions controlled by the branch is equal to or less than the threshold value (four or seven instructions) [27]. Therefore, by using the CUDA framework, we do not need to manually optimize for the `sign()` function, as we did on the Intel Harpertown, AMD Barcelona, and Cell/B.E. platforms. Optimization for the Tesla C1060 is more involved at a high level but simpler to program than the Cell/B.E. for these kernels.

Also, the proper use of on-chip cache memory significantly reduces the bandwidth requirement, and Kernel2 and Kernel3 become compute-bound even without explicit blocking.

9.4.4   Quantitative Comparison of Implementation Costs

Table 9.2 summarizes the comparison. For the Cell/B.E., a blocking approach fixes the local-store-space requirement regardless of input data size and simplifies the coding in addition to the improved performance. We can also identify that the code size for the Tesla C1060 is significantly smaller than the other architectures. For the Tesla C1060, the challenge is in extracting additional parallelism and best exploiting the memory subsystem (based on data access coalescing and broadcasting mechanism and efficient use of the shared memory).

Criteria	x86 (initial)	x86 (blocking)	Cell/B.E. (initial)	Cell/B.E. (blocking)	Tesla (initial)	Tesla (blocking)
Source Lines of Code (SLOC)	335	419	1620	1004	52 (Kernel1) + 88 (Kernel1) 97 (Kernel2/3)	

TABLE 9.2: Quantitative comparison of implementation costs in terms of code size. This excludes the code for the kernel invocation and the residual part computation.

9.5 Experimental Results for the Computational Statistics Kernels

We ran experiments on five platforms out of the seven platforms in Table 9.1 for the kernels from computational statistics—this excludes the 2P Intel Nehalem system and the Tesla C870 GPU-based system. For direct n-body computations, we use five platforms including the 2P Intel Nehalem and Tesla C870 GPU-based systems but excluding the 2P Intel Harpertown system and the QS20 blade. To measure the sustained bandwidth of the 2P Harpertown system, we use PAPI [37] and count the number of memory bus transactions. For the 4P Barcelona system, we use AMD CodeAnalyst [38] and count the number of DRAM accesses. For the systems with the Cell/B.E. processors, we attach counter variables to every DMA memory request and ignore the PPE initiated traffic. For the Tesla C1060, we estimate the total bandwidth requirement using the following equations: $n_X \times n_Y \times n \times$ sizeof(float or double) $\times 2$ for Kernel1 and $n_X \times n_Y \times n \times n \times$ sizeof(float or double) $\times 2 \times \frac{1}{16}$ (a half-warp width, owing to data sharing) for Kernel2 and Kernel3, where sizeof(float) $= 4$ and sizeof(double) $= 8$. For the NUMA-based AMD Barcelona and Cell/B.E. architectures, we replicate matrix X and Y, which are read multiple times, for higher bandwidth utilization. This replication cost is included for the timing, and the off-loading overhead to the device memory in the Tesla C1060 is also included.

9.5.1 Kernel1

The left half of Figure 9.2 depicts the sustained Gflop/s and bandwidth for Kernel1 in single precision with the initial implementation. Although the algorithm is computationally intensive, the performance is bounded by memory bandwidth since the entire data set does not fit into the on-chip memory. The Tesla C1060 benefits from its high bandwidth to onboard DRAM, but the sustained bandwidth is lower than the theoretical peak and varies significantly for different input matrix sizes. The off-loading overhead accounts for 18% (for the smallest matrix) to 2.4% (for the largest matrix) of the total execution time. The QS20 and QS22 blades achieve the highest bandwidth utilization on average across the different values of n owing to their DMA-based data transfer mechanism with a large chunk size. The QS20 blade (with Rambus XDR) achieves higher bandwidth utilization than the QS22 blade (with DDR2). The 4P Barcelona system achieves significantly higher sustained bandwidth and bandwidth utilization than the 2P Harpertown system. This exemplifies the scalability benefit of NUMA architecture. The AMD Barcelona has optimized the memory-access scheduling algorithms for interleaved streaming accesses, and this also contributes to higher bandwidth utilization. However, the 2P

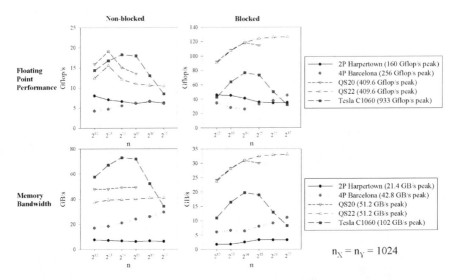

FIGURE 9.2: Sustained Gflop/s (top) and bandwidth utilization (GB/s) (bottom) for the initial (left) and the blocking-based (right) implementations of Kernel1 (single precision). Missing points for the QS20 are due to memory-allocation failure. Here, $n_X = n_Y = 1024$.

Harpertown system delivers higher flop rates per unit bandwidth consumption owing to the large shared cache memory.

The blocked implementations yield significantly better results than the initial implementation, as shown in the right half of Figure 9.2, but still deliver significantly lower performance than the theoretical peak. In the case of the x86 architectures, there are a number of possible explanations. First, we need to tune the blocking with respect to the different sizes and associativities at all levels of the cache hierarchy to achieve higher performance. This task is daunting task even for skilled programmers. Secondly, the blocked implementation may interact, and may even interfere, with the various hardware mechanisms in an unintentionally negative way. For example, the blocked version has a more complex memory-access pattern, which may reduce the effectiveness of the hardware prefetchers. Thirdly, the behavior of the memory system mechanisms are complex and challenging to reason about. For instance, on the Harpertown, data in DRAM are first read into the lower level (L2) cache, whereas there are read into the highest level (L1) cache first and moved to the noninclusive L2 and L3 caches (when the cache line is evicted) on the Barcelona. All these differences can affect the performance in nonintuitive ways, and this imposes challenges to a programmer if one wishes to extract the highest achievable flop rates out of the chip.

In contrast, for the Cell/B.E.–based systems, it is significantly easier to understand the data transfer related performance issues owing to its simple

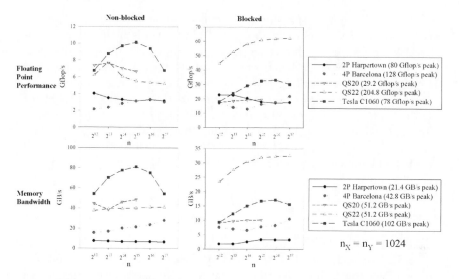

FIGURE 9.3: Sustained Gflop/s (top) and bandwidth (GB/s) (bottom) for the initial (left) and the blocking based (right) implementations of Kernel1 (double precision). Missing points for the QS20 are due to memory allocation failure. Here, $n_X = n_Y = 1024$.

architecture. Still, to achieve the highest flop rate, programmers need to consider their code at the assembly level. Each iteration of the innermost loop in Kernel1 requires two vector loads, one vector store, two address increments, and one vector of FMA (fused-multiply-and-add) instructions. As a result, the fixed-point instructions and load/store instructions can become a performance bottleneck. Extensive low-level tuning is required to balance the even and odd pipeline and minimize the address-calculation overhead.

For the Tesla C1060, the delivered performance is lower than 10% of its theoretical peak even in the best case. The off-loading overhead (11%–70% of the total execution time, which increases as n decreases), integer instructions for address increments, and the lack of additional multiply instructions to feed the SFU all lower the deliverable performance. Also, to load data to the block array in the shared memory, the Tesla C1060 needs to calculate an address and issue a load instruction for every single real number even though DRAM access latency can be, in principle, efficiently hidden with the hardware-multithreading mechanism; thus, the device memory to the shared memory traffic cannot be perfectly overlapped with the computation as is the case of the Cell/B.E. The Tesla C1060 architecture is less transparent than the Cell/B.E.'s, and so the performance impact of the tunable parameters (e.g., thread block size, block width and height in a blocking approach, and loop unrolling factors) are more difficult to predict, thereby requiring explicit search and tuning.

Figure 9.3 shows the results for double-precision. Interestingly, we can see that the sustained bandwidth for the Tesla C1060 is higher than the single-precision case, largely due to the increased granularity of data transfer from 64 byte (16 threads in a half-warp × 4 byte floating-point number) to 128 byte (16 × 8 byte floating-point number). In the case of double precision, the QS22 blade delivers higher performance than the QS20 blade, and the QS20 blade becomes compute-bound. We can also note that QS20 and the Tesla C1060 achieve a significantly higher fraction of its theoretical peak than the single-precision case. This shows that the QS20 and the Tesla C1060's peak double-precision performance is easier to achieve than its single-precision counterpart.

9.5.2   Kernel2

Figures 9.4 and 9.5 summarize the results for Kernel2 in single- and double-precision, respectively. This kernel is highly compute-intensive, and if a pair of two rows can fit into the on-chip memory, Kernel2 will be compute-bound even without blocking. For single precision, the Tesla C1060 delivers nearly 100 Gflop/s. Still, this number is significantly lower than the advertised peak numbers for similar reasons discussed in the case of Kernel1 in addition to `sign()` function. Kernel2's invocation of `sign()` is replaced with predicated instructions; however, computing predication also requires additional cycles. In addition, as the SP has only one execution pipeline and does not have

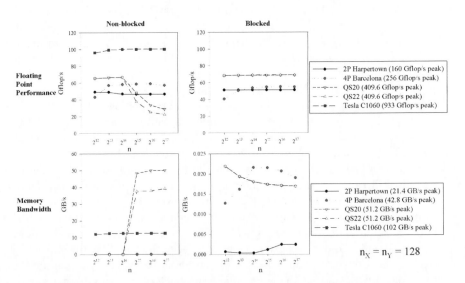

FIGURE 9.4: Sustained Gflop/s (top) and bandwidth (GB/s) (bottom) for the initial (left) and the blocking-based (right) implementations of Kernel2 (single precision). Here, $n_X = n_Y = 128$.

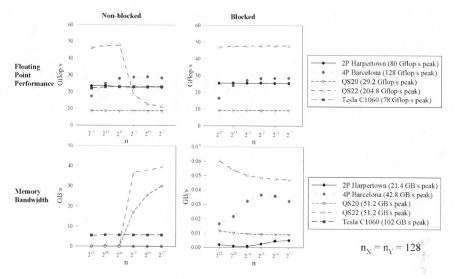

FIGURE 9.5: Sustained Gflop/s (top) and bandwidth (GB/s) (bottom) for the initial (left) and the blocking-based (right) implementations of Kernel2 (double precision). Here, $n_X = n_Y = 128$.

separate integer execution units, performance is more susceptible to the mix of integer instructions than on other architectures. For the QS20 and QS22 blades, the kernel is compute-bound for small n and bandwidth-bound for large n without blocking. In contrast, the QS20 and QS22 blades become compute-bound even with large n with blocking. In the bandwidth-limited cases, the QS20 blade achieves nearly full bandwidth utilization while the QS22 blade delivers approximately 80% of the theoretical peak bandwidth. The QS20 blade uses XDR DRAM and the QS22 blade uses DDR2 DRAM, which explains the reason for the different sustained bandwidth.

For the Tesla C1060, the kernel becomes compute-bound even without blocking because of the efficient use of the shared memory. Indeed, blocking does not improve performance, and so we need not consider it further.

9.5.3 Kernel3

Kernel3 is highly floating-point intensive, and its performance is given in Figures 9.6 and 9.7. For single precision, the QS20 blade, the QS22 blade, and the Tesla C1060 card deliver comparable performance for small n. The QS20 and QS22 blades' performances drop sharply with large n without blocking, but the flop rates remain nearly constant with blocking. Considering that our kernel has more adds and subtracts than multiplies, the QS20 and QS22 blades demonstrate near the maximum achievable floating-point performance, yet the performance for the Tesla C1060 is only one-quarter of its advertised

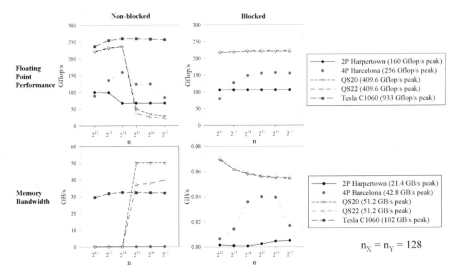

FIGURE 9.6: Sustained Gflop/s (top) and bandwidth (GB/s) (bottom) for the initial (left) and the blocking-based (right) implementations of Kernel3 (single precision). Here, $n_X = n_Y = 128$.

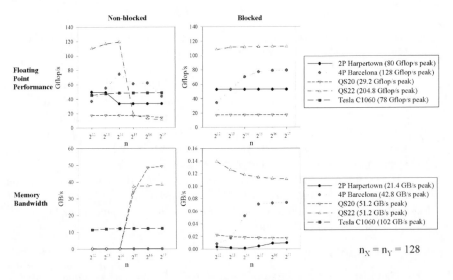

FIGURE 9.7: Sustained Gflop/s (top) and bandwidth (GB/s) (bottom) for the initial (left) and the blocking-based (right) implementations of Kernel3 (double precision). Here, $n_X = n_Y = 128$.

peak performance. For double-precision, the Tesla C1060 delivers over half of its peak flop rates in comparison to one-quarter in the single-precision case, as double-precision SP can be solely used for floating-point operations, and single-precision SPs can be exploited for integer instructions (e.g., address calculation).

9.6 Descriptions and Qualitative Analysis of Direct n-Body Kernels

Parallelization of direct n-body problems is well studied [39, 40] and has a long history that includes custom hardware (e.g., GRAPE systems). Most recently, there has been one thorough study for x86-based CPU tuning [41], as well as several studies on GPUs both prior to [42, 43] and following [44, 45] the introduction of the high-level CUDA model, which we use in the present study. The post-CUDA direct n-body GPU applications show particularly impressive performance, with comparison to existing custom hardware used for n-body simulations, like the GRAPE-6AF. These prior GPU studies focused primarily on single-precision kernels, as that was the only precision available in hardware at the time. To our knowledge, there have not been any published n-body implementation performance studies for the STI Cell/B.E. processor family. This work tries to fill these comparison gaps by including both GPUs and STI Cell/B.E. platforms, as well as multicore CPU systems.

We implement various parallel versions of a simple n-body gravitational simulation using direct (particle-particle) $O(n^2)$ force evaluation. In particular, we numerically integrate the equations of motion for each particle i,

$$\vec{F}_i = -Gm_i \sum_{\substack{1 \leq j \leq n \\ j \neq i}} m_j \frac{\vec{r}_i - \vec{r}_j}{||\vec{r}_i - \vec{r}_j||^3} \tag{9.2}$$

where particle i has mass m_i, is located at the three-dimensional position $\vec{r}_i$, and experiences a force $\vec{F}_i$ from all other particles; G is the universal gravitational constant, which we normalize to 1. We use the Verlet algorithm for our numerical integration scheme.

9.6.1 Characteristics, Costs, and Parallelization

The bottleneck is the $O(n^2)$ force evaluation, which computes the acceleration for all bodies i (with $G = 1$):

$$\vec{a}_i \approx \sum_{1 \leq j \leq n} m_j \frac{\vec{r}_i - \vec{r}_j}{\left(||\vec{r}_i - \vec{r}_j||^2 + \epsilon^2\right)^{\frac{3}{2}}} \tag{9.3}$$

Here, we adopt the common convention of a "softening parameter" term ϵ^2 in the denominator to improve the overall stability of the numerical integration scheme [17]. Computing the acceleration dominates the overall simulation cost so that, in this paper, we focus on its parallelization and tuning for accelerators and may effectively ignore the cost of the numerical integrator.

The scalar pseudocode for Equation (9.3) can be written as follows (comments prefixed by "//"):

1: **for** all bodies i **do**
2:     // Load (x_i, y_i, z_i) here
3:     $\vec{a}_i \equiv (a_x, a_y, a_z) \leftarrow (0, 0, 0)$ // Init acceleration
4:     **for** all bodies $j \neq i$ **do**
5:         // Load (x_j, y_j, z_j) here
6:         $(\Delta x, \Delta y, \Delta z) \leftarrow (x_i - x_j, y_i - y_j, z_i - z_j)$
7:         $\gamma \leftarrow (\Delta x)^2 + (\Delta y)^2 + (\Delta z)^2 + \epsilon^2$
8:         $s \leftarrow m_j / (\gamma \cdot \sqrt{\gamma})$ // $\gamma^{\frac{3}{2}} = \gamma\sqrt{\gamma}$
9:         $(a_x, a_y, a_z) \leftarrow (a_x + s \cdot \Delta x, a_y + s \cdot \Delta y, a_z + s \cdot \Delta z)$
10:    **end for**
11:    // Store acceleration
12:    $\vec{a}_i \leftarrow (a_x, a_y, a_z)$
13: **end for**

Lines 5–9 compute a single pairwise interaction. As written, there are 18 floating-point operations: 3 subtractions (line 6); 3 multiplies and 3 adds (line 7); 1 multiply, 1 square root, and 1 divide (line 8); and 3 multiplies and 3 adds (line 9). For consistency in comparing flop rates to other papers, we consider the cost of lines 5–8 to be 20 flops, i.e., we count 20 flops per pairwise interaction, or $20n(n-1)$ flops total.

There are $4n^2 + 6n$ loads and stores, with the dominant term coming from the loads of the three coordinates on lines 5–6 and mass on line 8. However, there is a significant amount of reuse, so that with appropriate cache blocking, we can incur close to the minimum of $4n$ compulsory misses so that the overall computation should be compute-bound with a computational intensity of $\approx 20n^2/4n = 5n$ flops per word.

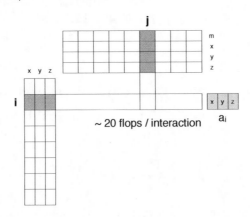

FIGURE 9.8: Computation and data access pattern for direct-force evaluation.

For all platforms, we adopt the standard approach to parallelization that exploits both coarse-gained parallelization of the outermost i loop followed by finer-grained data-parallelism across the i and j loops. We may visualize the general computation and data access pattern as shown in Figure 9.8. The n three-dimensional points appear as the $n \times 3$ matrix on the left. The i loop iterates over these points (rows); for each i, we stream over the same points j, shown mirrored (transposed) along the top, and accumulate an acceleration vector $\vec{a}_i$. The simplest coarse-grained parallelization is an owner-computes approach that partitions the i loop into n/p chunks among p threads. This approach requires no synchronization on writes to the final acceleration matrix.

9.7 Direct n-Body Implementations

9.7.1 x86 Implementations

For the general-purpose multicore platforms ($2 \times$ 4-core $\times$ 2 threads/core $= 16$ thread Intel Nehalem and $4 \times 4 = 16$-core AMD Barcelona systems), we consider several standard optimization techniques. These techniques are well known but essential to implement if we wish to fairly evaluate the accelerator-based GPU and PowerXCell8i platforms.

First, we exploit coarse-grained owner-computes parallelization at the outermost loop of the interaction calculation. Each thread computes the forces on a subset of $\frac{n}{p}$ particles. In this decomposition, all writes are independent.

We consider both OpenMP and Pthreads programming models, though we expect that OpenMP parallelization will be sufficient since the kernel is largely compute-bound. In the case of Pthreads, we create a team of threads at the beginning of program execution. These threads busy-wait at a barrier for a signal from the master thread to begin the interaction calculation. Upon completion, the threads again meet at a barrier. (Reported performance includes the time of these barriers.)

Secondly, we apply cache-level blocking of both loops. Since the memory footprint scales like $O(n)$ while flops scale as $O(n^2)$, we do not expect this technique to provide much benefit until n becomes very large since (a) typical L1 latencies for data that reside in L2 are relatively small, and (b) 2-MB or larger L2 and L3 caches today can easily accommodate $n \approx 100,000$ points.

Thirdly, we manually combine data alignment, unroll-and-jam, and SIMD vectorization transformations to improve register usage and exploit fine-grained data parallelism. In particular, we store the x, y, z coordinates and mass values in separate arrays aligned on an SIMD-friendly boundary. To enable SIMD vectorization, we then unroll-and-jam the outer loop and unroll the inner loop. In the single-precision code, we use the reciprocal and combined reciprocal square-root instruction (_mm_rcp_ps and _mm_rsqrt_ps). We also use SSE3-based reductions (e.g., _mm_hadd_ps) for some parts of the computation. Although the most aggressive compiler considered in this study, the Intel C v11.0, does perform vectorization, we observe significant benefits from manual vectorization ($\approx 2\times$).

9.7.2   PowerXCell8i Implementation

As discussed in Section 9.2, the STI PowerXCell8i processor architecture consists of a single general-purpose core, the PowerPC Element (PPE), coupled to eight Synergistic Processing Elements (SPEs), each of which has short-vector (SIMD) processors (SPUs) and a small 256-KB software-managed local store. The primary purpose of the PPE is to run the operating system and supervise the working of the SPEs by dispatching jobs and facilitating synchronization. The bulk of achievable performance for the direct n-body kernel comes from fully utilizing the SPEs, and a naive implementation, properly vectorized by the compiler, gives reasonable baseline performance. However, we find that additional optimization, some very trivial, enables an increase of almost ten times over the naive code. In this section, we discuss the most significant of these optimizations.

9.7.2.1   Parallelization strategy

We rely on the SPEs for acceleration, using the PPE only for dispatching the SPE threads, performing a single synchronization barrier, and coordinating the overall integrator. We follow the general owner-computes parallelization strategy outlined above. However, since the local-store on each SPE is purely software-managed, we must explicitly coordinate the sending and receiving of all other bodies. Since the SPEs themselves update the contents in PPE's main storage (to return the computed accelerations, velocities, and positions), a single synchronization barrier is necessary to enforce consistency. The barrier synchronization is done using mailboxes and is nonblocking in the SPEs. The SPEs may thereby overlap any potential delay with useful computation.

9.7.2.2   Data organization and vectorization

To best exploit the SIMD capabilities of the SPUs, we use the same storage scheme discussed in the CPU case, in which each coordinate of the positions is stored in its own SIMD-aligned array. Where possible, we manually coalesce squaring and addition operations into fused multiply-add (spu_madd) instructions, which execute with half the latency of separate multiply and add (6 versus 12 cycles). Finally, we write our code to enable all combinations of coordinate data alignments between source and target points (16 combinations in single precision, 4 in double precision). We can do this compactly using the SPU hardware rotate instructions to align the outer vector with all possible configurations of the inner vector. These rotate instructions do not execute in the same pipelines as the floating-point arithmetic instructions and thus incur virtually no cost.

9.7.2.3 Double buffering the DMA

As already mentioned, data for all the bodies have to be transferred from the PPU to all other SPUs before they can process the information. The STI PowerXCell8i allows up to 16 KB of data to be transferred using the available direct-memory access (DMA) engine between the various local stores. An SPE-initiated DMA was preferred here as the DMA would start only when (and as soon as) the SPE needed the data. DMA requests are asynchronous, which allows us to fetch data in small chunks and overlap data transfer with processing. This technique is well known among STI PowerXCell8i developers as *double buffering.*

Data organization plays a vital role in how the DMA is organized. Our choice of separately stroring the x, y, and z coordinates (and mass) in their own arrays (four in all) makes them easy to prefetch in a unit-stride streaming fashion. A single DMA can be dispatched to bring a chunk of bodies from the PPU to the SPU (and vice versa). Every DMA is now split into four smaller DMAs to fetch the chunks from four different arrays. We use the scatter-gather DMA support to fetch from each of the four elements in the lists in 16-KB chunks, bringing the total data transferred by a single list operation to 64 KB, allowing us to bring in four times as many bodies as before.

As expected, the code is compute-bound and highly tolerant of the relatively small DMA latencies. In principle, it might seem that double buffering could be eliminated altogether. However, the small 256-KB local store on each SPU cannot store all the data at once, making double buffering necessary.

9.7.2.4 SPU pipelines

The STI PowerXCell8i SPUs have two pipelines, with each pipeline executing a different subset of the possible instructions. As a result, achieving peak performance is only possible if the instruction mix is such that the pipelines can be perfectly balanced. The n-body kernel has a particularly bad instruction mix, as it is made up mostly of floating-point arithmetic instructions that can only execute in *pipeline 0*. The remaining and relatively fewer number of integer instructions are relegated to *pipeline 1*. In our case, these include the rotate instruction already mentioned along with floating-point reciprocal and reciprocal square-root estimate instructions.

9.7.3   NVIDIA GPU Implementation

9.7.3.1   Parallelization strategy

We follow the general owner-computes parallelization strategy outlined at the top of Section 9.6 with some refinements.

Our overall strategy follows the one used in the *GPU Gems 3* example [46]. In particular, we first partition the bodies in chunks of size q. Since we use an owner-computes strategy, this partitions the $O(n^2)$ total work into chunks of size $O(q \times n)$ each, each of which we assign to a thread block. Within the thread block, we will assign 1 thread to each of the q points and make that thread responsible for computing the acceleration (force) due to the other $n-1$ points. We will iterate over the other interaction points q at a time so that, at each step, the entire thread block is simultaneously computing q^2 interactions. We synchronize these threads (using _syncthreads()) every q^2 interactions.

There are several constraints on q. First, we need q to be a positive integer multiple of the so-called half-warp size (on current NVIDIA GPUs, 16). The warp size is, effectively, a vector length unit. Secondly, we need the $2q$ points to fit in the thread block's shared-memory (local store). In single-precision, that means 3 coordinates + 1 mass at 4 bytes each, or $2 \cdot q \cdot 4 \cdot 4$ bytes $\leq$ shared-memory capacity (e.g., a current typical value is 16 KB). Thirdly, we must satisfy the register-capacity constraint. For our n-body gravity kernel, each thread has a register working set of approximately 31 so that $q \times 31 \leq$ number of registers (on C870, ≤ 8192, and on the C1070, ≤ 16384). Finally, we cannot have more than a certain number of warps per thread block. With current warp sizes of 32 and the max number of warps at 8, there can be at most $q \leq 32 \cdot 8 = 256$. Within these constraints, we did a limited search to find a good value for q on a given platform.

Finally, we design the kernel to expose as many explicit FMA instructions as possible and to expose opportunities to use the hardware reciprocal square-root function, rsqrtf(). We show such a kernel in Listing 9.3.

The unoptimized implementation of the above GPU algorithm was able to reach approximately 65% of the final Gflop/s performance.

9.7.3.2   Optimizing the implementation

We tried a number of optimization techniques, which we enumerate below roughly in decreasing order of effectiveness.

Tuning the optimal block size (number of threads per block): We tried to choose the block size so that $n/q \geq 100$ to ensure that enough independent thread blocks were ready for scheduling. On the Tesla C1060, the number of independent blocks that worked well was 120, suggesting that each of the 30 multiprocessors context switched among four thread blocks. The number of threads in each block varied from 64 to 512 depending upon the number of bodies being simulated.

```
for(i = 1 ; i < blockDim.x + 1 ; i++ ) {
  indx = i + blockDim.x;// Each block has its own posvector
                        instead from 1
  r_vec.x = posvector[indx].x - body_current.x;
  r_vec.y = posvector[indx].y - body_current.y;
  r_vec.z = posvector[indx].z - body_current.z;
  r_vec.w = posvector[indx].w;
  float dSqr = softSquared;
  // For double precision the commented statement is used
  to find dSqr
  // dSqr += (r_vec.x * r_vec.x) + (r_vec.y * r_vec.y)
  + (r_vec.z * r_vec.z);
  dSqr += (r_vec.x * r_vec.x);// Fused-multiply-and-add
  dSqr += (r_vec.y * r_vec.y);// ..
  dSqr += (r_vec.y * r_vec.z);// ..
  float invr = rsqrtf(dSqr);
  // Hardware-implemented inverse square root
  float iinvr = invr * invr * invr;
  float u = r_vec.w * iinvr;
  acc.x += r_vec.x * u;// Fused-multiply-and-add
  acc.y += r_vec.y * u;// ..
  acc.z += r_vec.z * u;// ..
}
```

LISTING 9.3: Compute kernel for pairwise interaction.

Coalescing memory accesses via padding: We aligned float4 data types and C type struct consisting of four double variables to achieve padding for the single-precision and double-precision implementation, respectively. Doing so ensured coalesced shared memory loads from device memory and also coalesced shared memory access within each thread block. The CUDA profiler tool verified this fact.

Loop unrolling: We used the #pragma unroll to perform loop unrolling of the compute kernel. This pragma accepts an unrolling depth, which we tuned manually. This technique was particularly effective when the total number of bodies was small. The unrolling factor was a multiple of q and had to be tuned manually.

Encouraging FMAs explicitly: Rather than long expression sequences consisting of multiplies and adds, we broke up these expressions as shown in Listing 9.3. This technique worked well in single precision. However, for double precision, this technique had the opposite effect.

Register-latency tuning: We found that there was actually some benefit to manually reordering statements in our code. We ordered the statements heuristically, keeping in mind instruction latencies. For the double-precision computations, we only have one double-precision execution unit per multiprocessor, which effectively serializes threads and reduces overall performance compared to single precision.

9.8   Experimental Results and Discussion for the Direct n-Body Implementations

9.8.1   Performance

We evaluate the implementations described in Section 9.7 on the hardware platforms given in Table 9.1 (excluding the 2P Intel Harpertown system). We consider single-precision and, where possible, double-precision implementations. When reporting Gflop/s, recall that we use 20 flops per iteration, counting one divide (reciprocal) and one square root as single flops each. Since these operations normally have much higher latencies, we should not expect to get close to peak on any platform. Note that we time the force evaluation *in context* of the integrator, which executes on the host CPU (x86 on GPU, PowerPC on PowerXCell8i) but which accounts for a negligible number of flops. Therefore, the Gflop/s rates *include* the time to transfer data between host and accelerator.

9.8.1.1   CPU performance

We begin with the baseline CPU performance to have a reference for comparing the subsequent GPU and PowerXCell8i performance.

First, we consider the effect of parallelization, cache-level blocking, and manual vectorization on the Intel Nehalem and AMD Barcelona platforms for single precision and double precision in Figures 9.9–9.12. The least useful optimization was blocking, which never helped. The presence of large caches on both systems helps to explain these results. As expected, the most effective optimization was parallelization.

With regard to manual SIMD vectorization, the results are qualitatively similar across the two platforms but differ markedly between single and double precision. For single precision, manual SIMD vectorization pays off significantly, boosting sequential performance by two times (compare white and medium gray bars) and boosting parallel performance by two times as well (compare dark gray and black bars).

However, in double precision, SIMD vectorization made no difference, and overall double-precision performance is a much smaller fraction of peak than is the case in single precision. When we simply removed the square-root and reciprocal operations from the double-precision code, we saw a big boost in

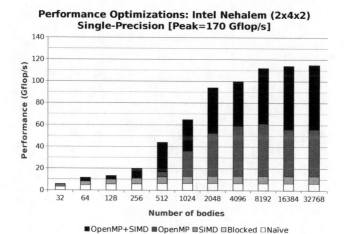

FIGURE 9.9: Direct *n*-body optimization breakdown: single-precision Intel Nehalem.

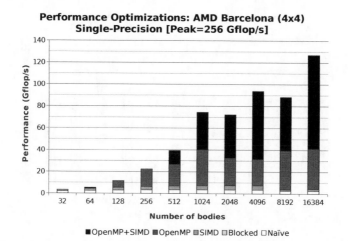

FIGURE 9.10: Direct *n*-body optimization breakdown: single-precision AMD Barcelona.

double-precision performance. Thus, the lack of good hardware support for these operations hampers performance.

9.8.1.2 GPU performance

The GPU implementation was the simplest of all the systems and required the least amount of tuning. Still, some degree of tuning was needed to get the best possible performance, with loop unrolling being a particularly effective

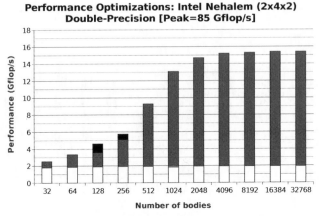

FIGURE 9.11: Direct n-body optimization breakdown: double-precision Intel Nehalem.

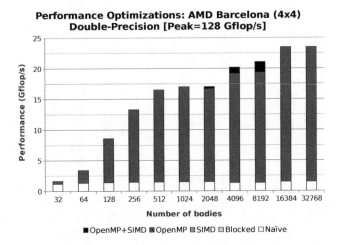

FIGURE 9.12: Direct n-body optimization breakdown: double-precision AMD Barcelona.

optimization. Figure 9.13 summarizes these results for the Tesla C1060 running in single precision.

The results for double precision (not shown) were similar. The main difference is that the `rsqrtf()` instruction does make a significant difference for the double-precision implementation as compared to the single-precision kernel.

Figure 9.14 shows how performance varies as the number of (independent) thread blocks increases. Not surprisingly, the number of thread blocks have to

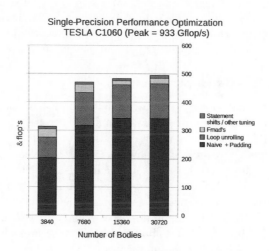

FIGURE 9.13: Breakdown of the effects of direct n-body performance tuning.

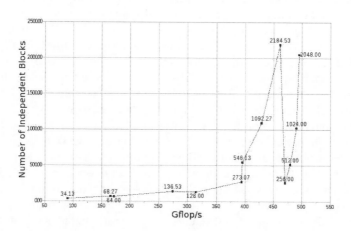

FIGURE 9.14: Performance versus number of independent thread blocks.

be some positive multiple of the number of multiprocessors for the architecture to perform at its best. The Tesla C1060 has 30 multiprocessors, while the Tesla C870 has 16. Hence, Tesla C870 performs best for a number of bodies in increasing powers of 2, while we need a slightly different number of bodies for the Tesla C1060 (for example, instead of 4096, we would use 3840 number of bodies).

9.8.1.3   PowerXCell8i performance

The performance of the STI PowerXCell8i implementations exhibits good scalability for a moderately large number of bodies, as shown in Figures 9.15 and 9.16. If the number of bodies is sufficiently large, we attain up to 61% of peak. The relatively low performance at small numbers of bodies is due in large part to the cost of scheduling the SPU threads. Once this overhead is overcome, we can expect the algorithm to perform linearly as the number of bodies go on increasing. Although one might guess that this overhead is due to the DMA that is done between the PPU and the SPU, our micro-benchmarking the double-buffered DMA indicates that we do in fact overlap DMA transfer with computation.

Apart from double buffering, some other techniques that tend to improve the overall throughput include loop unrolling, manual vectorization and FMA insertion, and choosing the correct data layout. Though we do not break down the impact of these optimizations due to technical limitations of doing so,[1] we observed about two times performance increase by switching the data layout using separate coordinate arrays, as compared to a layout that packs coordinates and mass together for each point. Furthermore, unroll-and-jam applied to the outer loop combined with using hardware rotate instructions to maintain the alignment gave additional improvement.

Curiously, performance decreases as the number of SPEs used increases, usually from 4 SPEs to 8 SPEs. We were able to attribute this phenomenon to the SPE thread dispatch overhead, which triples for this increment but only doubles for every other increment. This causes a sharp fall in the Gflop/s measured at these points (4 to 8). However, as the number of interactions increases, this fixed overhead is hidden by the computation.

The double-precision implementation builds upon the single-precision algorithm. The performance is very low—6 Gflop/s in the best case—compared to the peak of over 200 Gflop/s. However, this relatively poor performance is due to the lack of hardware support for square roots and reciprocals. Note that we are using IBM's vectorized software implementations from the SIMDMath library provided with the SDK. To estimate the latency of these operations, we replaced the reciprocal and square root by simple multiplies. Omitting only one of either reciprocal or square root boosted performance from 6 Gflop/s to approximately 10–11 Gflop/s, and removing both boosted performance to

---

[1] The program wouldn't run at all in absence of double buffering; loop unrolling is mandatory to perform SPU SIMD-ization and also depends on the data layout.

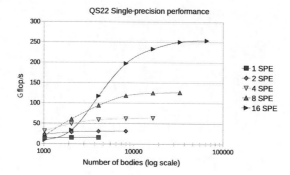

FIGURE 9.15: Performance of the direct n-body implementation on a QS22 blade (single precision, with two IBM PowerXCell8i processors).

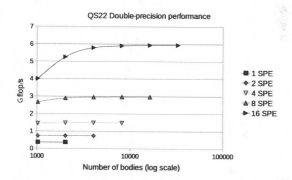

FIGURE 9.16: Performance of the direct n-body implementation on a QS22 blade (double precision, with two IBM PowerXCell8i processors).

over 100 Gflop/s. This demonstration shows the severity of a lack of hardware support for these operations.

Both the implementations were compiled with the IBM *xlc* (v 10.1) cross-compiler for the IBM PowerXCell8i platform, with optimization level −05. The single-precision program benefited from the compiler optimizations, with the IBM *xlc* adding a few tens of Gflop/s over the GNU *gcc* compiled code. However, we found that compiler optimizations, or the compiler itself for that matter, made almost no difference for the double-precision implementation.

9.8.1.4 Overall performance comparison

To summarize, we compare all implementations across all platforms in Figures 9.17 and 9.18. In absolute performance, the NVIDIA Tesla C1060 achieves the best single-precision performance by a factor of 2. Perhaps some-

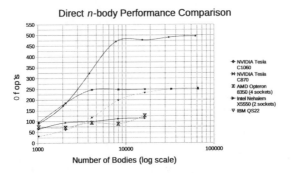

FIGURE 9.17: Cross-architecture performance comparison (single precision).

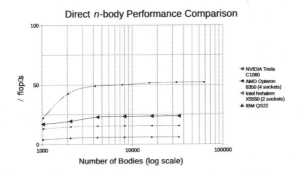

FIGURE 9.18: Cross-architecture performance comparison (double precision).

what surprisingly, the two GPU systems "win" even at very small particle sizes.

The CPU platforms achieve large fractions of peak as well in single precision (67% on Intel Nehalem and 50% on the AMD Barcelona). In double precision, the lack of hardware support for double-precision hampers performance; however, it is possible that more extensive and careful tuning of the double-precision implementation could make these platforms more competitive.

The Tesla C1060 and Tesla C870 both achieve approximately 50% of their theoretical peak in single precision, delivering a near constant performance of 500 Gflop/s and 250 Gflop/s respectively. For the double-precision implementation, the Tesla C1060 was able to reach 67% of its theoretical peak, which is impressive.

The Cell implementation was able to reach about 60% of its advertised peak of 410 Gflop/s of an entire blade for a single-precision implementation. The double-precision implementation, on the other hand, failed to achieve a substantial fraction of peak, due largely to the lack of good hardware support for square-root and reciprocal operations.

9.8.2   Productivity and Ease of Implementation

Criteria	Tesla C870	Tesla C1060	QS22	Nehalem	Barcelona
Source Lines of Code (SLOC)	380	390	850	500	500

TABLE 9.3: Summary of implementation complexity for various platforms.

We estimate programming complexity in terms of the lines of code required (see Table 9.3). However, we also want to emphasize that this is only a rough measure, and there are other factors that affect programming complexity. For example, the major challenge in the GPU implementation was in designing a high-level algorithm due to the uniqueness of the GPU architecture, and the actual implementation was straightforward. For the GPU implementation, the double-precision implementation was quite similar to the single-precision one, except that we had to take into account the absence float4 intrinsic type counterpart in double. However, we did not split the 64-bit double-precision number to store as two 32-bit values even though this technique is known to reduce bank conflicts while accessing data in shared memory (local store). We assume that the flops intensity is sufficient to hide such conflicts.

In the case of IBM's QS22, the situation was different. Given the simplicity of the PowerXCell8i architecture, the algorithm development was relatively easy. However, the actual coding required much more fine tuning and nontrivial optimization to reach the documented performance than expected. Moreover, the lack of hardware-implemented function in double-precision (reciprocal and reciprocal square root) limited performance.

The CPU implementation was moderately complex to code compared to the other platforms, and its double-precision performance suffers for the same reasons as PowerXCell8i.

9.9   Conclusions

This work joins previous cross-platform multicore performance studies [47, 48] with the computational statistics kernels and the direct force evaluation component from n-body simulations. The computational statistics kernels are simple to analyze but also have varying computational characteristics, and it is ideal to study the impact of different architectural design trade-offs. Direct n-body has two distinguishing characteristics from the kernels considered in the prior efforts as well. The direct n-body force evaluation is more heavily compute-bound and thus stresses different aspects of the architectures and programming models. This computation also includes square root and divide, whose latencies have large impacts in double precision across a number of our evaluation platforms. We have also tried in this paper to assess, at least anecdotally, a measure of end-user programmer productivity on these platforms.

Based on our detailed comparison of multicore processors and accelerators, we can draw several general conclusions. First, a hardware-multithreading approach lowers programming complexity by freeing programmers from latency issues at the expense of increased requirements for parallelism. Second, NUMA has benefits over UMA for data-intensive algorithms if the underlying system or programmer can properly handle data placement and thread binding. Third, an explicit DMA-based data transfer mechanism can effectively achieve high bandwidth utilization at the cost of higher coding complexity. Fourth, in the case of the Cell/B.E., blocking can also reduce the programming complexity as the programmer is forced to optimize the program with the fixed-size on-chip memory. Fifth, providing hardware support for special-purpose functions is essential for a certain class of computations—e.g., square root and divide for direct n-body computations. Sixth, increasing raw floating-point performance only may not lead to overall system performance enhancement in many cases, and the off-chip memory bandwidth is critical even when the algorithm appears to be compute-bound in asymptotic notation. Integer execution units are also important to achieve the peak flop rates as well. Seventh, the Cell/B.E.'s simple memory subsystem increases the coding complexity to implement the first correct version of the code with proper data-transfer mechanisms, yet the subsequent optimization process is easier and more intuitive. The opposite holds for the x86 based architectures. In the case of Tesla GPU, the key challenge is in designing high-level algorithms to best exploit the architecture, and the implementation of such high-level algorithms is more straightforward. By our proxy measures, GPUs and CUDA prove to be the most power-efficient and the simplest to implement in many cases. This paper provides quantitative support for these conclusions, presented in a tutorial-style discussion that we hope can be used to guide future tuning and architecture-selection efforts.

Looking forward, we intend to evaluate different hardware systems with more complex and irregular kernels, as well as real-world applications. In particular, we are interested in higher-level hierarchical tree-based codes for the n-body problem, not only in physics but also in statistical data analysis and mining [21]. Since the direct n-body calculation appears as the leaf-leaf interaction of those problems, we expect our results to be useful in those contexts.

Acknowledgements

This work was supported in part by an IBM Shared University Research (SUR) award, the National Science Foundation (NSF) under award number 0833136, NSF TeraGrid allocation CCR-090024, NSF Grants CNS-0614915 and IIP-0934114, and a grant from the Defense Advanced Research Projects Agency (DARPA). Any opinions, findings and conclusions or recommendations expressed in this material are those of the authors and do not necessarily reflect those of NSF or DARPA.

We acknowledge Georgia Institute of Technology, its Sony-Toshiba-IBM Center of Competence and the National Science Foundation for the use of Cell Broadband Engine resources that have contributed to this research. We also acknowledge Dr. Hyesoon Kim for her advice during this project. We thank the various readers of the IBM Cell/B.E. and the NVIDIA CUDA community forums for additional advice and assistance throughout.

Bibliography

[1] O. Bockenbach, M. Knaup, and M. Kachelriess. Implementation of a cone-beam backprojection algorithm on the Cell Broadband Engine processor. In *Proc. SPIE Medical Imaging*, San Diego, CA, Feb. 2007.

[2] D. Jimenez-Gonzalez, X. Martorell, and A. Ramirez. Performance analysis of Cell Broadband Engine for high memory bandwidth applications. In *Proc. 7th IEEE Int'l Symp. on Performance Analysis of Systems and Software (ISPASS)*, San Jose, CA, Apr. 2007.

[3] T.d Saidani, S. Piskorski, L. Lacassagne, and S. Bouaziz. Parallelization schemes for memory optimization on the cell processor: A case study of image processing algorithm. In *Proc. Workshop on memory performance (MEDEA)*, Brasov, Romania, Sep. 2007.

[4] J. D. Owens, D. Luebke, N. Govindaraju, M. Harris, J. Kruger, A. E. Lefohn, and T. J. Purcell. A survey of general-purpose computation on graphics hardware. *Computer Graphics Forum*, 26(1):80–113, Mar. 2007.

[5] Vasily Volkov and James W. Demmel. Benchmarking GPUs to tune dense linear algebra. In *Proc. Int'l Conf. on High Performance Computing and Networking (SC)*, Austin, TX, 2008.

[6] S. Ryoo, C. I. Rodrigues, S. S. Baghsorkhi, S. S. Stone, D. B. Kirk, and W. W. Hwu. Optimization principles and application performance

evaluation of a multithreaded GPU using CUDA. Salt Lake City, UT, Feb. 2008.

[7] S. Ryoo, C. I. Rodrigues, S. S. Stone, J. A. Stratton, S. Ueng, S. S. Baghsorkhi, and W. W. Hwu. Program optimization carving for GPU computing. *Journal of Parallel and Distributed Computing*, 2008.

[8] J. Nickolls, I. Buck, M. Garland, and K. Skadron. Scalable parallel programming with CUDA. *ACM Queue*, 6(2):40–53, Mar. 2008.

[9] S. Williams, L. Oliker, R. Vuduc, J. Shalf, K. Yelick, and J. Demmel. Optimization of sparse matrix-vector multiplication on emerging multicore platforms. In *Proc. Int'l Conf. on High Performance Computing and Networking (SC)*, 2007.

[10] J. S. Meredith, S. R. Alam, and J. S. Vetter. Analysis of a computational biology simulation technique on emerging processing architectures. In *Proc. 6th IEEE Int'l Workshop on High Performance Computational Biology (HICOMB)*, Long Beach, CA, 2007.

[11] S. Che, J. Li, J. W. Sheaffer, K. Skadron, and J. Lach. Accelerating compute-intensive applications with GPUs and FPGAs. In *Proc. 6th IEEE Symp. on Application Specific Processors (SASP)*, Anaheim, CA, Jun. 2008.

[12] Seunghwa Kang, David Bader, and Richard Vuduc. Understanding the design trade-offs among current multicore systems for numerical computations. In *Proc. IEEE Int'l. Parallel and Distributed Processing Symp. (IPDPS)*, Rome, Italy, May 2009.

[13] N. Arora, A. Shringarpure, and R. W. Vuduc. Direct n-body kernels for multicore platforms. In *Proc. Int'l Conf. on Parallel Processing (ICPP)*, Vienna, Austria, 2009.

[14] The R project for statistical computing. http://www.r-project.org/, 2009.

[15] M. G. Kendall. A new measure of rank correlation. *Biometrika Trust*, 1938.

[16] Simon Portegies Zwart, Stephen McMillan, Derek Groen, Alessia Gualandris, Michael Sipior, and Willem Vermin. A parallel gravitational n-body kernel. *New Astronomy*, 2007.

[17] S. Aarseth. In *Gravitational N-Body Simulations*. Cambridge University Press, 1st edition, 2003.

[18] Joshua Barnes and Piet Hut. A hierarchical O($n \log n$) force calculation algorithm. *Nature*, 1986.

[19] L. Greengard. The rapid evaluation of potential fields in particle systems. *The MIT Press*, 1987.

[20] Tsuyoshi Hamada and Toshiaki Iitaka. The chamomile scheme: An optimized algorithm for n-body simulations on programmable graphics processing units. *New Astronomy*, 2008.

[21] Alexander G. Gray and Andrew W. Moore. 'n-body' problems in statistical learning. In *Proc. Advances in Neural Information Processing Systems (NIPS)*, Vancouver, British Columbia, Canada, Dec. 2000.

[22] P. Gepner, D. L. Fraser, and M. F. Kowalik. Second generation quad-core Intel Xeon processors bring 45 nm technology and a new level of performance to HPC applications. *Lecture Notes in Computer Science*, 5101:417–426, 2008.

[23] T. Chen, R. Raghavan, J. Dale, and E. Iwata. Cell Broadband Engine Architecture and its first implementation—A performance view. *IBM Journal of Research and Developments*, 51(5):559–572, Sep. 2007.

[24] E. Lindholm, J. Nickolls, S. Oberman, and J. Montrym. NVIDIA Tesla: A unified graphics and computing architecture. *IEEE Micro*, 28(2):39–55, Mar. 2008.

[25] AMD Corporation. *Software Optimization Guide for AMD Family 10h Processors*, 3.06 edition, Apr. 2008.

[26] Intel Corporation. *Intel 64 and IA-32 Architectures Optimization Reference Manual*, Nov. 2007.

[27] NVIDIA Corporation. *NVIDIA CUDA Compute Unified Device Architecture Programming Guide*, 2.0 edition, Jun. 2008.

[28] S. Che, M. Boyer, J. Meng, D. Tarjan, J. W. Sheaffer, and K. Skadron. A performance study of general-purpose applications on graphics processors using CUDA. *Journal of Parallel and Distributed Computing*, 2008.

[29] O. Schenk, M. Christen, and H. Burkhart. Algorithmic performance studies on graphics processing units. *Journal of Parallel and Distributed Computing*, 2008.

[30] L. A. Polka, H. Kalyanam, G. Hu, and S. Krishnamoorthy. Package technology to address the memory bandwidth challenge for tera-scale computing. *Intel Technology Journal*, 11(3), 2007.

[31] E. Ipek, O. Mutlu, J. F. Martinez, and R. Caruana. Self-optimizing memory controllers: A reinforcement learning approach. *ACM SIGARCH Computer Architecture News*, 36(3):39–50, Jun. 2008.

[32] C. Natarajan, B. Christenson, and F. Briggs. A study of performance impact of memory controller features in multi-processor server environment. In *Proc. 3rd Workshop on Memory Performance Issues (WMPI)*, Munich, Germany, Jun. 2004.

[33] O. Mutlu and T. Moscibroda. Enhancing the performance and fairness of shared DRAM systems with parallelism-aware batch scheduling. In *Proc. 35th Ann. Int'l Symp. on Computer Architecture (ISCA)*, Beijing, China, Jun. 2008.

[34] N. Rafique, W. T. Lim, and M. Thottethodi. Effective management of DRAM bandwidth in multicore processors. In *Proc. 16th Int'l Conf. on Parallel Architectures and Compilation Techniques (PACT)*, Brasov, Romania, Sep. 2007.

[35] S. Rixner, W. J. Dally, U. J. Kapasi, P. Mattson, and J. D. Owens. Memory access scheduling. In *Proc. 27th Ann. Int'l Symp. on Computer Architecture (ISCA)*, Vancouver, Canada, Jun. 2000.

[36] S. A. McKee, W. A. Wulf, J. H. Aylor, R. H. Klenke, M. H. Salinas, S. I. Hong, and D. A. B. Weikle. Dynamic access ordering for streamed computations. *IEEE Transactions on Computers*, 49(11):1255–1271, Nov. 2000.

[37] PAPI. http://icl.cs.utk.edu/papi, 2009.

[38] AMD CodeAnalyst. http://developer.amd.com/cpu/CodeAnalyst, 2009.

[39] M.S. Warren and J.K. Salmon. A parallel hashed oct-tree n-body algorithm. *Supercomputing 93*, 15(19), 1993.

[40] Shang-Hua Teng. Provably good partitioning and load balancing algorithms for parallel adaptive n-body simulation. *SIAM Journal on Scientific Computing*, 19(2), 1998.

[41] Keigo Nitadori, Junichiro Makino, and Piet Hut. Performance tuning of n-body codes on modern microprocessors: I. Direct integration with a Hermite scheme on x86_64 architecture. *New Astronomy*, 12:169–181, 2006.

[42] Francisco Chinchilla, Todd Gamblin, Morten Sommervoll, and Jan F. Prins. Parallel n-body simulation using GPUs. *Technical Report TR04-032, University of North Carolina*, 2004.

[43] Erich Elsen, V. Vishal, Mike Houston, Vijay Pande, Pat Hanrahan, and Eric Darve. n-body simulations on GPUs. Stanford University, 2007.

[44] Mark J. Stock and Adrin Gharakhani. Toward efficient GPU-accelerated n-body simulations. *AIAA*, 2008.

[45] Robert G. Belleman, Jeroen Bedorf, and Simon F. Portegies Zwart. High performance direct gravitational n-body simulations on graphics processing units II: An implementation in CUDA. *New Astronomy*, 2008.

[46] L. Nyland, M. Harris, and J. Prins. Fast n-body simulation with CUDA. In *GPU Gems 3*. NVIDIA, 2007. Chapter 31.

[47] Sam Williams, Richard Vuduc, Leonid Oliker, John Shalf, Katherine Yelick, and James Demmel. Optimizing sparse matrix-vector multiply on emerging multicore platforms. *Journal of Parallel Computing*, 35:178–194, 2009.

[48] Kaushik Datta, Mark Murphy, Vasily Volkov, Samuel Williams, Jonathan Carter, Leonid Oliker, David A. Patterson, John Shalf, and Katherine A. Yelick. Stencil computation optimization and auto-tuning on state-of-the-art multicore architectures. In *Proc. ACM/IEEE Conf. on Supercomputing (SC)*, Austin, TX, November 2008.

Chapter 10

Sorting on the Cell Broadband Engine

Shibdas Bandyopadhyay

University of Florida

Dolly Sharma

University of Connecticut

Reda A. Ammar

University of Connecticut

Sanguthevar Rajasekaran

University of Connecticut

Sartaj Sahni

University of Florida

10.1   The Cell Broadband Engine

The Cell Broadband Engine (CBE) is a heterogeneous multicore architecture developed by IBM, Sony, and Toshiba. A CBE (Figure 10.1) consists of a Power PC (PPU) core, eight Synergistic Processing Elements or Units (SPEs or SPUs), and associated memory-transfer mechanisms [14]. The SPUs are connected in a ring topology, and each SPU has its own local store. However, SPUs have no local cache and no branch-prediction logic. Data may be moved between an SPU's local store and central memory via a DMA transfer, which is handled by a Memory Flow Control (MFC). Since the MFC runs independent of the SPUs, data transfer can be done concurrently with computation. The absence of branch-prediction logic in an SPU and the availability of SIMD instructions that can operate on vectors that are comprised of four numbers poses a challenge when developing high-performance CBE algorithms.

10.2   High-Level Strategies for Sorting

As noted in [11], a logical way to develop a sorting algorithm for a heterogeneous multicore computer such as the CBE is to (1) begin with a sorting algorithm for a single SPU; then (2) using this as a core, develop a sort algorithm for the case when data fit in all available cores; then (3) use this

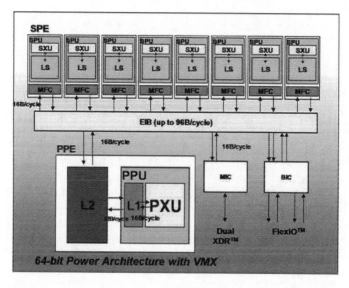

FIGURE 10.1: Architecture of the Cell Broadband Engine [9].

multi-SPU algorithm to develop a sort algorithm for the case when the data to be sorted do not fit in the local stores of all available SPEs but fit in main memory. The strategy would be to extend this hierarchical plan to the case where data to be sorted are so large that they are distributed over the main memories of a cluster of CBEs.

An alternative strategy to sort is to use the master-slave model in which the PPU serves as the master processor, and the SPUs are the slave processors. The PPU partitions the data to be sorted and sends each partition to a different SPU; the SPUs sort their partition using a single SPU sort; the PPU merges the sorted data from the SPUs so as to complete the sort of the entire data set. This strategy is used in [32] to sort on the nCube hypercube and in [29] to sort on the CBE.

Regardless of whether we sort large data sets using the hierarchical strategy of [11] or the master-slave strategy of [29, 32], it is important to have a fast algorithm to sort within a single SPU. The absence of any branch-prediction capability and the availability of vector instructions that support SIMD parallelism on an SPU make the development of efficient SPU sort algorithms a challenge. SPUs also have two instruction pipelines, which make them capable of issuing two instructions in the same cycle if they fall in different pipelines. It pays to hand-tune the generated assembly code so that two instructions are issued in as many of the cycles as possible.

10.3   SPU Vector and Memory Operations

We use several SIMD functions that operate on a vector of four numbers to describe the SPU adaptation of sorting algorithms. We describe these in this section. In the following, v1, v2, min, max, and temp are vectors, each comprised of four numbers, and p, p1, and p2 are bit patterns. Also, dest (destination) and src (source) are addresses in the local store of an SPU and bufferA is a buffer in local store while streamA is a data stream in main memory. Furthermore, Function names that begin with spu are standard C/C++ Cell SPU intrinsics while those that begin with mySpu are defined by us. Our description of these functions is tailored to the sorting application.

1. spu_shuffle(v1,v2,p) ⋯ This function returns a vector comprised of a subset of the eight numbers in v1 and v2. The returned subset is determined by the bit pattern p. Let W, X, Y, and Z denote the four numbers (left to right) of v1, and let A, B, C, and D denote those of v2. The bit pattern p = XCCW, for example, returns a vector comprised of the second number in v1 followed by two copies of the third number of v2 followed by the first number in v1. In the following, we assume that constant patterns such as XYZD have been predefined.

2. spu_cmpgt(v1,v2) ⋯ A 128-bit vector representing the pairwise comparison of the four numbers of v1 with those of v2 is returned. If an element of v1 is greater than the corresponding element of v2, the corresponding 32 bits of the returned vector are 1; otherwise, these bits are 0.

3. spu_add(v1,v2) ⋯ Returns the vector obtained by pairwise adding the numbers of v1 with the corresponding numbers of v2.

4. spu_sub(v1,v2) ⋯ Returns the vector obtained by pairwise subtracting the numbers of v2 from the corresponding numbers of v1.

5. spu_and(p1,p2) ⋯ Returns the vector obtained by pairwise adding the bits of p1 and p2.

6. mySpu_not(p) ⋯ Returns the vector obtained by complementing each of the bits of p. Although the CBE does not have a *not* instruction, we can perform this operation using the *nor* function that is supported by the CBE and which computes the complement of the bitwise *or* of two vectors. It is easy to see that spu_nor(p, v0), where v0 is an all-zero vector, correctly computes the complement of the bits of p.

7. spu_select(v1,v2,p) ⋯ Returns a vector whose *i*th bit comes from v1 (v2) when the *i*th bit of p is 0 (1).

8. spu_slqwbyte(v1,n) ··· Returns a vector obtained by shifting the bytes of v1 m bytes to the left, where m is the number represented by the five least significant bits of n. The left shift is done with zero fill, so the rightmost m bytes of the returned vector are 0.

9. spu_splat(s) ··· Returns a vector comprised of four copies of the number s.

10. mySpu_cmpswap(v1,v2) ··· Pairwise compares the numbers of v1 and v2 and swaps so that v1 has the smaller number of each compare and v2 has the larger number. Specifically, the following instructions are executed:

```
p = spu_cmpgt(v1,v2);
min = spu_select(v1,v2,p);
v2 = spu_select(v2,v1,p);
v1 = min;
```

11. mySpu_cmpswap_skew(v1,v2) ··· Performs the comparisons and swaps shown in Figure 10.2. Specifically, the following instructions are executed:

```
temp = spu_slqwbyte(v2,4);
p = spu_cmpgt(v1,temp);
min = spu_select(v1,temp,p);
v1 = spu_shuffle(min,v1,WXYD);
max = spu_select(temp,v1,p);
v2 = spu_shuffle(max,v2,AWXY);
```

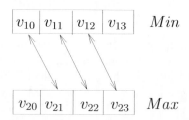

FIGURE 10.2: Comparisons for mySpu_cmpswap_skew.

12. mySpu_gather(vArray,v1) ··· Here, vArray is an array of vectors. Let W, X, Y, and Z be the numbers of v1. The function returns a vector whose first number is the first number of vArray[W], its second number is the second number of vArray[X], its third number is the third number of vArray[Y], and its fourth number is the fourth number of vArray[Z]. One implementation of this function first extracts W, X, Y, and Z from v1 using the function spu_extract and then executes the code

```
temp = spu_shuffle(vArray[W],vArray[X],WBWW);
temp = spu_shuffle(temp,vArray[Y],WXCC);
return spu_shuffle(temp,vArray[Z],WXYD);
```

13. mySpu_gather12(vArray,v1) ⋯ This function, which is similar to mySpu_gather, returns a vector whose first number is the first number of vArray[W] and whose second number is the second number of vArray[X]. The third and fourth numbers of the returned vector are set arbitrarily. Its code is
return spu_shuffle(vArray[W],vArray[X],WBWW);

14. mySpu_gather34(vArray,v1) ⋯ This function, which is similar to mySpu_gather12, returns a vector whose first number is the third number of vArray[W] and whose second number is the fourth number of vArray[X]. The third and fourth numbers of the returned vector are set arbitrarily. Its code is
return spu_shuffle(vArray[W],vArray[X],YDYY);

15. mySpu_gatherA(vArray,v1) ⋯ This function is similar to mySpu_gather and returns a vector whose first number is the first number of vArray[W], its second number is the third number of vArray[X], its third number is the first number of vArray[Y], and its fourth number is the third number of vArray[Z]. The code is
```
temp = spu_shuffle(vArray[W],vArray[X],WCWW);
temp = spu_shuffle(temp,vArray[Y],WXAA);
```
return spu_shuffle(temp,vArray[Z],WXYC);

16. mySpu_gatherB(vArray,v1) ⋯ This too is similar to mySpu_gather. The function returns a vector whose first number is the second number of vArray[W], its second number is the fourth number of vArray[X], its third number is the second number of vArray[Y], and its fourth number is the fourth number of vArray[Z]. The code is
```
temp = spu_shuffle(vArray[W],vArray[X],XDXX);
temp = spu_shuffle(temp,vArray[Y],WXBB);
```
return spu_shuffle(temp,vArray[Z],WXYD);

17. memcpy(dest, src, size) ⋯ This function copies the size number of bytes from the local store location beginning at src to dest.

18. dmaIn(bufferA, streamA) ⋯ This function triggers a DMA transfer of the next buffer load of data from streamA in main memory into bufferA in the local store. This is done asynchronously and concurrently with SPU execution.

19. `dmaOut(bufferA, streamA)` $\cdots$ This function is similar to `dmaIn` except that a buffer load of data is transferred asynchronously from `bufferA` in the local store to `streamA` in main memory.

10.4 Sorting Numbers

10.4.1 Single SPU Sort

Recently, three sorting algorithms—AA-sort [16], CellSort [11], and Merge sort [3]—were proposed for the CBE. AA-sort is an adaptation of comb sort, which was originally proposed by Knuth [17] and rediscovered by Dobosiewicz [8] and Box and Lacey [6]. CellSort is an adaptation of bitonic sort (e.g., [17]). Both AA-sort and CellSort are based on sorting algorithms that are inefficient on a single processor. Hence, parallelizing these algorithms begins with a handicap relative to the fastest serial sorting algorithms–merge sort for worst-case behavior and quick sort for average behavior. Comb sort is known to have a worst-case complexity that is $O(n^2)$ [10]. Although the best upper bound known for its average complexity is also $O(n^2)$, experimental results indicate an average complexity of $O(n \log n)$ [6, 10]. On the other hand, the average complexity of quick sort is known to be $O(n \log n)$. Since experiments indicate that comb sort runs in about twice as much time on a single processor as does quick sort [10], attempts such as [16] to develop a fast average-case sort for a single SPU of the CBE that begin with comb sort are handicapped by a factor of two compared to attempts that begin with quick sort. This handicap is overcome by the CBE adaptation of the Merge sort described in [3].

For integers and floats, the CBE supports 4-way parallelism within a single SPU as four integers (floats) may be stored in each of the SPU's 128-bit vector registers. Hence, we expect at best two-fold speedup over a conventional implementation of quick sort. However, due to possible anomalous behavior resulting from such factors as the absence of branch prediction, we may actually observe a greater speedup [18]. Similarly, attempts such as [11] to develop a fast worst-case sort for a single SPU starting with bitonic sort are handicapped relative to starting with merge sort because the worst-case complexity of bitonic sort is $O(n \log^2 n)$, while that of merge sort is $O(n \log n)$.

10.4.2 Shellsort Variants

Shellsort [17] sorts a sequence of n numbers in m passes, employing a decreasing increment sequence $i_1 > i_2 > \cdots > i_m = 1$. In the jth pass, increment $h = i_j$ is used; the sequence is viewed as comprised of h subsequences with the kth subsequence comprised of the numbers in positions k, $k+h$, $k+2h$, $\cdots$, of the overall sequence, $0 \leq k < h$; and each subsequence is sorted. The sorting of the subsequences done in each pass is called an h-sort. While an h-sort is typically accomplished using insertion sort, other simple sorting algorithms such as bubble sort may also be used. With the proper choice of increments, the complexity of Shellsort is $O(n \log^2 n)$ [17]. Shellsort variants replace the h-sort used in each pass of Shellsort with an h-pass that only partially sorts the subsequences. For example, in an h-bubble pass we make only the first pass of bubble sort on each subsequence. Since replacing h-sort by h-pass in Shellsort no longer guarantees a complete sort, we follow with some simple sorting algorithm such as bubble sort to complete the sort. So, the h-passes may be viewed as preprocessing passes done so as to improve the performance of the ensuing sort algorithm. In Shellsort, $i_m = 1$ is used to assure that the sequence is sorted following the final h-sort. However, in a Shellsort variant, this assurance comes from the sort algorithm run following the preprocessing h-passes. So, the h-pass with $h = 1$ is typically skipped. The general structure of a Shellsort variant is

Step 1 [Preprocess] Perform h-passes for $h = i_j$, $1 \leq j < m$.

Step 2 [Sort] Sort the preprocessed sequence.

10.4.2.1   Comb and AA sort

Knuth [17] proposed a Shellsort variant in which each h-pass is a bubble pass (Listing 10.1). This variant was rediscovered later by Dobosiewicz [8], and Box and Lacey [6] named this variant *comb sort*. The increment sequence used by comb sort is geometric with factor s. Dobosiewicz [8] has shown that the preprocessing step sorts $a[0 : n - 1]$ with very high probability whenever $s < 1.33$. As a result, $s = 1.3$ is recommended in practice (note that a larger s requires a smaller number of h-passes). With this choice, the outer **for** loop of the second step (bubble sort) is entered only once with high probability, and the complexity of comb sort is $O(n \log n)$ with high probability. Experiments indicate that the algorithm's average runtime is close to that of quick sort [10]. However, the worst-case complexity of comb sort is $O(n^2)$ [10].

Inoue et al. [16] have adapted comb sort to the CBE to obtain the sort method AA-sort, which efficiently sorts numbers using all eight SPUs of a CBE. The single SPU version begins with a vector array $d[0 : r-1]$ of numbers; each vector $d[i]$ has four numbers. Hence, d is an $r \times 4$ matrix of numbers. This matrix is first sorted into column-major order, and then the numbers permuted so as to be sorted in row-major order. Listing 10.2 gives the algorithm for the column-major sort, and Listing 10.3 gives the column-major to row-major reordering algorithm.

The column-major to row-major reordering is done in two steps. In the first step, the numbers in each 4×4 submatrix of the $r \times 4$ matrix of numbers are transposed so that each vector now has the four numbers in some row of the result. For simplicity, we assume that r is a multiple of 4. In the second

```
Algorithm combsort(a,n)
{// Sort a[0:n-1]
   // Step 1: Preprocessing
   for (h = n/s; h > 1; h /= s) {
       // h-bubble pass
       for (i = 0; i < n-h; i++)
          if (a[i] > a[i+h]) swap(a[i],a[i+h]);
   }
   sorted = false;
   // Step 2: Bubble sort
   for (pass = 1; pass < n && !sorted; pass++) {
     sorted = true;
     for (i = 0; i < n-pass; i++)
       if (a[i] > a[i+1]) {swap(a[i],a[i+1]);
                                    sorted = false;}
   }
}
```

LISTING 10.1: Comb sort.

```
Algorithm AA(d,r)
{// Sort d[0:r-1] into column-major order
   // Step 1: Preprocessing
   for (i = 0; i < r; i++) sort(d[i]);
   for (h = r; h > 1; h /= s) {
       // h-bubble pass
       for (i = 0; i < r-h; i++)
         mySpu_cmpswap(d[i],d[i+h]);
       for (i = r-h; i < r; i++)
         mySpu_cmpswap_skew(d[i],d[i+h-r]);
   }
   sorted = false;
   // Step 2: Bubble sort
   do {
     for (i = 0; i < r-1; i++)
       mySpu_cmpswap(d[i],d[i+1]);
     mySpu_cmpswap_skew(d[r-1],d[0]);
   } while (not sorted)
}
```

LISTING 10.2: Single SPU column-major AA-sort [16].

```
Algorithm transpose(d,r)
{// Column-major to row-major reordering
    // Step 1: Transpose 4 x 4 submatrices
    for (i = 0; i < r; i += 4) {
        // Compute row02A, row02B, row13A, and row13B
        row02A = spu_shuffle(d[i], d[i+2], WAXB);
        row02B = spu_shuffle(d[i], d[i+2], YCZD);
        row13A = spu_shuffle(d[i+1], d[i+3], WAXB);
        row13B = spu_shuffle(d[i+1], d[i+3], YCZD);
        // Complete the transpose
        d[i]   = spu_shuffle(row02A, row13A, WAXB);
        d[i+1] = spu_shuffle(row02A, row13A, YCZD);
        d[i+2] = spu_shuffle(row02B, row13B, WAXB);
        d[i+3] = spu_shuffle(row02B, row13B, YCZD);
    }
    // Step 2: Reorder vectors
    for (i = 0; i < r; i++)
        if (!inPlace[i]) {
            current = i;
            next = i/(r/4) + (i mod (r/4))*4;
            temp = d[i];
            while (next != i) {// Follow cycle
                d[current] = d[next];
                inPlace[current] = true;
                current = next;
                next = current/(r/4) + (current mod (r/4))*4;
            }
            d[current] = temp;
            inPlace[current] = true;
        }
}
```

LISTING 10.3: Column-major to row-major.

step, the vectors are permuted into the correct order. For the first step, we collect the first and second numbers in rows 0 and 2 of the 4×4 matrix being transposed into the vector row02A. The third and fourth numbers of these two rows are collected into row02b. The same is done for rows 1 and 3 using vectors row13A and row13B. Figure 10.3 shows this rearrangement. Then, the transpose is constructed from the just-computed four vectors.

10.4.2.2 Brick sort

In brick sort, we replace the h-bubble pass of comb sort by an h-brick pass [21, 24] in which we first compare-exchange positions $i, i+2h, i+4h, \cdots$ with positions $i+h, i+3h, i+5h, \cdots, 0 \leq i < h$, and then we compare-exchange positions $i+h, i+3h, i+5h, \cdots$ with positions $i+2h, i+4h, i+6h, \cdots$,

a	b	c	d
e	f	g	h
i	j	k	l
m	n	o	p

(a) Initial

a	i	b	j
c	k	d	l
e	m	f	n
g	o	h	p

(b) Rows 02A, 02B, 03A, 03B, top to bottom

a	e	i	m
b	f	j	n
c	g	k	o
d	h	l	p

(c) Transpose

FIGURE 10.3: Collecting numbers from a 4×4 matrix.

$0 \leq i < h$. Listing 10.4 gives our CBE adaptation of the preprocessing step (Step 1) for brick sort. Step 2 is a bubble sort, as was the case for AA-sort. The bubble sort needs to be followed by a column-major to row-major reordering step (Listing 10.3). It is known that the preprocessing step of brick sort nearly always does a complete sort when the increment sequence is geometric with shrink factor (i.e., s) less than 1.22 [21, 24]. Hence, when we use $s < 1.22$, the **do-while** loop of Step 2 (bubble sort) is entered only once (to verify the data are sorted) with high probability.

```
Algorithm Brick(d,r)
{// Sort d[0:r-1] into column-major order
    // Step 1: Preprocessing
    for (i = 0; i < r; i++) sort(d[i]);
    for (h = r; h > 1; h /= s) {
        // h-brick pass
        // Compare-exchange even:odd bricks
        for (i = 0; i < r-2*h; i += 2*h)
            for (j = i; j < i + h; j++)
                mySpu_cmpswap(d[j],d[j+h]);
        // Handle end conditions
        if (j < n - h) {// More than 1 brick remains
        end = j + h;
        for (; j < n - h; j++)
            mySpu_cmpswap(d[j],d[j+h]);
        }
        else end = r;
        while (j < end) {
            mySpu_cmpswap_skew(d[j],d[j+h-n]);
            j++;
        }
        // Compare-exchange odd:even bricks beginning with
        // i = h. Similar to even:odd bricks
    // Step 2: Bubble sort
    // Same as for AA-sort
    }
}
```

LISTING 10.4: Column-major brick sort.

```
Algorithm Shaker(d,r)
{// Sort d[0:r-1] into column-major order
    // Step 1: Preprocessing
    for (i = 0; i < r; i++) sort(d[i]);
    for (h = r; h > 1; h /= s) {
        // h-shake pass
        // Left-to-right bubble pass
        for (i = 0; i < r-h; i++)
            mySpu_cmpswap(d[i],d[i+h]);
        for (i = r-h; i < r; i++)
            mySpu_cmpswap_skew(d[i],d[i+h-r]);
        // Right-to-left bubble pass
        for (i = r-h-1; i > 0; i--)
            mySpu_cmpswap(d[i],d[i+h]);
        for (i = r-1; i >= r - h; i--)
            mySpu_cmpswap_skew(d[i],d[i+h-r]);
    }
    // Step 2: Bubble sort
    // Same as for AA-sort
}
```

LISTING 10.5: Column-major shaker sort.

10.4.2.3   Shaker sort

Shaker sort differs from comb sort in that h-bubble passes are replaced by h-shake passes. An h-shake pass is a left-to-right bubble pass as in comb sort followed by a right-to-left bubble pass. Listing 10.5 gives our CBE adaptation of shaker sort. The preprocessing step of shaker sort almost always sorts the data when the shrink factor s is less than 1.7.

10.4.3   Merge Sort

Unlike the Shellsort variants, comb, brick, and shaker sort of Section 10.4.2 whose complexity is $O(n \log n)$ with high probability, the worst-case complexity of merge sort is $O(n \log n)$. Further, merge sort is a stable sort (i.e., the relative order of elements that have the same key is preserved). While this property of merge sort isn't relevant when we are simply sorting numbers (as you can't tell two equal numbers apart), this property is useful in some applications where each element has several fields, only one of which is the sort key. The Shellsort variants of Section 10.4.2 are not stable sorts. On the down side, efficient implementations of merge sort require added space. When sorting numbers in the vector array $d[0:r-1]$, we need an additional vector array $t[0:r-1]$ to support the merge. CBE merge-sort adaptation is presented as a stable sort, and later we point out the simplifications that are possible when

22	30	5	17		22	25	19	15		5	2	1	3
14	26	32	9		30	6	13	23		9	6	4	8
25	6	20	10		5	20	29	8		14	10	7	12
2	28	16	11		17	10	1	31		17	11	13	15
19	13	29	1		14	2	21	12		22	16	19	18
21	4	24	7		26	28	4	18		26	20	21	23
15	23	8	31		32	16	24	27		30	25	24	27
12	18	27	3		9	11	7	3		32	28	29	31
(a) Initial					(b) Phase 1					(c) Phase 2			

2	17		1	18		1	2	3	4
5	20		3	19		5	6	7	8
6	22		4	21		9	10	11	12
9	25		7	23		13	14	15	16
10	26		8	24		17	18	19	20
11	28		12	27		21	22	23	24
14	30		13	29		25	26	27	28
16	32		15	31		29	30	31	32
(d) Phase 3						(e) Phase 4			

FIGURE 10.4: Merge sort example.

we wish to sort numbers rather than elements that have multiple fields. We again assume that the numbers are in the vector array $d[0 : r - 1]$.

There are four phases to our stable merge-sort adaptation:

Phase 1: Transpose the elements of $d[0 : r - 1]$, which represents an $r \times 4$ matrix, from row-major to column-major order.

Phase 2: Sort the four columns of the $r \times 4$ matrix independently and in parallel.

Phase 3: In parallel, merge the first two columns together and the last two columns together to get two sorted sequences of length $2r$ each.

Phase 4: Merge the two sorted sequences of length $2r$ each into a row-major sorted sequence of length $4r$.

10.4.3.1   Merge Sort Phase 1—Transpose

We note that Phase 1 is needed only when we desire a stable sort. Figure 10.4 shows an initial 8×4 matrix of numbers, and the result following each of the four phases of our merge-sort adaptation.

The Phase-1 transformation is the inverse of the column-major to row-major transformation done in Listing 10.3 and we do not provide its details. Details for the remaining three phases are provided in the following subsections.

```
Algorithm Phase2(d,r)
{// Sort the 4 columns of d[0:r-1],
 // use additional array t[0:r-1]
   for (s = 1; s < r; s *= 2)
      for (i = 0; i < r; i += 2*s) {
          A = spu_splats(i); // Initialize a counters
          B = spu_splats(i+s); // Initialize b counters
          for (k = i; k < i + 2*s; k++) {// Merge the segments
              // One round of compares
              aData = mySpu_gather(d,A);
              bData = mySpu_gather(d,B);
              p = spu_cmpgt(aData,bData);
              t[k] = spu_select(aData,bData,p);
              // Update counters
              notP = mySpu_not(p);
              A = spu_sub(A,notP);
              B = spu_sub(B,p);
          }
          swap(d,t); // Swap roles
      }
}
```

LISTING 10.6: Phase 2 of merge sort.

10.4.3.2   Merge Sort Phase 2—Sort columns

Phase 2 operates in $\log r$ subphases characterized by the size of the sorted segments being merged. For instance, in the first subphase, we merge together pairs of sorted segments of size 1 each; in the next subphase the segment size is 2, in the third it is 4, and so forth. At any time, the two segments being merged have the same physical locations in all four columns. So, for our 8×4 example, when merging together segments of size 2, we shall first merge, in parallel, four pairs of segments, one pair from each column. The first segment of a pair is in rows 0 and 1 of the $r \times 4$ matrix and the second in rows 2 and 3. Then, we shall merge together the segments of size 2 that are in rows 4 through 7. Following this, the segment size becomes 4.

To merge four pairs of segments in parallel, we employ eight counters to keep track of where we are in the eight segments being merged. The counters are called $a_0, \cdots, a_3, b_0, \cdots, b_3$. (a_i, b_i) are the counters for the segments of column i, $0 \leq i \leq 3$ that are being merged. When the segment size is s and the segments being merged occupy rows i through $i + 2s - 1$, the a counters are initialized to i and the b counters to $i + s$. Although all a counters have the same value initially as do all b counters, as merging progresses, these counters have different values. Listing 10.6 gives the Phase 2 algorithm. For simplicity, we assume that r is a power of 2.

10.4.3.3 Merge Sort Phase 3—Merge pairs of columns

In Phase 3 we merge the first two and last two columns of the $r \times 4$ matrix together to obtain two sorted sequences, each of size $2r$. The first sequence is in columns 0 and 1 and the second in columns 2 and 3 of an output matrix. We do this merging using eight counters. Counters a_0, b_0, a_1, b_1 start at the top of the four columns of our matrix and move downwards, while counters a_2, b_2, a_3, b_3 start at the bottom and move up (see Figure 10.5(a)). Let $e(c)$ be the matrix element that counter c is at. The comparisons $e(a_i) : e(b_i)$, $0 \leq i \leq 3$ are done in parallel, and depending on the outcome of these comparisons, four of the eight elements compared are moved to the output matrix. When $e(a_0) \leq e(b_0)$ $(e(a_1) \leq e(b_1))$, $e(a_0)$ $(e(a_1))$ is moved to the output and a_0 (a_1) incremented by 1; otherwise, $e(b_0)$ $(e(b_1))$ is moved to the output and b_0 (b_1) incremented by 1. Similarly, when $e(a_2) \leq e(b_2)$ $(e(a_3) \leq e(b_3))$, $e(b_2)$ $(e(b_3))$ is moved to the output and b_2 (b_3) decremented by 1; otherwise, $e(a_2)$ $(e(a_3))$ is moved to the output and a_2 (a_3) decremented by 1. The merge is complete when we have done r rounds of comparisons. Listing 10.7 gives the algorithm for Phase 3.

Theorem 1 *Algorithm* `Phase3` *correctly merges four sorted columns into two.*

Proof 1 *To prove the correctness of Algorithm* `Phase3`, *we need to show that each element of the first (last) two columns of the input $r \times 4$ matrix is copied into the first (last) two columns of the output matrix exactly once and that the elements of the first (third) output column followed by those of the second (fourth) are in sorted order. It is sufficient to show this for the first two columns of the input and output matrices. First, observe that when $a_0 \leq a_2$ $(b_0 \leq b_2)$, these counters are at input elements that have yet to be copied to the output. Further, when $a_0 > a_2$ $(b_0 > b_2)$, all elements of the respective column have been copied from the input to the output (note that a counter is updated only when its element has been copied to the output matrix). We consider four cases: $a_0 < a_2$, $a_0 = a_2$, $a_0 = a_2 + 1$, and $a_0 > a_2 + 1$.*

Case $a_0 < a_2$: *When $b_0 < b_2$ (Figure 10.5(a)), exactly one of $e(a_0)$ and $e(b_0)$ and one of $e(a_2)$ and $e(b_2)$ are copied to the output, and the corresponding counters are advanced. No element is copied to the output twice.*

Next, consider the case $b_0 = b_2$ (Figure 10.5(b)). If $e(a_0) \leq e(b_0)$, $e(a_0)$ and one of $e(a_2)$ and $e(b_2)$ are copied to the output, and the corresponding counters advanced. Again no element is copied to the output twice. If $e(a_0) > e(b_0) = e(b_2)$, $e(b_2) < e(a_0) \leq e(a_2)$ and $e(b_0)$ and $e(a_2)$ are copied to the output, and their counters advanced. Again, no element is copied twice.

The next case we consider has $b_0 = b_2 + 1$. Let the values of b_0 and b_2 be b_0' and b_2' just before the update(s) that resulted in $b_0 = b_2 + 1$, and let a_0' and a_2' be the values of the a counters at this time. One of the

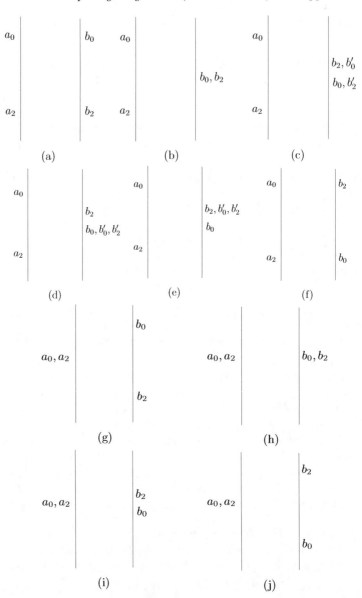

FIGURE 10.5: Phase 3 counters.

following must be true: (a) $b_2' = b_0' + 1$ (both b_0 and b_2 were advanced, Figure 10.5(c)), (b) $b_0' = b_2' = b_0$ (only b_2 was advanced, Figure 10.5(d)), or (c) $b_0' = b_2' = b_2$ (only b_0 was advanced, Figure 10.5(e)). In (a), it must be that $b_2 = b_0'$ and $b_0 = b_2'$. So, $e(a_0) > e(b_0')$ and $e(a_2) \leq e(b_2')$. Hence,

```
Algorithm Phase3(d,r)
{// Merge the 4 sorted columns of d[0:r-1] into 2 sorted
 // sequences. Use additional array t[0:r-1]
    A = {0, 0, r-1, r-1}; // Initialize a counters
    B = {0, 0, r-1, r-1}; // Initialize b counters
    for (k = 0; k < r; i++) {
        aData = mySpu_gatherA(d,A);
        bData = mySpu_gatherB(d,B);
        p = spu_cmpgt(aData,bData);
        e = spu_equal(aData,bData);
        e = spu_and(e, vector(0,0,-1,-1));
        p = spu_or(p, e);
        min = spu_select(aData, bData, p);
        max = spu_select(bData, aData, p);
        t[k] = spu_shuffle(min,t[k],WBXD);
        t[r-k-1] = spu_shuffle(max,t[r-k-1],AYCZ);
        // Update counters
        notP = mySpu_not(p);
        f1 = spu_and(p,vector(-1,-1,0,0));
        s1 = spu_and(p,vector(0,0,-1,-1));
        f2 = spu_and(notP,vector(-1,-1,0,0));
        s2 = spu_and(notP,vector(0,0,-1,-1));
        A = spu_sub(A,f2);
        A = spu_add(A,s2);
        B = spu_sub(B,f1);
        B = spu_add(B,s1);
    }
}
```

LISTING 10.7: Phase 3 of merge sort.

$e(a_0) \le e(a_2) \le e(b_2') = e(b_0)$ *and* $e(a_2) \ge e(a_0) > e(b_0') = e(b_2)$. *There-fore,* $e(a_0)$ *and* $e(a_2)$ *are copied to the output, and* a_0 *and* a_2 *advanced. Again, only previously uncopied elements are copied to the output, and each is copied once. For subcase (b), when* b_2' *was decremented to* b_2, a_0' *was incremented to* a_0, $e(b_2') \ge e(a_2)$ *and* $e(a_0') \le a(b_0')$. *Since* $b_0 > b_2$, *all elements of the second column have been copied to the output. We see that* $e(a_0) \le e(a_2) \le e(b_2') = e(b_0)$. *So,* $e(a_0)$ *is copied, and* a_0 *is advanced. Further, as a result of some previous comparison,* b_0 *was advanced to its current position from the present position of* b_2. *So, there is an* $a_0'' \le a_0$ *such that* $e(b_2) < e(a_0'') \le e(a_0) \le e(a_2)$. *Therefore,* $e(a_2)$ *is copied and* a_2 *advanced. Again, no previously copied element is copied to the output, and no element is copied twice. Subcase (c) is symmetric to subcase (b).*

The final case has $b_0 > b_2 + 1$ *(Figure 10.5(f)). From the proof of subcases* $b_0 = b_2$ *and* $b_0 = b_2 + 1$, *it follows that this case cannot arise.*

Case $a_0 = a_2$: *There are four subcases to consider: (a) $b_0 < b_2$, (b) $b_0 = b_2$, (c) $b_0 = b_2 + 1$, and (d) $b_0 > b_2 + 1$ (Figures 10.5(g–j)). Subcase (a) is symmetric to the case $a_0 < a_2$ and $b_0 = b_2$ considered earlier. In subcase (b), independent of the outcome of the comparison $e(a_0) : e(b_0)$, which is the same as the comparison $e(a_2) : e(b_2)$, $e(a_0)$ (equivalently $e(a_2)$) and $e(b_0)$ (equivalently $e(b_2)$) are copied to the output. For subcase (c), we notice that when $a_0 = a_2$, these two counters have had a cumulative advance of $r - 1$ from their initial values, and when $b_0 = b_2 + 1$ these two counters have together advanced by r. So, the four counters together have advanced by $2r - 1$ from their initial values. This isn't possible as the four counters advance by a total of 2 in each iteration of the for loop. So, subcase (c) cannot arise. Next, consider subcase (d). From the proof for the case $a_0 < a_2$, we know that we cannot have $b_0 > b_2 + 1$ while $a_0 < a_2$. So, we must have gotten into this state from a state in which $a_0 = a_2$ and $b_0 \le b_2$. It isn't possible to get into this state from subcase (a) as subcase (a), at worst, increases b_0 by 1 and decreases b_2 by 1 each time we are in this subcase. So, it is possible to get into this subcase only from subcase (b). However, subcase (b) only arises at the last iteration of the* for *loop. Even otherwise, subcase (b) either increments b_0 by 1 or decrements b_2 by 1 and so cannot result in $b_0 > b_2 + 1$.*

Case $a_0 > a_2 + 1$: *From the proofs of the remaining cases, it follows that this case cannot arise.*

From the proof of Theorem 1, it follows that when we are sorting numbers rather than records with numeric keys, algorithm Phase3 works correctly even with the statements

```
e = spu_equal(aData,bData);
e = spu_and(e, vector(0,0,-1,-1));
p = spu_or(p, e);
```

omitted.

10.4.3.4   Merge Sort Phase 4—Final merge

For the Phase-4 merge, we employ four counters. Counters a_0 and a_1, respectively, begin at the first and last element of the first sorted sequence (i.e., at the top of the first column and bottom of the second column, respectively), while b_0 and b_1 begin at the first and last elements of the second sequence (Figure 10.6). In each round, the comparisons $a_0 : b_0$ and $a_1 : b_1$ are done in parallel; $e(a_0)$ $(e(b_1))$ is moved to the output if $e(a_0) \le e(b_0)$ $(e(b_1) \ge e(a_1))$; otherwise, $e(b_0)$ $(e(a_1))$ is moved to the output. The sorted output is assembled in row-major order in the vector array t. We use the variables k and pos to keep track of the row and column in t in which we must place the output element from the comparison $e(a_0) : e(b_0)$. The output element from $e(a_1) : e(b_1)$ goes into row $(r - k - 1)$ and column $(3 - pos)$ of t. Listing 10.8 gives the algorithm for the case when the counters remain within the bounds of their respective

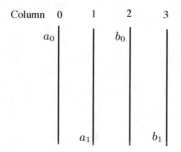

FIGURE 10.6: Phase-4 counters.

```
Algorithm Phase4(d,r)
{// Partial algorithm to merge 2 sorted sequences of
 // d[0:r-1] into 1 sorted sequence
 // Use additional array t[0:r-1]
    A = {0, r-1, 0, 0}; // Initialize a counters
    B = {0, r-1, 0, 0}; // Initialize b counters
    k = 0; pos = 0;
    while (no column is exhausted) {
      aData = mySpu_gather12(d,A);
      bData = mySpu_gather34(d,B);
      p = spu_cmpgt(aData,bData);
      e = spu_equal(aData,bData);
      e = spu_and(e, vector(0,-1,0,0));
      p = spu_or(p, e);
      min = spu_select(aData, bData, p);
      max = spu_select(bData, aData, p);
      max = spu_slqwbyte(max, 4);
      t[k] = spu_shuffle(min,t[k],mask[pos]);
      t[r-k-1] = spu_shuffle(max,t[r-k-1],mask[3-pos]);
      // Update counters
      notP = mySpu_not(p);
      f1 = spu_and(p,vector(-1,0,0,0));
      s1 = spu_and(p,vector(0,-1,0,0));
      f2 = spu_and(notP,vector(-1,0,0,0));
      s2 = spu_and(notP,vector(0,-1,0,0));
      A = spu_sub(A,f2);
      A = spu_add(A,s1);
      B = spu_sub(B,f1);
      B = spu_add(B,s2);
      k += (pos+1)/4;
      pos = (pos+1)%4;
    }
}
```

LISTING 10.8: Phase 4 of merge sort with counters in different columns.

# Integers	AASort	Shaker sort	Brick Sort	Bitonic Sort	Merge Sort	Merge Sort (Sequential)	Quick Sort
128	52	53.6	53	47.8	50.8	146.6	145.6
256	62.4	65.4	63.4	65.6	57.6	178.6	206.8
512	81.8	86.4	81.4	72.6	70.4	272.2	332
1024	123.8	142.2	116.8	125.4	97	315.4	605.6
2048	222.8	262	190.2	165.8	142	543	1164
4096	438.6	494.8	332.6	297.8	268.4	989.8	2416.6
8192	912.4	1033.6	663.8	609.6	508	2011.2	4686.6
16384	1906.4	2228	1361	1331.2	1017	4103	9485.2

TABLE 10.1: Comparison of various SPU sorting algorithms.

columns. $mask[pos]$, $0 \leq pos \leq 3$ is defined so as to change only the number in position pos of a $t[]$ vector.

As was the case in Phase 3, the statements

```
e = spu_equal(aData,bData);
e = spu_and(e, vector(0,0,-1,-1));
p = spu_or(p, e);
```

may be omitted when we are sorting numbers rather than records with numeric keys.

10.4.4   Comparison of Single-SPU Sorting Algorithms

We programmed our merge-sort, brick-sort, and shaker-sort adaptations using the CBE SDK Version 3.0. For comparison purposes, we used an AA sort code developed by us, the Cell sort code of [11], a nonvectorized merge-sort code developed by us, and the quick-sort routine available in the CBE SDK. The codes were first debugged and optimized using the CBE simulator that is available with the SDK. The optimized codes were run on the Georgia Tech-STI Cellbuzz cluster to obtain actual runtimes. Table 10.1 gives the average time required to sort n 4-byte integers for various values of n. The average for each n is taken over five randomly generated sequences. The variance in the sort time from one sequence to the next is rather small, and so the reported average is not much affected by taking the average of a larger number of

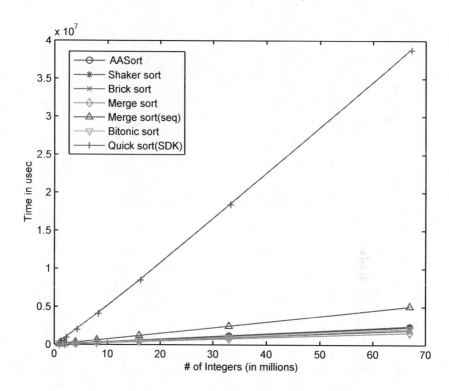

FIGURE 10.7: Plot of average time to sort 1 to 67 million integers.

random input sequences. Figure 10.7 is a plot of the average times reported in Figure 10.1. The shown runtimes include the time required to fetch the data to be sorted from main memory and to store the sorted results back to main memory.

Our experiments reveal that a standard nonvectorized textbook implementation of merge sort takes about four times the time taken by the vectorized merge-sort adaptation. Further, the quick-sort method that is part of the CBE SDK takes about nine times the time taken by our merge-sort adaptation. Brick sort is the fastest of the shellsort-like algorithms—AA sort, shaker sort, and brick sort—considered in this chapter, taking about 71% the time taken by AA sort to sort 16,384 integers. Although cell (bitonic) sort is slightly faster than brick sort, it takes about 31% more time to sort 16,384 integers than taken by merge sort.

10.4.5 Hierarchical Sort

For sorting a large set of numbers, a hierarchical approach similar to external sort is employed where first each SPU sorts a local memory load of

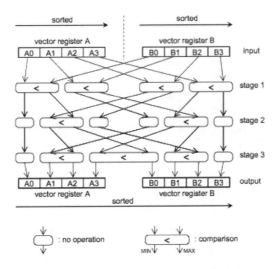

FIGURE 10.8: SIMD odd-even merge of two vectors [16].

data to generate sorted sequences called runs. These runs are then merged by the SPUs to produce the final sorted sequence. Depending on how many runs are merged at a time, there will be multiple rounds of merging before generating the final sorted sequence. The PPU dispatches the runs to the SPUs, which do the merging and return the results back to the PPU. In the run-merging phase, each SPU independently merges a different set of runs. So, one needs to develop only a run-merging algorithm for a single SPU. Inoue et al. [16] propose a single-SPU merging algorithm that merges runs in pairs (i.e., a 2-way merge) using an adaptation of odd-even merge. Odd-even merge of two four-number sorted sequences is done in three stages. First, the two sorted sequences are concatenated to get an eight-number sequence where each half is in sorted order. During the first stage, numbers that are four positions apart are compare-exchanged.[1] In the second stage, numbers that are two positions apart are compare-exchanged, and in the last stage, alternate numbers are compare-exchanged if needed. This scheme can be effectively SIMD-ized by beginning with two vectors, each containing one of the two sorted four-number sequences. Vector compare instructions are used so that four compare-exchange may be done at a time. Figure 10.8 shows the process with two sorted vectors A and B, and Listing 10.9 gives the pseudocode for this adaptation.

As the runs are too long to fit in the local memory of an SPU, buffers are used to hold a portion of each of the runs currently being merged. Multi-buffering techniques are employed to overlap the computation with the data

---

[1]In a compare-exchange, two numbers are compared and swapped if the first number is larger than the second.

```
Algorithm oddEvenMerge(v1, v2)
{// Merge two vectors v1 and v2
   vector f1, f2;
   vetor f3, f4;
   vector p; // For storing pattern
   p = spu_cmpgt(v1, v2);
   f1 = spu_select(v1, v2, pattern);
   f2 = spu_select(v2, v1, pattern);
   // Stage 2
   f3 = spu_rotate(f1, 8);
   f1 = spu_select(f3, f2, p);
   f4 = spu_select(f2, f3, p);
   f2 = spu_shuffle(f1, f4, WACY);
   f3 = spu_shuffle(f1, f4, ZXBD);
   // Stage 3
   p = spu_cmpgt(f2, f3);
   p = spu_shuffle(p, vZero, WXYA);
   f1 = spu_select(f2, f3, p);
   f4 = spu_select(f3, f2, p);
   // Output
   v1 = spu_shuffle(f1, f4 ZWAX);
   v2 = spu_shuffle(f1, f4, BYCD);
}
```

LISTING 10.9: SIMD 2-way merge of two vectors v1 and v2.

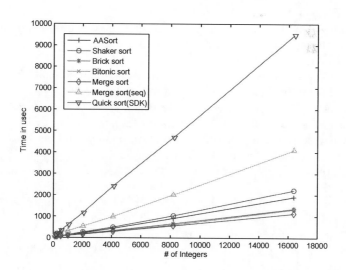

FIGURE 10.9: Plot of average time to sort 4-byte integers.

transfer. Figure 10.9 is the plot of the average times taken for sorting 1 to 67 million integers with different SPU sorting algorithms followed by SIMD odd-even 2-way merge to merge the runs except in the case of Bitonic Sort [11], where bitonic merge is used to merge the runs, and in the case of sequential merge sort, where a textbook merge is done to combine the runs. Similar to single-SPU sorting, Brick sort is the fastest among shellsort-like algorithms, taking 82% of the time taken by AA Sort for sorting 67 million numbers, while Shaker sort is the slowest of the bunch, running 21% slower than Brick sort for sorting 67 million numbers. Merge sort is the fastest of all algorithms tested, taking 84% of the time taken by Brick sort and runs 24% faster than Bitonic sort for sorting 67 million numbers. Compared to sequential sorts, the SIMD version of merge sort runs three times faster than the textbook merge sort and 24 times faster than the SDK quick sort.

10.4.6   Master-Slave Sort

10.4.6.1   Algorithm SQMA

In this section, we propose a sorting algorithm that can be thought of as a hybrid between Quick Sort and Merge Sort. We call this algorithm SQMA (Algorithm 10.1). In this algorithm and the follow-up algorithms, each SPU receives a chunk of data. We use a double-buffering technique to communicate

---

Algorithm 10.1 Algorithm SQMA (A, ArraySize)

{ // Sort A[0 : ArraySize-1]

Summary: Divide the data in PPE into equal sized chunks and send them to SPEs; perform Quick Sort, and send the partially sorted data back to PPE. Perform Merge operation on PPE.

Input: Unsorted Array A[ArraySize] of integers, number of SPUs = p.

Output: Sorted Array A[ArraySize], of integers.

Step 1: In the PPU, generate a 1 dimensional array, A[ArraySize] of randomly generated integers, where size of the array is ArraySize.

Step 2: Divide the array into ArraySize/p parts and send one part to each SPE using double-buffered DMA data transfer.

Repeat Steps 3, 4, and 5 until all the data in ArraySize/p parts are sent to the SPEs

       Step 3: In each SPE, receive a chunk of 4096 elements, X[4096] at a time via a double-buffered DMA call from the memory.

       Step 4: Perform a Quick Sort on these elements.

       Step 5: Send the result back to the PPE by the double-buffered DMA call to the memory.

Step 6: In the PPE, perform a merge operation on the partially sorted array chunks received in A[ArraySize] and output the sorted array.

}

---

data between the PPU and SPUs in all our implementations. Since the SPUs can store a limited amount of data at a time, they pick 4096 integers at a time from the chunk assigned to it and sort it. After the SPUs finish, we get the results back in the PPE. The main idea behind SQML is to divide the data in PPE into equal-sized chunks, send them to SPEs, perform Quick Sort, send the partially sorted data back to PPE, and perform merge operation on the PPE.

10.4.6.2 Random Input Integer Sorting with Single Sampling & Quick Sort (RISSSQS)

SQMA did not offer a good speedup. To rectify this, RISSSQS (Algorithm 10.2) preprocesses the data before sending them to the SPEs. The input is generated randomly with a range of integers [1, R]. We create a new array

---

Algorithm 10.2 Algorithm RISSSQS(A, ArraySize)

{ // Sort A[0 : ArraySize-1]

Summary: Perform single sampling on the data in PPE by choosing keys from the Range [1, R] and rearranging data based on it. Divide data into equal sized chunks, and send to the SPEs. Perform Quick Sort on the chunks, and send result back to PPE.

Input: Unsorted Array A[ArraySize] of integers, number of SPUs = p, scalingFactor, Range of Integers =[1,R].

Output: Sorted Array A[ArraySize] of integers.

Step 1: In the PPE, Generate a one-dimensional array A[ArraySize] of randomly generated integers in the range [1, R], where the size of the array is ArraySize.

Step 2: Compute the numberOfBuckets to be
ArraySize*scalingFactor/4096.

Expected Number of integers with a specific value, E = ArraySize/R.

Number of distinct elements in any bucket, n = 4096/E* scalingFactor.

Step 3: Compute the range of each bucket as $[1..n]$, $[n+1..2n]$, ...

Step 4: Partition the elements in Array A[] according to the range calculated above for each bucket.

Step 5: Now divide the array into ArraySize/p parts and send one part to each SPE.

Repeat Steps 6 and 7 until all the data have been sent to the SPEs.

    Step 6: In each SPE, receive an array X[4096] elements at a time using the double-buffered DMA data transfer.

    Step 7: Perform Quick Sort on these elements, and send the result back to the PPE.

Step 8: In the PPE, receive the data back from the SPE, and output the sorted data.

}

---

of size (scaling factor × the size of the original array) and initialize all the elements to zero. SPE sorts 4096 integers at a time. We apply bucket sort on the data in the PPE, where the buckets are of size 4096 each and hence calculate the number of buckets. So, we can now calculate the range of integers for each bucket. After applying Bucket Sort, the new array has a number of buckets, and each bucket has some zeros and some integers. Now we divide this array into chunks similar to SQMA and send the chunks to SPEs. The SPEs then perform Quick Sort on the chunks. When the result comes back to the PPE from the SPEs, it is completely sorted. In our implementation, we used different values of scaling factors.

10.4.6.3   Random Input Integer Sorting with Single Sampling using Bucket Sort (RISSSBS)

Steps involved in Algorithm 10.3 are as follows: perform a single level of sampling on the data in PPE by choosing keys from the Range [1, R] and

---

Algorithm 10.3 Algorithm RISSSBS(A, ArraySize)

{ // Sort A[0 : ArraySize-1]

Summary: Perform single sampling on the data in PPE by choosing keys from the Range [1, R] and rearranging data based on them. Divide data into equal sized chunks, and send to the SPEs. Perform Bucket Sort on the chunks, and send sorted result back to PPE.

Input: Unsorted Array A[ArraySize] of integers, number of SPUs = p, scalingFactor, Range of Integers = [1, R].

Output: Sorted Array A[ArraySize], of integers.

Step 1: In the PPE, Generate a one-dimensional array A[ArraySize] of randomly generated integers in the range [1, R], where size of array is ArraySize.

Step 2: Compute the numberOfBuckets to be ArraySize*scalingFactor/4096. Expected Number of each element, $E = ArraySize/R$. Number of distinct elements in any bucket, $n = 4096*E/2*ArraySize$

Step 3: Compute the range of each bucket as $[1..n], [n+1..2n]...$

Step 4: Partition the elements in Array A[] according to the range calculated above for each bucket.

Step 5: Now divide the array into ArraySize/p parts, and send one part to each SPE using double-buffered DMA data transfer.

Repeat Steps 6 and 7 until all the data have been sent to the SPEs.

Step 6: In each SPE, receive an array X[4096] elements at a time using the double-buffered DMA data transfer.

Step 7: Perform Bucket Sort on these elements, and send the result back into the PPE using double-buffered DMA data transfer.

Step 8: In the PPE, receive the data back from the SPE, and output the sorted data.

}

---

rearranging the data based on it; divide the data into equal-sized chunks and send to the SPEs; perform bucket sort on the chunks and send sorted result back to PPE.

10.4.6.4 Algorithm RSSSQS

Algorithm 10.4 is a randomized sorting algorithm. The data is generated randomly and has no range like the previous algorithms. We do single sampling on the data in the PPE. Here, too, we create a new array of size (scaling factor × the size of the original array). The new array is then divided into chunks and sent to the PPE for Quick Sort. We receive fully sorted data in the PPE.

Algorithm 10.4 Algorithm RSSSQS(A, ArraySize)

{ // Sort A[0 : ArraySize-1]

Summary: Perform single sampling on the data in PPE by picking Random keys and rearranging data based on them. Divide data into chunks, and send to the SPEs. Perform Quick Sort on the chunks, and send result back to PPE.
Input: Unsorted Array A[ArraySize] of integers, number of SPUs = p, scalingFactor, Range of Integers = [1, R].
Output: Sorted Array A[ArraySize], of integers.
 Step 1: In the PPE, Generate a one-dimensional array A[ArraySize] of randomly generated integers, where size of array is ArraySize.
 Step 2: Compute the numberOfBuckets to be ArraySize*scalingFactor/4096.
 Step 3: Randomly pick numberOfBuckets keys.
 Step 4: Sort these keys, and use these as keys to partition the data elements in Array A[] into elements falling within these keys.
 Step 5: Now divide Array A into ArraySize/p parts, and send one part to each SPE.
 Repeat Steps 5 and 6 until all the data have been sent to the SPEs.
  Step 5: In each SPE, receive an array X[4096] elements at a time using the double-buffered DMA data transfer.
  Step 6: Perform Quick Sort on these elements, and send the result back into the PPE using double-buffered DMA data transfer.
 Step 7: In the PPE, receive the data back from the SPE, and output the sorted data.
}

10.4.6.5 Randomized Sorting with Double Sampling using Quick Sort (RSDSQS)

RSDSQS (Algorithm 10.5) is also a randomized sorting algorithm where the data is generated randomly with no range. In this case, we do double sampling of the data. Double Sampling gives us a better sample of the data, and that reduces the risk of overflow in any bucket. We create a new array of larger size like in the previous algorithms. We then divide the data into chunks and send them to the SPEs for sorting. After that, we get the sorted result back in the PPE.

Algorithm 10.5 Algorithm RSDSQS(A, ArraySize)

{ // Sort A[0 : ArraySize-1]

Summary: Perform double sampling on the data in PPE, and rearrange data based on it. Divide data into chunks and send to the SPEs. Perform Quick Sort on the chunks, and send results back to PPE.

   Step 1: In the PPE, Generate a one-dimensional array A[ArraySize] of randomly generated integers, where size of array is ArraySize.

   Step 2: Compute the numberOfBuckets to be ArraySize*scalingFactor/4096.

   Step 3: Randomly pick n keys (where n is any integer multiple of 2 between 4096 and 8192).

   Step 4: Sort these keys, and sequentially pick numberOfBuckets keys (equidistant from one another) from the array of n keys.

Use these as keys to partition the data elements in Array A[] into elements falling within these keys.

   Step 5: Now divide Array A into ArraySize/p parts and send one part to each.

Repeat Steps 6 and 7 until all the data have been sent to the SPEs.

   Step 6: In each SPE, receive an array X[4096] elements at a time.

   Step 7: Perform Quick Sort on these elements, and send the result back into the PPE using double-buffered DMA data transfer.

   Step 8: In the PPE receive the data back from the SPE, and output the sorted data.

}

10.4.6.6   Randomized Sorting with Double Sampling using Merge Sort (SDSMS)

Algorithm 10.6 is similar to RSDSQS except that we used merge sort in the SPEs to do the sorting.

10.4.6.7   Evaluation of SQMA, RISSSQS, RISSSBS, RSSSQS, RSDSQS, and SDSMS

Each of the algorithms was run 100 times for each reading, and an average was taken to get the final score. IBM SDK Version 1.0 was used, and the coding for all the algorithms was done in C using IBM's SDK. The data were transferred using using the double-buffering scheme [35]. For the comparison between the Cell/B.E., Altix 350, and the Pentium 4, we used a Pentium 4 processor with 1 GB of RAM (1.99 GHz) and a clock resolution of 100 clock ticks per second and Altix 350 with 64 nodes each with 1 GB of RAM

Algorithm 10.6 Algorithm SDSMS(A, ArraySize)

{ // Sort A[0 : ArraySize-1]

Summary: Perform double sampling on the data in PPE, and rearrange data based on it. Divide data into chunks and send to the SPEs. Perform merge sort on the chunks, and send result back to PPE.

Input: Unsorted Array A[ArraySize] of integers, Num of SPUs = p, scalingFactor.

Output: Sorted Array A[ArraySize], of integers.

    Step 1: In the PPE, Generate a one-dimensional array A[ArraySize] of randomly generated integers, where size of array is ArraySize.

    Step 2: Compute the numberOfBuckets to be ArraySize*scalingFactor/4096.

    Step 3: Randomly pick n keys (where n is any integer multiple of 2 between 4096 and 8192).

    Step 4: Sort these keys, and sequentially pick numberOfBuckets keys (equidistant from one another) from the array of n keys. Use these as keys to partition the data elements in Array A[] into elements falling within these keys.

    Step 5: Now divide Array A into ArraySize/p parts, and send one part to each SPE using double-buffered DMA data transfer.

    Repeat Steps 6 and 7 until all the data has been sent to the SPEs.

        Step 6: In each SPE, receive an array X[4096] elements at a time using the double-buffered DMA data transfer.

        Step 7: Perform Merge Sort on these elements, and send the result back into the PPE using double-buffered DMA data transfer.

    Step 8: In the PPE, receive the data back from the SPE and output the sorted data.

}

(1.5 GHz) and a clock resolution of 1024 clock ticks per second. Cell has 1 GB of RAM on the PPE with a clock resolution of 100 clock ticks per second and 256 KB of local memory on each SPE.

In the following sections, we give the results of each algorithm when run on the Cell/B.E. followed by a graph depicting the speedup as well as, in the case of sampling, the effect of changing the scaling factor of the data. Scaling factor is used to account for the random data sampling. Thus, when we say a scaling factor of 2, it means that we allocate twice the space for each sample to account for the randomness.

10.4.6.8 Results

Tables 10.2 and 10.3 and Figure 10.10 show data obtained using Algorithm SQMA (Simple Quick and Merge Algorithm).

Data Size	Time (s)
16k	0.026
64k	0.09
128k	0.19
1M	1.23

TABLE 10.2: Change the data size, and observe the time taken.

Data Size	No. of SPE	Time (s)	Speedup	% Efficiency
1M	1	1.34	1	100
1M	2	1.3	1.03	51.5
1M	4	1.2	1.11	27.75
1M	8	1.2	1.11	13.875

TABLE 10.3: Efficiency achieved on increasing the number of SPEs.

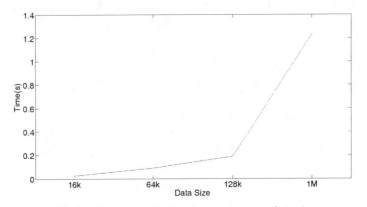

(a) Graph to study timing when we increase data size.

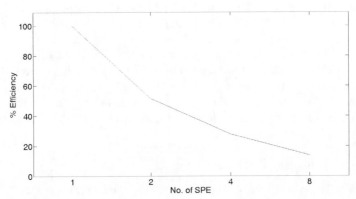

(b) Graph to study percent efficiency when increasing the number of SPEs.

FIGURE 10.10: Algorithm SQMA.

Tables 10.4–10.10 show data obtained using Algorithm RISSSQS (Random Input Integer Sorting with Single Sampling using Quick Sort). The range of the input data is from 1 to 6000.

# of SPU	Time	Speedup	% Efficiency
1	0.022	1	100
2	0.019	1.15789	57.8
4	0.013	1.69230	42.3
8	0.008	2.75	34.3
16	*NoResult*		

TABLE 10.4: Data size 16,000.

# of SPU	Time	Speedup	% Efficiency
1	0.032	1	
2	0.021	1.523809	76.190476
4	0.016	2	50
8	0.014	2.285714	28.571428
16	0.01	3.2	20

TABLE 10.5: Data size 32,000.

# of SPU	Time	Speedup	% Efficiency
1	0.058	1	100
2	0.036	1.6	80
4	0.026	2.23	55.75
8	0.021	2.76	34.5
16	0.016	3.62	22

TABLE 10.6: Data size 64,000.

# of SPU	Time	Speedup	% Efficiency
1	0.108	1	100
2	0.059	1.8	90
4	0.034	3.17	79.25
8	0.022	4.9	61.25
16	0.018	6	37.5

TABLE 10.7: Data size 128,000.

# of SPU	Time	Speedup	% Efficiency
1	0.216	1	100
2	0.11	1.96	98
4	0.06	3.6	90
8	0.034	6.35	79.375
16	0.023	9.39	58.6

TABLE 10.8: Data size 256,000.

# of SPU	Time	Speedup	% Efficiency
1	0.45	1	100
2	0.23	1.95	97.5
4	0.12	3.75	93.75
8	0.065	6.9	86.25
16	0.039	11.5	72.1

TABLE 10.9: Data size 512,000.

# of SPU	Time	Speedup	% Efficiency
1	0.99	1	100
2	0.5	1.98	99
4	0.26	3.8	95
8	0.137	7.22	90.25
16	0.079	12.5	78.3

TABLE 10.10: Data size one million.

Tables 10.11–10.16 and Figure 10.11 show data obtained using Algorithm RISSSQS (Random Input Integer Sorting with Single Sampling using Quick Sort). We now see the effect of changing the range from 1024 to 6×1024.

# of SPU	Time	Speedup	% Efficiency
1	1.95	1	100
2	0.98	1.99	99.5
4	0.488	3.99	99.75
8	0.25	7.8	97.5
16	0.137	14.2	88.75

TABLE 10.11: Range 1–1024.

# of SPU	Time	Speedup	% Efficiency
1	1.405	1	100
2	0.709	1.98	99
4	0.36	3.9	97.5
8	0.189	7.4	92.5
16	0.105	13.3	83.125

TABLE 10.12: Range 1–2×1024.

# of SPU	Time	Speedup	% Efficiency
1	1.19	1	100
2	0.605	1.96	98
4	0.31	3.8	95
8	0.163	7.4	92.5
16	0.09	13.2	82.5

TABLE 10.13: Range 1–3×1024.

# of SPU	Time	Speedup	% Efficiency
1	1.09	1	100
2	0.55	1.98	99
4	0.28	3.89	97.25
8	0.15	7.2	90
16	0.085	12.82	80.125

TABLE 10.14: Range 1–4×1024.

# of SPU	Time	Speedup	% Efficiency
1	1.03	1	100
2	0.526	1.96	98
4	0.27	3.81	95.25
8	0.144	7.35	91.875
16	0.082	12.5	78.125

TABLE 10.15: Range 1–5×1024.

# of SPU	Time	Speedup	% Efficiency
1	0.99	1	100
2	0.5	1.98	99
4	0.26	3.8	95
8	0.137	7.22	90.25
16	0.079	12.5	78.3

TABLE 10.16: Range 1–6×1024.

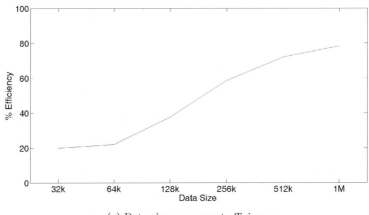

(a) Data size vs. percent efficiency.

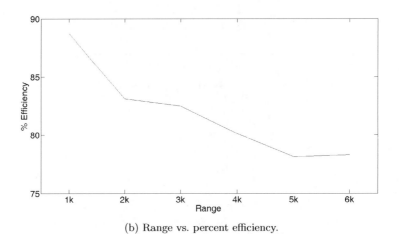

(b) Range vs. percent efficiency.

FIGURE 10.11: Algorithm RISSQS.

Table 10.17 and Figure 10.12 show data obtained using Algorithm RISSSBS (Random Input Integer Sorting with Single Sampling using Bucket Sort).

# Of SPU	Time (s)	Speedup	% Efficiency
1	0.0113	1	100
2	0.011	1.07	53.5
4	0.009	1.25	31.5
8	0.011	1.05	13.12
16	0.013	0.88	5.4

TABLE 10.17: Performance results for one million data points.

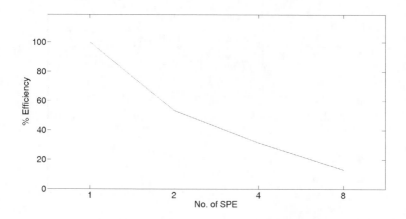

FIGURE 10.12: Speedup versus number of SPUs.

Tables 10.18–10.24 and Figure 10.13 show data obtained using Algorithm RSSQS (Randomized Single Sampling using Quick Sort).

# of SPU	Time	Speedup	% Efficiency
1	0.08096	1	100
2	0.0505	1.603	80.15
4	0.02797	2.89	72.25
8	0.01845	4.38	54.75
16	0.01466	5.52	34.60

TABLE 10.18: For data size 64,000 elements.

# of SPU	Time	Speedup	% Efficiency
1	0.31447	1	100
2	0.1713	1.83	91.5
4	0.09898	3.17	79.25
8	0.05478	5.74	71.75
16	0.03457	9.09	56.80

TABLE 10.19: For data size 256,000 elements.

# of SPU	Time	Speedup	% Efficiency
1	0.666461	1	100
2	0.368733	1.8	90
4	0.213127	3.12	78
8	0.126389	5.27	65.875
16	0.076181	8.74	54.60

TABLE 10.20: For data size of 512,000 elements.

# of SPU	Time	Speedup	% Efficiency
1	1.29906	1	100
2	0.662404	1.96	98
4	0.347023	3.74	93.5
8	0.193637	6.7	83.75
16	0.123963	10.47	64.5

TABLE 10.21: For data size of one million elements.

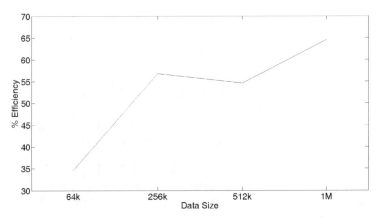

(a) Data size vs. percent efficiency.

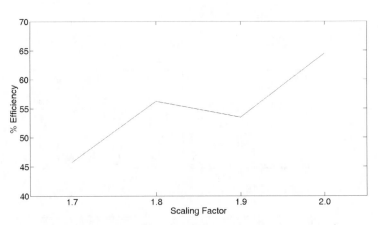

(b) Scaling factor vs. percent efficiency (data size one million).

FIGURE 10.13: Algorithm RSSQS.

For single sampling and varying the scaling factor (1.9 to 1.7) on a data size of one million elements, see Tables 10.22–10.24.

Scaling factor	# of SPU	Time	Speedup	% Efficiency
1.9	1	1.190	1	100
1.9	2	0.835	1.42	71
1.9	4	0.414	2.87	71.75
1.9	8	0.224	5.3	66.25
1.9	16	0.139	8.57	53.50

TABLE 10.22: Scaling factor 1.9.

Scaling factor	# of SPU	Time	Speedup	% Efficiency
1.8	1	1.073	1	100
1.8	2	0.748	1.43	71.5
1.8	4	0.392	2.73	68.25
1.8	8	0.206	5.21	65.13
1.8	16	0.119	9	56.25

TABLE 10.23: Scaling factor 1.8.

# of SPU	Time	Speedup	% Efficiency
1	0.958	1	100
2	0.492	1.94	97
4	0.274	3.5	87.5
8	0.195	4.9	61.25
16	0.131	7.32	45.75

TABLE 10.24: Scaling factor 1.7.

Tables 10.25–10.33 and Figure 10.14 show data obtained using Algorithm RSDSQS (Randomized Sorting with Double Sampling using Quick Sort).

# of SPU	Time	Speedup	% Efficiency
1	0.032	1	100
2	0.017	1.86	93.37
4	0.011	2.79	69.80
8	0.0088	3.62	45.30
16	0.0107	3.00	18.75

TABLE 10.25: Data size 32,000.

# of SPU	Time	Speedup	% Efficiency
1	0.115	1	100
2	0.068	1.69	84.71
4	0.037	3.10	77.54
8	0.023	4.91	61.45
16	0.015	7.71	48.18

TABLE 10.26: Data size 128,000.

# of SPU	Time	Speedup	% Efficiency
1	0.228	1	100
2	0.117	1.95	97.5
4	0.063	3.63	90.75
8	0.037	6.24	78
16	0.025	9.5	59.375

TABLE 10.27: Data size 256,000.

# of SPU	Time	Speedup	% Efficiency
1	0.928	1	100
2	0.480	1.931	96.5
4	0.265	3.5	87.5
8	0.147	6.30	78.88
16	0.0897	10.344	64.63

TABLE 10.28: Data size one million.

The effect of change in scaling factor for data size of one million is shown in Tables 10.29–10.33.

# of SPU	Time	Speedup	% Efficiency
1	0.820	1	100
2	0.716	1.146	57.27
4	0.401	2.049	51.21
8	0.21	3.91	48.85
16	0.115	7.142	44.64

TABLE 10.29: Scaling factor 1.9.

# of SPU	Time	Speedup	% Efficiency
1	0.711	1	100
2	0.629	1.13	56.49
4	0.369	1.92	48.09
8	0.190	3.74	46.74
16	0.103	6.92	43.24

TABLE 10.30: Scaling factor 1.8.

# of SPU	Time	Speedup	% Efficiency
1	0.603	1	100
2	0.310	1.94	97.05
4	0.178	3.38	84.38
8	0.158	3.81	47.65
16	0.105	5.75	35.97

TABLE 10.31: Scaling factor 1.7.

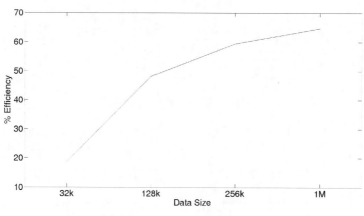

(a) Data size vs. percent efficiency.

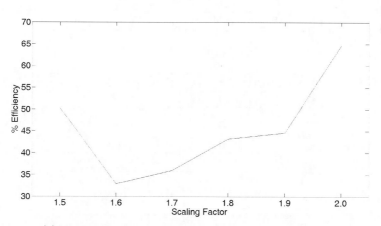

(b) Scaling factor vs. percent efficiency (data size one million).

FIGURE 10.14: Algorithm RSDSQS.

# of SPU	Time	Speedup	% Efficiency
1	0.506	1	100
2	0.270	1.872	93.61
4	0.236	2.146	53.66
8	0.146	3.480	43.51
16	0.096	5.267	32.92

TABLE 10.32: Scaling factor 1.6.

# of SPU	Time	Speedup	% Efficiency
1	0.426	1	100
2	0.225	1.90	94.90
4	0.126	3.39	84.81
8	0.073	5.88	73.44
16	0.053	8.04	50.27

TABLE 10.33: Scaling factor 1.5.

Tables 10.34–10.41 and Figure 10.15 show data obtained using Algorithm RSDSMS (Randomized Sorting with Double Sampling using Merge Sort).

# Of SPU	Time (s)	Speedup	% Efficiency
1	0.058	1	100
2	0.03	1.93	96.67
4	0.017	3.41	85.29
8	0.01	5.8	72.5
16	0.008	7.25	45.31

TABLE 10.34: Data size 32,000.

# Of SPU	Time (s)	Speedup	% Efficiency
1	0.477	1	100
2	0.242	1.97	98.55
4	0.123	3.88	96.95
8	0.066	7.23	90.34
16	0.038	12.55	78.45

TABLE 10.35: Data size 256,000.

# Of SPU	Time (s)	Speedup	% Efficiency
1	1.91	1	100
2	0.964	1.98	99.07
4	0.496	3.85	96.27
8	0.258	7.40	92.54
16	0.141	13.55	84.66

TABLE 10.36: Data size one million.

The effect of scaling factor on time and percent efficiency for data size of on million elements is shown in Tables 10.37–10.41.

# of SPU	Time	Speedup	% Efficiency
1	1.79	1	100
2	0.89	2.01	100.56
4	0.45	3.98	99.44
8	0.232	7.72	96.44
16	0.12	14.92	93.23

TABLE 10.37: Scaling factor 1.9.

# of SPU	Time	Speedup	% Efficiency
1	1.66	1	100
2	0.83	2	100
4	0.42	3.95	98.81
8	0.21	7.90	98.81
16	0.117	14.19	88.67

TABLE 10.38: Scaling factor 1.8.

# of SPU	Time	Speedup	% Efficiency
1	1.54	1	100
2	0.78	1.97	98.72
4	0.4	3.85	96.25
8	0.208	7.40	92.55
16	0.111	13.87	86.71

TABLE 10.39: Scaling factor 1.7.

# of SPU	Time	Speedup	% Efficiency
1	1.411	1	100
2	0.712	1.98	99.09
4	0.371	3.80	95.08
8	0.19	7.43	92.83
16	0.101	13.97	87.31

TABLE 10.40: Scaling factor 1.6.

# of SPU	Time	Speedup	% Efficiency
1	1.27	1	100
2	0.64	1.98	99.22
4	0.33	3.85	96.21
8	0.178	7.13	89.19
16	0.102	12.45	77.82

TABLE 10.41: Scaling factor 1.5.

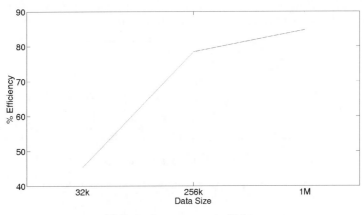

(a) Data size vs. percent efficiency.

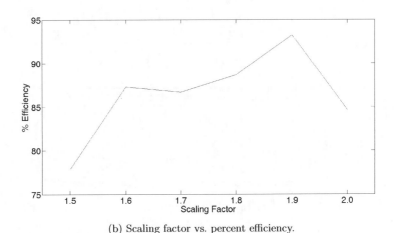

(b) Scaling factor vs. percent efficiency.

FIGURE 10.15: Algorithm RSDSMS.

10.4.6.9   Analysis

For the Cell/B.E., we can see that for Algorithm 10.1 (SQMA, Simple Quick and Merge Sort), we did not get any speedup. Both Quick sort and Merge Sort are $O(\log n)$ algorithms. It seems like whatever speedup we are able to generate from the Quick Sort in the SPEs is negligible compared to the time it takes for merging in the PPE. Hence, the overall time for the algorithm shows no speed up. For all other algorithms, except for Algorithm 10.3 (RISSSBS, Random Input Integer Sorting with Single Sampling using Bucket Sort), we can see that we have a distinct speedup. This is because the entire sorting is being done in the SPEs. The PPE just acts as a preprocessor that divides the input based on some keys while the SPEs do the sorting. This leads to a consistent speedup.

Algorithm 10.2 (RISSSQS, Random Input Integer Sorting using Quick Sort) shows the best speedup. We studied the results for different data sizes, and we observed that the speedup increases as we increase the data size. We also generated data randomly with the range varying from [1, 1024] to [1, 6*1024] to see the effect of range variation on the speedup results, and the results show that the speedup drops down when we increase the range.

Algorithm 10.3 (RISSSBS), however, does not show any speedup. The reason for this could be that bucket sort is an $O(n)$ algorithm and probably takes less time than the communication time between the PPE and SPEs. Notice that the time for Bucket Sort (Algorithm RISSSBS) is much faster than that for Quick Sort (Algorithm RISSSQS).

Algorithm 10.6 (SDSMS, Randomized Sorting with Double Sampling using Merge Sort) shows a very good speedup, but Algorithm 10.5 (RSDSQS, Randomized Sorting with Double Sampling using Quick Sort) takes less time to sort.

Algorithm 10.4 (RSSSQS, Randomized Sorting with Single Sampling using Quick Sort) and Algorithm 10.6 (SDSMS) show a speedup of about 65% for one million elements. In all algorithms (except Algorithm SQMA), we use a scaling factor of 2. We reduced the scaling factor in both Algorithm RSSSQS and Algorithm RSDSQS and studied the results. Algorithm RSSSQS does not work for scaling factors less than 1.7, and Algorithm RSDSQS goes down up to 1.5. With double sampling, we get a better sample, which gives us a better distribution of data; this explains the fact that we are able to use a smaller value of the scaling factor for Algorithm RSDSQS. Although the speedup drops when we decrease the scaling factor, it gives us better performance.

We also tested the performance of Randomized Sorting with Single Sampling and Randomized Sorting with Double Sampling on Pentium 4 processor with 1 GB of RAM and Altix 350 with 64 nodes each with 1 GB of RAM. Cell has 1 GB of RAM on the PPE and 256 KB of local memory on each SPE. The Cell outperformed the others in both the cases.

10.4.6.10   Conclusion

We designed six new algorithms to perform integer sorting on Cell/B.E. Our first algorithm is to send chunks of data to the SPE and sort the integers in parallel, then send the results back to PPE and apply a merge operation on the partially sorted list. This is not very efficient as merging in the PPE takes a lot of time and thus nullifies the speedup. In the other five algorithms, we use single and double sampling techniques in the PPE before we send the data to the SPEs. We also try different sorting methods in the SPEs and observe their effect on timing and speedup. The preprocessing takes much less time compared to the merge operation at the end. We observe better performance results for sorting using single sampling, but we prefer sorting using double sampling, as it gives us a better sample and thus a better distribution. In this case, it is highly likely that the data will not overflow the boundaries of buckets, and we will get the correct results. Cell also performs much better than Pentium or Altix.

10.5   Sorting Records

10.5.1   Record Layout

A record R is comprised of a key k and m other fields $f_1, f_2, \cdots, f_m$. For simplicity, we assume that the key and each other field occupies 32 bits. Hence, a 128-bit CBE vector may hold up to four of these 32-bit values. Although the development in this section relies heavily on storing four keys in a vector (each key occupying 32 bits), the size of the other fields isn't very significant. Let k_i be the key of record R_i, and let f_{ij}, $1 \leq j \leq m$ be this record's other fields. With our simplifying assumption of uniform-size fields, we may view the n records to be sorted as a two-dimensional array `fieldsArray[][]` with `fieldsArray[i][0]` $= k_i$ and `fieldsArray[i][j]` $= f_{ij}$, $1 \leq j \leq m$, $1 \leq i \leq n$. When this array is mapped to memory in column-major order, we get the first layout considered in [16]. We call this layout the `ByField` layout as, in this layout, the n keys come first. Next, we have the n values for the first field of the records followed by the n second fields, and so on. When the field array is mapped to memory in row-major order, we get the second layout considered in [16]. This layout, which is a more common layout for records, is called the `ByRecord` layout as, in this layout, all the fields of R_1 come first, then we have all the fields of R_2 and so on. When the sort begins with data in the `ByField` (`ByRecord`) layout, the result of the sort must also be in the `ByField` (`ByRecord`) layout.

10.5.2   High-Level Strategies for Sorting Records

There are two high-level strategies to sort multifield records. In the first, we strip the keys from the records and create n tuples of the form (k_i, i). We then sort the tuples by their first component. The second component of the tuples in the sorted sequence defines a permutation of the record indexes that corresponds to the sorted order for the initial records. The records are rearranged into this permutation by either copying from `fieldsArray` to a new space or in place using a cycle-chasing algorithm as described for a table sort in [15]. This strategy has the advantage of requiring only a linear number of record moves. So, if the size of each record is s and if the time to sort the tuples is $O(n \log n)$, the entire sort of the n records can be completed in $O(n \log n + ns)$ time. The second high-level strategy is to move all the fields of a record each time its key is moved by the sort algorithm. In this case, if the time to sort the keys alone is $O(n \log n)$, the time to sort the records is $O(ns \log n)$. For relatively small s, the first strategy outperforms the second when the records are stored in uniform access memory. However, since reordering records according to a prescribed permutation with a linear number of moves makes random accesses to memory, the second scheme outperforms the first (unless s is very large) when the records to be rearranged are in relatively slow memory such as disk or the main memory of the CBE. For this reason, we focus, in this chapter, on using the second strategy. That is, the sort algorithm moves all the fields of a record whenever its key is moved.

10.5.3   Single-SPU Record Sorting

Two SIMD vector operations used frequently in number-sorting algorithms are `findmin` and `shuffle`. The `findmin` operation compares corresponding elements in two vectors and returns a vector `min` that contains, for each compared pair, the smaller. For example, when the two vectors being compared are (4, 6, 2, 9) and (1, 8, 5, 3), the `min` is (1, 6, 2, 3). Suppose that v_i and v_j are vectors that, respectively, contain the keys for records $R_{i:i+3}$ and $R_{j:j+3}$. Listing 10.10 shows how we may move the records with the smaller keys to a block of memory beginning at `minRecords`.

```
pattern = spu_cmpgt(v_i, v_j);
minRecords = fields_select(v_i, v_j, pattern);
```
LISTING 10.10: The `findmin` operation for records.

When the `ByField` layout is used, `fields_select` takes the form given in Listing 10.11.

```
for(p = 1; p <= m; p++) {
  minRecords[p] = spu_select(fieldsArray[i][p],
                             fieldsArray[j][p], pattern);
}
```
LISTING 10.11: `fields_select` operation in `ByField` layout.

Notice that in the `ByField` layout, the elements `fieldsArray[i:i+3][p]` are contiguous and define a four-element vector. However, in the `ByRecord` layout, these elements are not contiguous and a different strategy (Listing 10.12) must be employed. The function `memcpy(dest, src, size)` moves `size` number of bytes from the local store location beginning at `src` to the local store location beginning at `dest`; `recSize` is the length of a record including its key (i.e., it is the number of bytes taken by a row of `fieldsArray`).

```
for(p = 0; p < 4; p++) {
  memcpy(minRecords + p * recSize, (pattern[p] == 1) ?
         fieldsArray[i+p] : fieldsArray[j+p], recSize);
}
```

LISTING 10.12: fields_select operation in ByRecord layout.

```
for(p = 0; p <= m; p++) {
  resultRecords[p] = spu_shuffle(fieldsArray[i][p],
                                 fieldsArray[j][p], WYAC);
}
```

LISTING 10.13: Shuffling two records in ByField layout.

```
memcpy(resultRecords, fieldsArray[i], recSize);
memcpy(resultRecords + recSize, fieldsArray[i + 2], recSize);
memcpy(resultRecords + 2 * recSize, fieldsArray[j], recSize);
memcpy(resultRecords + 3 * recSize, fieldsArray[j + 2],
       recSize);
```

LISTING 10.14: Shuffling two records in ByRecord layout.

The shuffle operation defined by spu_shuffle for the case of sorting numbers may be extended to the case of multifield records using the code of Listing 10.13 for the ByField layout and that of Listing 10.14 for the ByRecord layout. Both codes are for the case when the shuffle pattern is WYAC. Other shuffle patterns are done in a similar way. This extension of spu_shuffle to records is referred to as fields_shuffle. Other vector operations like spu_slqwbyte can be thought of as a spu_shuffle operation with a certain pattern, and one can define a similar operation fields_rotate along those lines for the records in both layouts; e.g., fields_rotate(v, 8) is equivalent to fields_shuffle(v, v, CDWX).

We observe that when the ByField layout is used, the findMin and shuffle operations perform $O(m)$ vector operations, and when the ByRecord layout is used, a constant number (4) of memcpy operations are done. However, the time for each memcpy operation increases with m.

10.5.4 Hierarchical Sorting for Records

10.5.4.1 4-way merge for records

In Section 10.5.3, we saw the adaptations needed to run the generation algorithms of [3, 11, 16] so that these may be used to generate runs for multifield records rather than for numbers. In the run-merging phase, each of the SPUs independently merges a different set of runs. Inoue et al. [16] propose a single-SPU merging algorithm that merges runs in pairs (i.e., a 2-way merge) using an adaptation of odd-even merge. It takes three `spu_cmpgt` instructions, six `spu_shuffle`, and six `spu_select` instructions to merge two SPU vectors using this scheme. The run-merging strategy of [16] may be adapted to the case of multifield records using the methods of Section 10.5.3. A higher-order merge in the run-merging phase reduces IO time and has little impact on computation time [15]. Moreover, when sorting multifield records, the IO time increases with the size of a record, and for suitably large records, this IO time will exceed the computation time and so cannot be effectively hidden using a 2-way merge and double buffering. So, for multifield records, there is merit to developing a higher-order merge. Correspondingly, two 4-way merge algorithms are proposed in [2]. One is a scalar algorithm and the other is a vectorized SIMD algorithm. Both algorithms are based on the high-level strategy shown in Figure 10.16.

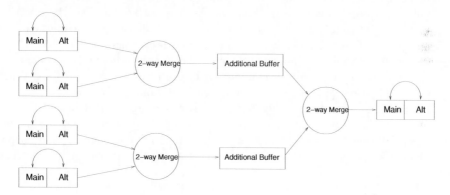

FIGURE 10.16: 4-way merge.

The 4-way merge strategy involves performing three 2-way merges in a single SPU using two buffers (main and alt) for each of the four input streams A, B, C, and D as well as two buffers for the output stream O. An additional buffer is used for the output (E and F, respectively) of each of the two left 2-way merge nodes of Figure 10.16. So, we employ a total of 12 buffers. Runs A and B are pairwise merged using the top-left 2-way merge node while runs C and D are pairwise merged using the bottom-left 2-way merge node. The former 2-way merge generates the intermediate run E, while the latter generates the intermediate run F. The intermediate runs E and F are merged by the right 2-way merge node to produce the output run O, which is written back to main memory. Run generation is done one block or buffer load at a time. Double buffering is employed for the input of A, B, C, and D from main memory and the output of O to main memory. By using double buffering and asynchronous DMA transfers to and from main memory, we are able to overlap much of the IO time with computation time.

10.5.4.2   Scalar 4-way merge

Listing 10.15 gives the pseudocode for the scalar 4-way merge algorithm. For simplicity, algorithm 4wayPipelinedMerge assumes that we have an integral number of blocks of data in each run. So, if each of the runs A, B, C, and D is (say) 10 blocks long, the output run O will be $n = 40$ blocks long. 4wayPipelinedMerge generates these output n blocks one block at a time. Even blocks are accumulated in one of the output buffers and odd blocks in the other. When an output buffer becomes full, we write the block to memory using an asynchronous DMA transfer (dmaOut) and continue output run generation using the other outbut buffer. So, other than when the first output block is being generated and the last being written to main memory, one of the output blocks is being written to main memory while the other one is being filled with records for the next block. At the end of each iteration of the outer **for** loop, we switch the roles of the two output buffers—the one that was being written to main memory becomes the buffer to place records for the next block, and the one that was being filled is written out. Of course, this switch may entail some delay as we must wait for the ongoing (if any) dmaOut to complete before we use this buffer for the records of the next block. When generating a block of the output run, we merge from the buffers bufferE and bufferF to the output buffer bufferO that is currently designated for this purpose. The number of records in a full buffer (i.e., the block size) is bSize. In case either bufferE or bufferF is empty, the generation of the output block is suspended, and we proceed to fill the empty buffer using the method

```
Algorithm 4wayPipelinedMerge(A, B, C, D, O, n)
{// Merge runs/streams A, B, C, and D to produce O
 // with n blocks of size bSize
   // bufferA is a buffer for A
   initiate a dmaIn for bufferA, bufferB,
   bufferC and bufferD;
   for (i = 0; i < n; i++) {
      for (j = 0; j < bSize; j++) { // Do block i
         if(bufferE is empty)
            mergeEF(A, B, E);
         if(bufferF is empty)
            mergeEF(C, D, F);
         move smaller record from front of bufferE
         and bufferF to bufferO
      }
      dmaOut(bufferO, O);
      switch the roles of the output buffers;
   }
}
```

LISTING 10.15: 4-way merge.

mergeEF, which merges from either input streams A and B to bufferE or from streams C and D to bufferF. The algorithm mergeEF merges for either the input streams A and B to bufferE or from E and F to bufferF. It uses double buffering on the streams A, B, C, and D and ensures that there is always an active dmaIn for these four input streams. Since the pseudocode is similar to that for 4wayPipelinedMerge, we do not provide this pseudocode here. Records are moved between buffers using the memcpy instructions when the ByRecord layout is used, and records are moved one field at a time when the layout is ByField.

10.5.4.3 SIMD 4-way merge

The SIMD version differs from the scalar version only in the way each of the three 2-way merges comprising a 4-way merge works. These 2-way merges move four records at a time from input buffers to the output buffer using an odd-even merge scheme on the keys of those records. Two sorted vectors each consisting of keys from four sorted records are merged using odd-even merge. The fields are also moved correspondingly using the fields operations introduced in the previous sections. The odd-even merge of two vectors is essentially the same process as in the case of merging numbers described in Section 10.4.5. Listing 10.16 gives the pseudocode of the adaptation for merging records.

```
Algorithm oddEvenMerge(v1, v2)
{// Merge records whose keys are in v1 and v2
  fields f1[], f2[];
  fields f3[], f4[];
  vector p; // For storing pattern
  p = spu_cmpgt(v1, v2);
  f1 = fields_select(v1, v2, pattern);
  f2 = fields_select(v2, v1, pattern);
  // Stage 2
  f3 = fields_rotate(f1, 8);
  f1 = fields_select(f3, f2, p);
  f4 = fields_select(f2, f3, p);
  f2 = fields_shuffle(f1, f4, WACY);
  f3 = fields_shuffle(f1, f4, ZXBD);
  // Stage 3
  p = spu_cmpgt(f2, f3);
  p = spu_shuffle(p, vZero, WXYA);
  f1 = fields_select(f2, f3, p);
  f4 = fields_select(f3, f2, p);
  // Output
  v1 = fields_shuffle(f1, f4 ZWAX);
  v2 = fields_shuffle(f1, f4, BYCD);
}
```

LISTING 10.16: SIMD 2-way merge of two vectors v1 and v2.

In Algorithm `oddEvenMerge`, v1 and v2 are two vectors each containing the keys of the next four records in the input buffers for the two streams being merged. It is easy to see that the next four records in the merged output are a subset of these eight records and, in fact, are the four records (of these eight) with the smallest keys. Algorithm `oddEvenMerge` determines these four smallest records and moves these to the output buffer.

10.5.5   Comparison of Record-Sorting Algorithms

We programmed several multifield record sorting algorithms using Cell/B.E. SDK 3.1. Specifically, the following algorithms were implemented and evaluated:

1. 2-way AA Sort ... this is the multifield record-sorting algorithm of Inoue et al. [16]. This uses a comb-sort variant for run generation and 2-way odd-even merge for run merging.

2. 4-way AA Sort ... this uses a comb sort variant for run generation as in [16] and our 4-way odd-even merge for run merging (Section 10.5.4.3).

3. 2-way Bitonic Sort ... this is an adaptation of the CellSort algorithm of Gedik et al. [11] to multifield records (Section 10.5.3). It uses bitonic sort for run generation and 2-way bitonic merge for run merging.

4. 4-way Bitonic Sort ... this uses bitonic sort for run generation as in [11] and our 4-way odd-even merge for run merging (Section 10.5.4.3).

5. 2-way Merge Sort ... this uses an adaptation of the SPU merge sort algorithm of Bandyopadhyay and Sahni [3] to multifield records (Section 10.5.3) for run generation and the 2-way odd-even merge of [16] for run merging.

6. 4-way Merge Sort ... this uses an adaptation of the SPU merge sort algorithm of Bandyopadhyay and Sahni [3] to multifield records (Section 10.5.3) for run generation and our 4-way odd-even merge for run merging (Section 10.5.4.3).

7. 2-way Scalar Merge Sort ... this uses an adaptation of the SPU merge sort algorithm of Bandyopadhyay and Sahni [3] to multifield records (Section 10.5.3) for run generation. Run merging is done using a 2-way scalar merging algorithm derived from the 4-way scalar merging algorithm of Section 10.5.4.2 by eliminating the bottom-left and the right 2-way merge nodes.

8. 4-way Scalar Merge Sort ... this uses an adaptation of the SPU merge sort algorithm of Bandyopadhyay and Sahni [3] to multifield records (Section 10.5.3) for run generation and our 4-way scalar merge for run merging (Section 10.5.4.2).

We experimented with the above eight multifield sorting algorithms using randomly generated input sequences. In our experiments, the number of 32-bit fields per record is varied from 5 to 15 (in addition to the key field), and the number of records varied from four thousand to one million. Also, we tried both layouts—ByField and ByRecord. For each combination of number of fields, number of records, and layout type, the time to sort 10 random sequences was obtained. The standard deviation in the observed runtimes was small, and we report only the average times.

10.5.5.1 Runtimes for ByField layout

Figures 10.17(a) through 10.18(d) give the average runtimes for our eight sorting algorithms using the ByField layout, and Figures 10.19(a)–(d) compare the average runtimes for the 2-way and 4-way versions of each of our sorting algorithms for the case when the number of records to be sorted is one million. For all our data, the 4-way version outperformed the 2-way version.

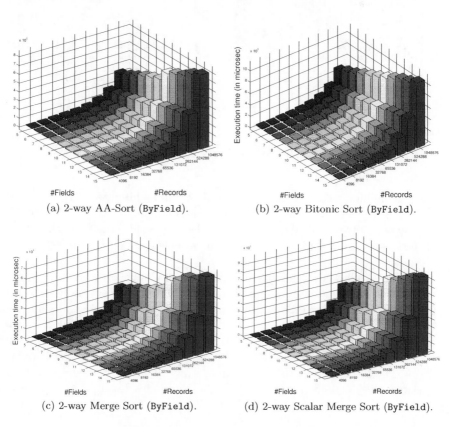

(a) 2-way AA-Sort (ByField).

(b) 2-way Bitonic Sort (ByField).

(c) 2-way Merge Sort (ByField).

(d) 2-way Scalar Merge Sort (ByField).

FIGURE 10.17: 2-way Sorts (ByField).

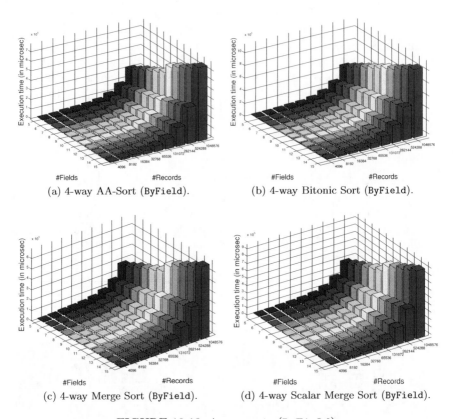

(a) 4-way AA-Sort (`ByField`). (b) 4-way Bitonic Sort (`ByField`).

(c) 4-way Merge Sort (`ByField`). (d) 4-way Scalar Merge Sort (`ByField`).

FIGURE 10.18: 4-way sorts (`ByField`).

For one million records with five 32-bit fields (in addition to a 32-bit key), the 4-way versions of AA Sort, Bitonic Sort, Merge Sort, and Scalar Merge Sort, respectively, took 5%, 4%, 7%, and 4% less time than taken by their 2-way counterparts, and these percentages for 15 fields were 9%, 6%, 9%, and 6% respectively.

Figure 10.20 shows the runtimes for the four 4-way sort algorithms for one million records. As can be seen, 4-way Bitonic Sort is the slowest, followed by 4-way Scalar Merge Sort, followed by 4-way AA Sort; 4-way Merge Sort was the fastest. In fact, across all our data sets, 4-way Bitonic Sort took between 17% and 23% more time than taken by 4-way Scalar Merge Sort, which in turn took between 18% and 19% more time than taken by 4-way AA Sort. The fastest 4-way sort algorithm, 4-way Merge Sort took, respectively, between 40% and 35%, 26% and 25%, 13% and 10% less time than taken by 4-way Bitonic Sort, 4-way Scalar Merge Sort, and 4-way AA Sort.

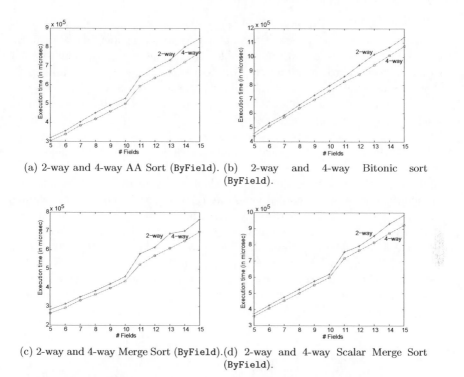

(a) 2-way and 4-way AA Sort (ByField). (b) 2-way and 4-way Bitonic sort (ByField).

(c) 2-way and 4-way Merge Sort (ByField).(d) 2-way and 4-way Scalar Merge Sort (ByField).

FIGURE 10.19: 2-way and 4-way sorts (ByField), one million records.

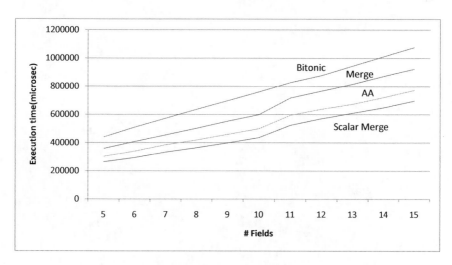

FIGURE 10.20: 4-way sorts (ByField), one million records.

10.5.5.2   Runtimes for `ByRecord` layout

Figures 10.21(a) through 10.22(d) give the average runtimes for sorting algorithms using the `ByRecord` layout and Figures 10.23(a)–(d) present the comparison of average runtimes for the 2-way and 4-way versions of each sorting algorithm when the number of records to be sorted is one million. In this layout as well, the 4-way version outperformed the 2-way version for all the data sets. For one million records with five 32-bit fields (in addition to a 32-bit key), the 4-way versions of AA Sort, Bitonic Sort, Merge Sort, and Scalar Merge Sort, respectively, took 5%, 4%, 7%, and 0.1% less time than taken by their 2-way counterparts, and these percentages for 15 fields were 9%, 6%, 9%, and 4%, respectively.

Figure 10.24 shows the runtimes for the four 4-way sort algorithms for one million records. As we can observe, 4-way Bitonic Sort is the slowest, followed by 4-way AA Sort, followed by 4-way Merge Sort; 4-way Scalar Merge Sort was the fastest. In fact, across all our data sets, 4-way Bitonic Sort took between 16% and 17% more time than taken by 4-way AA Sort, which in turn took between 24% and 35% more time than taken by 4-way Merge Sort. The fastest of them in `ByRecord` format, 4-way Scalar Merge Sort took, respectively, 88%, 86%, and between 81% and 88% less time than taken by 4-way Bitonic Sort, 4-way AA Sort, and 4-way Merge Sort.

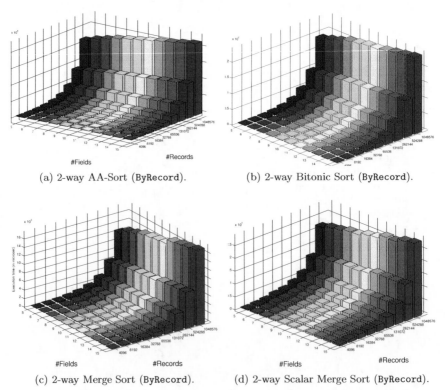

(a) 2-way AA-Sort (`ByRecord`).

(b) 2-way Bitonic Sort (`ByRecord`).

(c) 2-way Merge Sort (`ByRecord`).

(d) 2-way Scalar Merge Sort (`ByRecord`).

FIGURE 10.21: 2-way Sorts (`ByRecord`).

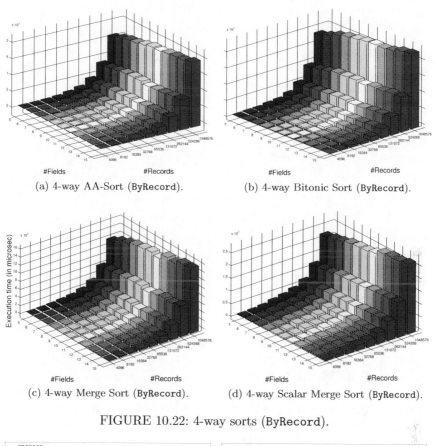

(a) 4-way AA-Sort (`ByRecord`). (b) 4-way Bitonic Sort (`ByRecord`).

(c) 4-way Merge Sort (`ByRecord`). (d) 4-way Scalar Merge Sort (`ByRecord`).

FIGURE 10.22: 4-way sorts (`ByRecord`).

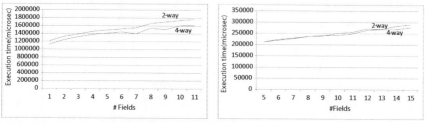

(a) 2-way and 4-way AA Sort (`ByRecord`). (b) 2-way and 4-way Bitonic sort (`ByRecord`).

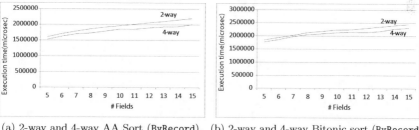

(c) 2-way and 4-way Merge Sort (`ByRecord`). (d) 2-way and 4-way Scalar Merge Sort (`ByRecord`).

FIGURE 10.23: 2-way and 4-way Sorts (`ByRecord`), one million records.

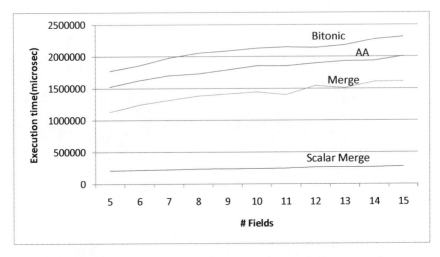

FIGURE 10.24: 4-way sorts (`ByRecord`), one million records.

10.5.5.3   Cross-layout comparison

Although in a real application, one may not be able to choose the layout format for the data to be sorted, it is worthwhile to compare the relative performance of the eight sort methods using the better layout for each. This means that we use the `ByField` layout for AA Sort and Merge Sort and the `ByRecord` layout for Merge Sort and Scalar Merge Sort. Figure 10.25 gives the runtimes for the 4-way versions using these formats for the case of one million records. Although Figure 10.25 is only for the case of one million records,

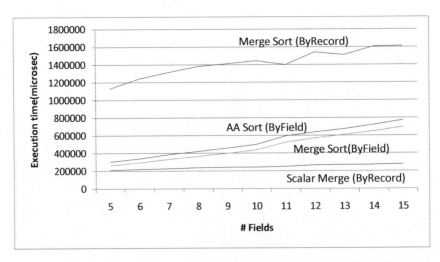

FIGURE 10.25: 4-way sorts using the best algorithms for different layouts.

4-way Scalar Merge Sort was the fastest for all of our data sets. For five 32-bit fields (in addition to the key field), 4-way Scalar Merge Sort (`ByRecord`) ran 81% faster than 4-way Merge Sort (`ByRecord`), 30% faster than 4-way AA Sort (`ByField`), and 20% faster than 4-way Merge Sort (`ByField`). When the number of fields was 15, these percentages were 88%, 64%, and 60%, respectively.

Experiments indicate that the 4-way Scalar Merge Sort is the fastest method (from among those tested) to sort multifield records on the CBE.

Acknowledgements

This research was supported in part by the National Science Foundation and National Institutes of Health under grants CNS-0963812 and CNS-1115184, CCF-0829916, and NIH R01-LM010101.

Bibliography

[1] S. Bandyopadhyay and S. Sahni. GRS—GPU radix sort for large multifield records, *International Conference on High Performance Computing* (HiPC), 2010.

[2] S. Bandyopadhyay and S. Sahni. Sorting large records on a Cell Broadband Engine, *IEEE International Symposium on Computers and Communications* (ISCC), 2010.

[3] S. Bandyopadhyay and S. Sahni. Sorting on a Cell Broadband Engine SPU, *IEEE International Symposium on Computers and Communications* (ISCC), 2009.

[4] K.E. Batcher. Sorting networks and their applications, *Proc. AFIPS Spring Joint Computing Conference*, vol. 32, 307–314, 1968.

[5] G.E. Blelloch. *Vector Models for Data-Parallel Computing.* Cambridge, MA: MIT Press, 1990.

[6] R. Box and S. Lacey. A fast, easy sort. *Byte*, vol. 4, pages 315–318. 1991.

[7] D. Cederman and P. Tsigas. GPU-quicksort: A practical quicksort algorithm for graphics processors, *ACM Journal of Experimental Algorithmics* (JEA), 14, 4, 2009.

[8] W. Dobosiewicz. An efficient variation of bubble sort. *Information Processing Letters*, vol. 11, pages 5–6, 1980.

[9] A. C. Chow, G. C. Fossum, and D. A. Brokenshire. A programming example: Large FFT on the Cell Broadband Engine. IBM White paper, 2005.

[10] A. Drozdek. Worst case for Comb Sort. *Informatyka Teoretyczna i Stosowana*, vol. 9, pages 23–27, 2005.

[11] B. Gedik, R. Bordawekar and P. Yu. CellSort: High performance sorting on the Cell processor. *VLDB*, pages 1286–1297, 2007.

[12] N. Govindaraju, J. Gray, R. Kumar and D. Manocha. GPUTeraSort: High performance graphics coprocessor sorting for large database management, *ACM SIGMOD International Conference on Management of Data*, 2006.

[13] A. Greb and G. Zachmann. GPU-ABiSort: Optimal parallel sorting on stream architectures, *IEEE International Parallel and Distributed Processing Symposium* (IPDPS), 2006.

[14] H. Hofstee. Power efficient processor architecture and the Cell Processor, In *Proceedings of 11th International Symposium on High Performance Computer Architecture*, 2005.

[15] E. Horowitz, S. Sahni and D. Mehta. *Fundamentals of data structures in C++*, Second Edition, San Jose, CA: Silicon Press, 2007.

[16] H. Inoue, T. Moriyama, H. Komatsu and T. Nakatani. AA-sort: A new parallel algorithm for multi-core SIMD processors, *16th International Conference on Parallel Architecture and Compilation (PACT)*, 2007.

[17] D. Knuth. *The Art of Computer Programming: Sorting and Searching* Volume 3, Second Edition, Reading, MA: Addition Wesley, 1998

[18] S. Lai and S. Sahni. Anomalies in parallel branch-and-bound algorithms, *Comm. of the ACM*, vol. 6, pages 594–602, 1984.

[19] S. Le Grand. Broad-phase collision detection with CUDA, In *GPU Gems 3*, Reading, MA: Addison-Wesley Professional, 2007.

[20] N. Leischner, V. Osipov and P. Sanders. GPU sample sort, *IEEE International Parallel and Distributed Processing Symposium (IPDPS)*, 2010.

[21] P. Lemke. The performance of randomized Shellsort-like network sorting algorithms *SCAMP working paper P18/94*, Institute for Defense Analysis, Princeton, NJ, 1994.

[22] E. Lindholm, J. Nickolls, S. Oberman and J. Montrym. NVIDIA Tesla: A unified graphics and computing architecture, *IEEE Micro*, 28, 3955, 2008.

[23] D Merrill and A. Grimshaw. Revisiting Sorting for GPGPU Stream Architectures, University of Virginia, Department of Computer Science, Technical Report CS2010-03, 2010.

[24] R. Sedgewick. Analysis of Shellsort and related algorithms, *4th European Symposium on Algorithms*, 1996.

[25] S. Sahni. Scheduling master-slave multiprocessor systems, *IEEE Trans. on Computers*, 45, 10, 1195–1199, 1996.

[26] N. Satish, M. Harris and M. Garland. Designing efficient sorting algorithms for manycore GPUs, *IEEE International Parallel and Distributed Processing Symposium* (IPDPS), 2009.

[27] S. Sengupta, M. Harris, Y. Zhang and J.D. Owens. Scan primitives for GPU computing, *Graphics Hardware 2007*, 97–106, 2007.

[28] E. Sintorn, and U. Assarsson. Fast parallel GPU-sorting using a hybrid algorithm, *Journal of Parallel and Distributed Computing*, 10, 1381–1388, 2008.

[29] D. Sharma, V. Thapar, R. Ammar, S. Rajasekaran and M. Ahmed. Efficient sorting algorithms for the Cell Broadband Engine, *IEEE International Symposium on Computers and Communications (ISCC)*, 2008.

[30] V. Volkov and J.W. Demmel. Benchmarking GPUs to Tune Dense Linear Algebra, *ACM/IEEE Conference on Supercomputing*, 2008.

[31] X. Ye, D. Fan, W. Lin, N. Yuan and P. Ienne. High performance comparison-based sorting algorithm on many-core GPUs, *IEEE International Parallel and Distributed Processing Symposium (IPDPS)*, 2010.

[32] Y. Won and S. Sahni. Hypercube-to-host sorting, *Journal of Supercomputing*, vol. 3, pages 41–61, 1989.

[33] CUDPP: CUDA Data-Parallel Primitives Library, http://www.gpgpu.org/developer/cudpp/, 2009.

[34] NVIDIA CUDA Programming Guide, *NVIDIA Corporation*, Version 3.0, Feb. 2010.

[35] Cell Broadband Engine, http://www-01.ibm.com/chips/techlib/techlib.nsf/products/Cell_Broadband_Engine, 2013.

Chapter 11

GPU Matrix Multiplication

Junjie Li

University of Florida, Gainesville

Sanjay Ranka

University of Florida, Gainesville

Sartaj Sahni

University of Florida, Gainesville

11.1   Introduction

Graphics Processing Units (GPUs) were developed originally to meet the computational needs of algorithms for rendering computer graphics. The rapid and enormous growth in sophistication of graphics applications such as computer games has resulted in the availability of GPUs that have hundreds of processors and peak performance near a teraflop and that sell for hundreds up to a few thousand dollars. Although GPUs are optimized for graphics calculations, their low cost per gigaflop has motivated significant research into their efficient use for non-graphics applications. The effort being expended in this direction has long-lasting potential because the widespread use of GPUs in the vibrant computer games industry almost ensures the longevity of GPUs. So, unlike traditional multimillion-dollar supercomputers whose development cost had to be borne entirely by a relatively small supercomputing community, GPUs are backed by a very large gaming industry. This makes it more likely that GPU architectures will remain economically viable and will continue to evolve.

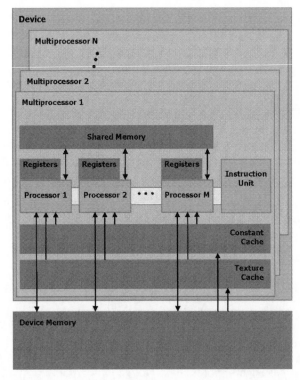

FIGURE 11.1: NVIDIA's GPU hardware model [1].

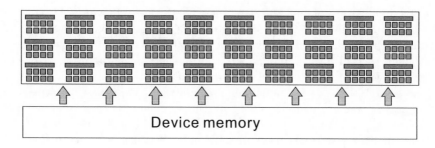

FIGURE 11.2: NVIDIA's Tesla C1060 GPU.

Although the cost of a GPU measured in dollars per peak gigaflop is very low, obtaining performance near the peak requires very careful programming of the application code. This programming is complicated by the availability of several different memories (e.g., device memory, shared memory, constant cache, texture cache, and registers), each with different latency, and the partitioning of the available scalar processors or cores into groups called streaming multiprocessors (Figure 11.1). In this chapter, we explore the intricacies of programming a GPU to obtain high performance for the multiplication of two single-precision square matrices. We focus our development on NVIDIA's Tesla series of GPUs of which the C1060 is an example (Figure 11.2). Our example programs are developed using CUDA.

11.2   GPU Architecture

NVIDIA's Tesla C1060 GPU, Figure 11.2, is an example of NVIDIA's general-purpose parallel computing architecture CUDA (Compute Unified Driver Architecture). Figure 11.2 is a simplified version of Figure 11.1 with $N = 30$ and $M = 8$. The C1060 is comprised of 30 streaming multiprocessors (SMs), and each SM is comprised of eight scalar processors (SPs), 16 KB of on-chip shared memory, and 16,384 32-bit registers. Each SP has its own integer and single-precision floating-point units. Each SM has one double-precision floating-point unit and two single-precision transcendental function (special function, SF) units that are shared by the eight SPs in the SM. The 240 SPs of a Tesla C1060 share 4 GB of off-chip memory referred to as *device* or *global* memory [1]. A C1060 has a peak performance of 933 GFlops of single-precision floating-point operations and 78 GFlops of double-precision operations. The peak of 933 GFlops is for the case when Multiply-Add (MADD) instructions are dual issued with SF instructions. In the absence of SF instructions, the peak is 622 GFlops (MADDs only) [2]. The C1060 consumes 188 W of power.

The architecture of the NVIDIA Tesla C2050 (also known as Fermi) corresponds to Figure 11.1 with $N = 14$ and $M = 32$. So, a C2050 has 14 SMs, and each SM has 32 SPs, giving the C2050 a total of 448 SPs or cores. Although each SP of a C2050 has its own integer, single- and double-precision units, the 32 SPs of an SM share four single-precision transcendental function units. An SM has 64 KB of on-chip memory that can be *"configured as 48 KB of shared memory with 16 KB of L1 cache (default setting) or as 16 KB of shared memory with 48 KB of L1 cache"* [1]. Additionally, there are 32 thousand 32-bit registers per SM and 3 GB of off-chip device/global memory that is shared by all 14 SMs. The peak performance of a C2050 is 1,288 GFlops (or 1.288 TFlops) of single-precision operations and 515 GFlops of double-precision operations, and the power consumption is 238 W [3]. Once again, the peak of 1,288 GFlops requires that MADDs and SF instructions be dual issued. When there are MADDs alone, the peak single-precision rate is 1.03 GFlops. Notice that the ratio of power consumption to peak single-precision GFlop rate is 0.2 W/GFlop for the C1060 and 0.18 W/GFlop for the C2050. The corresponding ratio for double-precision operations is 2.4 W/GFlops for the C1060 and 0.46 W/GFlop for the C2050. In NVIDIA parlance, the C1060 has compute capability 1.3, while the compute capability of the C2050 is 2.0.

A Tesla GPU is packaged as a double-wide PCIe card (Figure 11.3), and using an appropriate motherboard and a sufficiently large power supply, one can install up to four GPUs on the same motherboard. In this chapter, we focus on single-GPU computation.

FIGURE 11.3: NVIDIA's Tesla PCIex16 Card (www.nvidia.com).

11.3 Programming Model

At a high-level, a GPU uses the master-slave programming model [4] in which the GPU operates as a slave under the control of a master or host processor. In our experimental setup, the master or host is a 2.8-GHz Xeon quadcore processor, and the GPU is the NVIDIA Tesla C1060. Programming in the master-slave model requires us to write a program that runs on the master processor (in our case the Xeon). This master program sends data to the slave(s) (in our case a single C1060 GPU), invokes a kernel or function that runs on the slave(s) and processes these sent data, and finally receives the results back from the slave. This process of sending data to the slave, executing a slave kernel, and receiving the computed results may be repeated several times by the master program. In CUDA, the host/master and GPU/slave codes may be written in C. CUDA provides extensions to C to allow for data transfer to/from device memory and for kernel/slave code to access registers, shared memory, and device memory.

At another level, GPUs use the SIMT (single instruction multiple thread) programming model in which the GPU accomplishes a computational task using thousands of lightweight threads. The threads are grouped into blocks, and the blocks are organized as a grid. While a block of threads may be one-, two-, or three-dimensional, the grid of blocks may only be one- or two-dimensional. Kernel invocation requires the specification of the block and grid dimensions along with any parameters the kernel may have. This is illustrated below for a matrix multiply kernel `MatrixMultiply` that has the parameters a, b, c, and n, where a, b, and c are pointers to the start of the row-major representation of $n \times n$ matrices, and the kernel computes $c = a * b$.

```
MatrixMultiply <<<GridDimensions , BlockDimensions >>>(a,b,c,n)
```

A GPU has a block scheduler that dynamically assigns thread blocks to SMs. Since all the threads of a thread block are assigned to the same SM, the threads of a block may communicate with one another via the shared memory of an SM. Further, the resources needed by a block of threads (e.g., registers and shared memory) should be sufficiently small that a block can be run on an SM. The block scheduler assigns more than one block to run concurrently on an SM when the combined resources needed by the assigned blocks do not exceed the resources available to an SM. However, since CUDA provides no mechanism to specify a subset of blocks that are to be coscheduled on an SM, threads of different blocks can communicate only via the device memory.

Once a block is assigned to an SM, its threads are scheduled to execute on the SM's SPs by the SM's warp scheduler. The warp scheduler divides the threads of the blocks assigned to an SM into warps of 32 consecutively indexed threads from the same block. Multidimensional thread indexes are serialized in row-major order for partitioning into warps. So, a block that has 128 threads is partitioned into four warps. Every thread currently assigned to an SM has its own instruction counter and set of registers. The warp scheduler selects a warp of ready threads for execution. If the instruction counters for all threads in the selected warp are the same, all 32 threads execute in one step. On a GPU with compute capability 2.0, each SM has 32 cores, so all 32 threads can perform their common instruction in parallel, provided, of course, this common instruction is an integer or floating-point operation. On a GPU with compute capability 1.3, each SM has eight SPs, so a warp can execute the common instruction for only eight threads in parallel. Hence, when the compute capability is 1.3, the GPU takes four rounds of parallel execution to execute the common instruction for all 32 threads of a warp. When the instruction counters of the threads of a warp are not the same, the GPU executes the different instructions serially. Note that the instruction counters may become different as the result of *"data dependent conditional branches"* in a kernel [1].

An SM's warp scheduler is able to hide much of the 400 to 600 cycle latency of a device-memory access by executing warps that are ready to do arithmetic while other warps wait for device-memory accesses to complete. So, the performance of code that makes many accesses to device memory can often be improved by optimizing it to increase the number of warps scheduled on an SM (i.e., increase the multiprocessor occupancy, see Section 11.4). This optimization could involve increasing the number of threads per block and/or reducing the shared memory and register utilization of a block to enable the scheduling of a larger number of blocks on an SM.

11.4   Occupancy

The *occupancy* of a multiprocessor (SM) is defined to be the ratio of the number of warps coscheduled on a multiprocessor and the GPU's limit, maxWarpsPerSM, on the number of warps that may be coscheduled on a multiprocessor. So,

$$\text{occupancy} \;=\; \frac{\text{number of warps coscheduled per SM}}{\texttt{maxWarpsPerSM}} \qquad (11.1)$$

In practice, the number of warps coscheduled on a multiprocessor is determined by several factors. These factors may be broadly classified as those that are characteristics of the GPU and those that specify resource requirements of the application kernel. Therefore, for a given GPU, the occupancy of a multiprocessor varies from one kernel to the next, and even when the GPU and kernel are fixed, the occupancy varies with the block size (i.e., number of threads per block) with which the kernel is invoked. Further, the same kernel and block size may result in different occupancy values on (say) a Tesla C1060 and a Tesla C2050 GPU. Since *applications with higher occupancy are better able to hide the latency of the GPU global/device memory,* increasing the occupancy of an application is one of the strategies used to improve performance.

In describing the factors that contribute to the occupancy of a multiprocessor, we use some of the variable names used in the CUDA occupancy calculator [4]. The GPU factors depend on the compute capability of the GPU.

1. GPU Factors

   (a) The maximum number, `maxThreadsPerWarp`, of threads in a warp.

   (b) The maximum number of blocks, `maxBlocksPerSM`, that may be coscheduled on a multiprocessor.

   (c) Warp allocation granularity, `warpAllocationUnit`. For purposes of allocating registers to a block, the number of warps in a block is rounded up (if necessary) to make it a multiple of the warp-allocation granularity. So, for example, when `warpAllocationUnit` $= 2$ and the number of warps in a block is 7, the number of registers assigned to the block is that for 8 warps.

   (d) Registers are allocated to a warp (compute capability 2.0) or block (compute capability 1.3) in units of size `regUnit`. So, for example, on a GPU with compute capability 2.0 and `regUnit` $= 64$, a warp that requires 96 registers would be allocated $\lceil 96/64 \rceil * 64 = 2 * 64 = 128$ registers. A block of 6 such warps would be assigned $6 * 128 = 768$ registers. The same block on a GPU with compute capability 1.3 and `regUnit` $= 512$ would be allocated $\lceil 6*96/512 \rceil * 512 = 1024$ registers.

   (e) Shared-memory granularity, `sharedMemUnit`. Shared memory is allocated in units whose size equals `sharedMemUnit`. So, when `sharedMemUnit` = 128, a block that requires 900 bytes of shared memory is allocated $\lceil 900/128 \rceil * 128 = 8 * 128 = 1024$ bytes of shared memory.

   (f) Number of registers, `numReg`, available per multiprocessor.

   (g) Shared memory, `sharedMemSize`, available per multiprocessor.

In addition to the explicit limit `maxBlocksPerSM`, the number of blocks that may be coscheduled on a multiprocessor is constrained by the limit, `maxWarpsPerSM`, on the number of warps as well as by the number of registers and the size of the shared memory. Table 11.1 gives the values of the GPU specific factors for GPUs with compute capabilities 1.3 (e.g., Tesla C1060) and 2.0 (e.g., Tesla C2050). From these values, we see that at most eight blocks may be coscheduled on a multiprocessor of a GPU with compute capability 1.3 or 2.0. Further, if a block is comprised of ten warps, the number of blocks that may be coscheduled on a multiprocessor is reduced from 8 to $\lfloor 32/10 \rfloor = 3$ when the compute capability is 1.3 and to $\lfloor 48/10 \rfloor = 4$ when the compute capability is 2.0.

variable↓/compute capability→	1.3	2.0
maxBlocksPerSM	8	8
maxThreadsPerWarp	32	32
maxWarpsPerSM	32	48
numReg	16,384	32,768
regUnit	512	64
sharedMemUnit	512	128
sharedMemSize	16,384	49,152
warpAllocationUnit	2	1

TABLE 11.1: GPU constraints for compute capabilities 1.3 and 2.0 [4].

2. Kernel Factors

   (a) Number, `myThreadsPerBlock`, of threads in a block.

   (b) Number, `myRegPerThread`, of registers required per thread.

   (c) Amount, `mySharedMemPerBlock`, of shared memory required per block.

To determine the occupancy of a kernel, we need to determine the number WPM of warps that will be coscheduled on a multiprocessor (Equation 11.1). This number is the product of the number BPM of blocks that get coscheduled on a multiprocessor and the number WPB of warps per block. From the definition of the GPU and kernel factors that contribute to occupancy, we see that

$$WPB \;=\; \left\lceil \frac{\texttt{myThreadsPerBlock}}{\texttt{maxThreadsPerWarp}} \right\rceil \tag{11.2}$$

$$BPM \;=\; \min \left\{ \texttt{maxBlocksPerSM}, \right.$$

$$\left\lfloor \frac{\texttt{maxWarpsPerSM}}{\texttt{WPB}} \right\rfloor,$$

$$\left\lfloor \frac{\texttt{numReg}}{\texttt{myRegPerBlock}} \right\rfloor,$$

$$\left. \left\lfloor \frac{\texttt{sharedMemSize}}{\texttt{myMemPerBlock}} \right\rfloor \right\} \tag{11.3}$$

where

$$
\mathtt{myRegPerBlock} = \begin{cases}
\left\lceil \dfrac{\mathtt{GWPB}*\mathtt{myRegPerThread}*\mathtt{maxThreadsPerWarp}}{\mathtt{regUnit}} \right\rceil \\
*\mathtt{regUnit} \\
\text{(capability 1.3)} \\[1em]
\left\lceil \dfrac{\mathtt{myRegPerThread}*\mathtt{maxThreadsPerWarp}}{\mathtt{regUnit}} \right\rceil \\
*\mathtt{regUnit} * \mathtt{WPB} \\
\text{(capability 2.0)}
\end{cases} \qquad (11.4)
$$

$$
\mathtt{GWPB} = \lceil \mathtt{WPB}/\mathtt{warpAllocationUnit} \rceil * \mathtt{warpAllocationUnit} \qquad (11.5)
$$

$$
\mathtt{myMemPerBlock} = \left\lceil \dfrac{\mathtt{mySharedMemPerBlock}}{\mathtt{sharedMemUnit}} \right\rceil * \mathtt{sharedMemUnit} \qquad (11.6)
$$

As an example, consider a kernel that requires 20 registers per thread and 3000 bytes of shared memory. Assume that the kernel is invoked on a C1060 (compute capability 1.3) with a block size of 128. So, $\mathtt{myThreadsPerBlock}$ = 128, $\mathtt{myRegPerThread}$ = 20, and $\mathtt{mySharedMemPerBlock}$ = 3000. From Equations 11.2–11.6, we obtain

$$
\begin{aligned}
\mathtt{WPB} &= \lceil 128/32 \rceil = 4 \\
\mathtt{GWPB} &= \lceil 4/2 \rceil * 2 = 4 \\
\mathtt{myRegPerBlock} &= \lceil 4*20*32/512 \rceil * 512 = 2560 \\
\mathtt{myMemPerBlock} &= \lceil 3000/512 \rceil * 512 = 3072 \\
\mathtt{BPM} &= \min\{8, \lfloor 32/4 \rfloor, \lfloor 16384/2560 \rfloor, \lfloor 16384/3072 \rfloor\} \\
&= \min\{8, 8, 6, 5\} = 5
\end{aligned}
$$

For our example kernel, we see that the number of blocks that may be co-scheduled on a multiprocessor is limited by the available shared memory. From Equation 11.1, we get

$$
\text{occupancy} = \frac{\mathtt{BPM} * \mathtt{WPB}}{\mathtt{maxWarpsPerSM}} = 5*4/32 = 0.63
$$

Since, $\mathtt{BPM} \leq 8$ for the C1060 (Equation 11.3) and $\mathtt{WPB} = 4$ for our example kernel, optimizing our example kernel to use less shared memory and fewer registers could raise occupancy to $8*4/32 = 1.0$. So, if we reduced the shared-memory requirement of a block to 1000 bytes while keeping register usage the same,

$$
BPM = \min\{8, 8, 6, \lfloor 16384/1024 \rfloor\} = \min\{8, 8, 6, 16\} = 6
$$

and occupancy $= 6 * 4/32 = 0.75$. Now, the occupancy is limited by the number of registers. Note that if we reduce the number of threads per block to 64, $WPB = 2$ and occupancy $\leq 8 * 2/32 = 0.5$.

11.5   Single-Core Matrix Multiply

Listing 11.1 gives the classical textbook algorithm to multiply two $n \times n$ matrices A and B stored in row-major order in the one-dimensional arrays a and b. The result matrix $C = A * B$ is returned, also in row-major order, in the one-dimensional array c. This classic algorithm is referred to as the ijk algorithm because the three nested **for** loops are in this order. It is well known that computing the product of two matrices in ikj order as in Listing 11.2 (see [5], for example) reduces cache misses and so results in better performance when n is large.

Table 11.2 gives the runtimes for the ijk and ikj versions on our Xeon quadcore processor. As expected, the ijk version is faster for small n because of its smaller overhead, but for large n, the cache effectiveness of the ikj version enables it to outperform the ijk version. For $n = 4096$, for example, the ijk

```
void SingleCoreIJK (float *a, float *b, float *c, int n)
{// Single-core algorithm to multiply two
  // n x n row-major matrices a and b.
   for (int i = 0; i < n; i++)
      for (int j = 0; j < n; j++)
      {
         float temp = 0;
         for (int k = 0; k < n; k++)
            temp += a[i*n+k]*b[k*n+j];
         c[i*n+j] = temp;
      }
}
```

LISTING 11.1: Single-core matrix multiply.

```
void SingleCoreIKJ (float *a, float *b, float *c, int n)
{// Single-core cache-aware algorithm to multiply two
  // n x n row-major matrices a and b
   for (int i = 0; i < n; i++)
   {
      for (int j = 0; j < n; j++)
         c[i*n+j] = 0;

      for (int k = 0; k < n; k++)
         for (int j = 0; j < n; j++)
            c[i*n+j] += a[i*n+k]*b[k*n+j];
   }
}
```

LISTING 11.2: Single-core cache-aware matrix multiply.

version takes almost seven times as much time as is taken by the ikj version! In computing the ratio ijk/ikj, we have used the measured runtimes with three decimal digits of precision. This explains the slight difference in the ratios reported in Table 11.2 and what you would get using the two-decimal-digit times given in this figure. In the remainder of this paper also, we computed derived results such as speedup using data with three-digit precision even though reported data have two-digit precision.

Method	256	512	1024	2048	4096
			n		
ijk	0.11	0.93	10.41	449.90	4026.17
ikj	0.14	1.12	8.98	72.69	581.28
ijk/ikj	0.80	0.83	1.16	6.19	6.93

TABLE 11.2: Runtimes (seconds) for the ijk and ikj versions of matrix multiply.

11.6 Multicore Matrix Multiply

A multicore version of the cache-efficient single-core algorithm `SingleCoreIKJ` is obtained quite easily using OpenMP [6]. When t threads are used, each thread computes an $(n/t) \times n$ submatrix of the result matrix C using the algorithm `multiply` (Listing 11.3). The similarity between `multiply` and `SingleCoreIKJ` is apparent.

Table 11.3 gives the time taken by `MultiCore` to multiply $n \times n$ matrices for various values of n using 1, 2, 4, and 8 threads on our Xeon quadcore processor. We note that, as expected, when $t = 1$, the runtimes for `MultiCore` and `SingleCoreIKJ` are virtually the same. As we did for Table 11.2, in computing the speedups reported in Table 11.3, we have used the measured runtimes with three decimal digits of precision. This explains the slight difference in the reported speedups and what you would get using the two-decimal-digit times given in Table 11.3. The speedup for $t = 2$, 4, and 8 relative to the case $t = 1$, is given in the last three lines of Table 11.3. The speedup relative to $t = 1$ is virtually the same as that relative to `SingleCoreIKJ`. It is interesting to note that the simple parallelization done to arrive at `MultiCore` achieves a speedup of 2.00 when $t = 2$ and a speedup very close to 4.00 when $t = 4$. Further increase in the number of threads does not result in a larger speedup as we have only four cores to execute these threads on. Actually, when we go from $t = 4$ to $t = 8$, there is a very small drop in performance for most values of n because of the overhead of scheduling eight threads rather than four.

```
void MultiCore (float *a, float *b, float *c, int n, int t)
{// Multicore algorithm to multiply two
 // n x n row-major matrices a and b.
 // t is the number of threads.
    #pragma omp parallel shared (a, b, c, n, t)
    {
        int tid = omp_get_thread_num (); // Thread ID
        multiply(a, b, c, n, t, tid);
    }
}


void multiply(float *a, float *b, float *c, int n, int t,
              int tid)
{// Compute an (n/t) x n submatrix of c; t is the number of
 // threads tid, 0 <= tid < t is the thread ID
    int height = n/t;
    int offset = height*n*tid;

    a += offset;
    c += offset;

    for (int i = 0; i < height; i++)
    {
        for (int j = 0; j < n; j++)
            c[i*n+j] = 0;

        for (int k = 0; k < n; k++)
            for (int j = 0; j < n; j++)
                c[i*n+j] += a[i*n+k]*b[k*n+j];
    }
}
```

LISTING 11.3: Multicore matrix multiply using OpenMP.

	n				
t	256	512	1024	2048	4096
1	0.14	1.12	8.98	72.67	581.24
2	0.07	0.56	4.50	36.39	291.14
4	0.04	0.28	2.26	18.21	146.37
8	0.04	0.28	2.28	18.42	146.98
$T1/T2$	2.00	2.00	2.00	2.00	2.00
$T1/T4$	4.00	3.97	3.98	3.99	3.97
$T1/T8$	3.89	4.00	3.94	3.95	3.95

t is the number of threads

Ti is the time using i threads, $i \in \{1, 2, 4, 8\}$

TABLE 11.3: Runtimes (seconds) for MultiCore.

11.7   GPU Matrix Multiply

Although `SingleCoreIKJ` (Listing 11.2) is faster than `SingleCoreIJK` when n is large, we do not expect a massively threaded adaptation of `SingleCoreIKJ` to outperform a similarly threaded version of `SingleCoreIJK` because `SingleCoreIKJ` makes $O(n^3)$ accesses to c while the `SingleCoreIJK` makes $O(n^2)$ accesses, and c resides in device memory. (Note that both versions make $O(n^3)$ accesses to a and b.) So, we focus on obtaining a GPU kernel based on `SingleCoreIJK`.

We begin by tiling an $n \times n$ matrix using $p \times q$ tiles as in Figure 11.4. For simplicity, we assume that p and q divide n so that an integral number of tiles is needed to cover the matrix. We may index the tiles using a two-dimensional index (u, v), $0 \le u < n/q$, $0 \le v < n/p$. We adopt the convention that the top-left tile is indexed $(0,0)$, the first coordinate of an index increases left to right, and the second coordinate increases top to bottom. Our GPU code will use a block of threads to compute a tile (more accurately, the submatrix corresponding to a tile) of the result matrix C. To compute the entire matrix C, we will use an $n/p \times n/q$ grid of thread blocks with thread block (u, v) computing tile (u, v) of C. In CUDA, a thread may determine the coordinates of the block (u, v) that it is part of using the variables `blockIdx.x` and `blockIdx.y`. So, $(u, v) = (\texttt{blockIdx.x}, \texttt{blockIdx.y})$.

(0, 0)	(1, 0)	(2, 0)	(3, 0)
(0, 1)	(1, 1)	(2, 1)	(3, 1)
(0, 2)	(1, 2)	(2, 2)	(3, 2)
(0, 3)	(1, 3)	(2, 3)	(3, 3)
(0, 4)	(1, 4)	(2, 4)	(3, 4)
(0, 5)	(1, 5)	(2, 5)	(3, 5)
(0, 6)	(1, 6)	(2, 6)	(3, 6)
(0, 7)	(1, 7)	(2, 7)	(3, 7)

FIGURE 11.4: Tiling a 16×16 matrix using 32 2×4 tiles.

Several different GPU adaptations of `SingleCoreIJK` are possible. These differ in their usage of registers, shared memory, number of C elements computed per thread, and so on. Different adaptations result in different performance. In the following, we examine some of the possible adaptations.

11.7.1   A Thread Computes a 1×1 Submatrix of C

11.7.1.1   Kernel code

In our first adaptation of `SingleCoreIJK`, each thread computes exactly one element of a tile. So, the number of threads in a block equals the number of elements $p * q$ in a tile. Threads of a block are indexed using the same convention as used to index blocks. That is, thread $(0,0)$ computes the top-left element of a tile, the first coordinate increases left to right, and the second coordinate increases top to bottom. With this convention, the index of a thread within a block is $(\texttt{threadIdx.x}, \texttt{threadIdx.y})$. Notice that $\texttt{threadIdx.x}$ is in the range $[0, q)$, and $\texttt{threadIdx.y}$ is in the range $[0, p)$. A CUDA kernel may determine the dimensions of a thread block using the variables $\texttt{blockDim.x}$ and $\texttt{blockDim.y}$. For our example, $p = \texttt{blockDim.y}$ and $q = \texttt{blockDim.x}$.

The convention to index elements of a matrix is different from that used to index blocks in a grid and threads in a block. $C(i,j)$ refers to element (i,j) of C, and i increases top to bottom while j increases left to right; $C(0,0)$ is the top-left element. Because of this difference in the indexing convention, overlaying a submatrix tile with a thread block results in thread (d,e) corresponding to element (e,d) of the tile/submatrix.

Figure 11.5 illustrates the work done by the thread that computes $C(5,6)$ of an 8×8 matrix C. Matrix C is tiled using 2×4 tiles. The total number of tiles is 8. The submatrix defined by a tile is computed by a block of threads. The dimensions of a block are $(4,2)$. That is, $\texttt{blockDim.x} = 4$ and $\texttt{blockDim.y} = 2$. The dimensions of the grid of blocks are $(2,4)$. $C(5,6)$ is in tile/block $(1,2)$ of the grid and is computed by thread $(2,1)$ (i.e., $\texttt{threadIdx.x} = 2$ and $\texttt{threadIdx.y} = 1$) of the block. When writing the code, we have to do the inverse computation. That is, a thread has to determine which element, $C(i,j)$, of the C matrix it is to compute. A thread does this inverse computation using

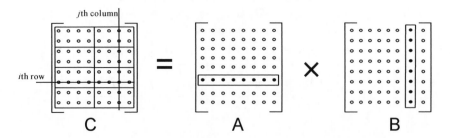

FIGURE 11.5: Thread $(2,1)$ of block $(1,2)$ computes $C(5,6)$.

```
__global__ void GPU1 (float *a, float *b, float *c, int n)
{// Thread code to compute one element of c
   // Element (i,j) of the product matrix is to be computed
   int i = blockIdx.y * blockDim.y + threadIdx.y;
   int j = blockIdx.x * blockDim.x + threadIdx.x;

   float temp = 0;
   for (int k = 0; k < n; k++)
      temp += a[i*n+k]*b[k*n+j];
   c[i*n+j] = temp;
}
```

LISTING 11.4: Thread code to compute one element of c.

its index as well as the index of the block it is part of. To compute $C(5,6)$, thread $(2,1)$ of block $(1,2)$ reads the pairs $(A(5,k), B(k,6))$ one pair at a time, multiplies the two components of the pair, and adds to a variable `temp`, whose final value will be $C(5,6)$. $A(5,0)$ and $B(0,6)$ are located in their row-major representations a and b using the standard row-major formula (i.e., $X(u,v)$ of and $n \times n$ matrix X is at position $u * n + v$ in its row-major representation x).

Listing 11.4 gives the kernel code for our first adaptation GPU1 of SingleCoreIJK. This code defines the computation done by a single thread. The first two lines determine the matrix element (i,j) to be computed by the thread. The remaining lines of code in this figure compute $C(i,j) = c[i*n+j]$. The variables i, j, k, and `temp` are assigned to registers, while a, b, and c are arrays that reside in device memory.

Notice the similarity between the kernel code of Listing 11.4 and the single-core code of Listing 11.1. The total number of threads employed to compute the product of two $n \times n$ matrices is n^2 and these threads are partitioned into $n^2/(pq)$ blocks of pq threads each. The global block scheduler assigns these blocks to the SMs for execution in some unspecified order and, resources permitting, several blocks may be coscheduled on an SM. The warp scheduler of an SM schedules the threads of the assigned blocks for execution in warps of 32 threads; each block has $\lceil pq/32 \rceil$ warps.

We note that GPU1 is identical to Matrix Multiplication without Shared Memory in [1].

11.7.1.2 Host code

Listing 11.5 gives a code fragment that could be used to compute the product of two $n \times n$ matrices using GPU1. In this fragment, d_A, d_B, and d_C refer to (previously allocated) device memory for the matrices A, B, and C, while h_A, h_B, and h_C refer to host memory for the same matrices. The code fragment begins by copying the matrices A and B from the host to the device. Then, the grid and block dimensions are defined and the kernel that defines a

```
// Copy from host to device
cudaMemcpy (d_A, h_A, memSize, cudaMemcpyHostToDevice);
cudaMemcpy (d_B, h_B, memSize, cudaMemcpyHostToDevice);

// Define grid and block dimensions
dim3 grid (n/q, n/p);
dim3 block (q, p);

// Invoke kernel
GPU1 <<<grid, block>>> (d_A, d_B, d_C, n);

// Wait for all threads to complete
cudaThreadSynchronize ();

// Copy result matrix from device to host
cudaMemcpy (h_C, d_C, memSize, cudaMemcpyDeviceToHost);
```

LISTING 11.5: Fragment to invoke GPU1.

thread computation invoked. Following the invocation of this kernel, we wait for all threads to complete and then copy the computed matrix product back to the host.

11.7.1.3   Tile/block dimensions

Before we can run our code on a GPU, we must decide the tile dimensions p and q, which in turn determine the block dimensions (q, p).[1] Since NVIDIA GPUs limit the number of threads in a block to be no more than 512, $pq \leq 512$. Since threads are scheduled in units of a warp, and a warp is 32 threads, we would like pq to be a multiple of 32. So, when using a square block, the only possible choices for the block dimension are 8×8 and 16×16. When using a rectangular block, additional choices (e.g., 1×64, 128×2, 8×64) become available. Some of these block sizes may be infeasible because they require more resources (e.g., registers and shared memory) than are available on an SM. An instrumentation of the kernel GPU1 reveals that it needs 10 registers per thread and 44 bytes of shared memory (this includes registers and shared memory required for any compiler-generated temporary variables and shared memory used to store the values of kernel parameters).[2] So, the maximum block size (i.e., pq) for GPU1 is not constrained by the number of available registers or by the size of shared memory but only by the NVIDIA-imposed limit of 512 threads per block. Using too small a block size will result in an occupancy less than 1 (Section 11.4). For example when the block size is 64

---

[1] By convention, the dimensions of a $p \times q$ block of threads are specified as (q, p).

[2] The register and shared-memory utilization as well as the occupancy of our GPU codes reported in this chapter were obtained using the ptx option to the CUDA compiler.

(say an 8×8 block is used), the number of blocks coscheduled on an SM is eight (this is the value of `maxBlocksPerSM`, Table 11.1). So, only 16 warps are co-scheduled on an SM and the occupancy is $16/32 = 0.5$ for compute capability 1.3 and $16/48 = 0.33$ for compute capability 2.0 (Equation 11.1).

11.7.1.4 Runtime

Table 11.4 gives the time taken by GPU1 for different values of n and different block dimensions. These times were obtained on a C1060, which is of compute capability 1.3. The shown times do not include the time taken to copy matrices A and B to device memory or that needed to copy C from device memory to host memory. This figure also gives the multiprocessor occupancy. Recall that when the block dimensions are (x, y), we use a $y \times x$ arrangement of threads and a $y \times x$ tile. So, each block computes a $y \times x$ submatrix of the product matrix. Hence, when $x = 1$ as is the case for the first five rows of Table 11.4, each thread block computes y contiguous elements of a column of C, and when $y = 1$ (last five rows of Table 11.4), each thread block computes x contiguous elements of a row of C. Observe that when $n = 16,384$ the best $((256,1))$ and worst $((1,512))$ choice for block dimensions result in performances that differ by a factor of almost 30. This is despite the fact that both choices result in an occupancy of 1.0. Interestingly, when 16×16 thread blocks are used, as is done in [1], the runtime for $n = 16383$ is 28% more than when the thread block dimensions have an area of $(256, 1)$.

We can explain some of the difference in the observed runtimes of GPU1 by the number of device-memory accesses that are made and the amount of data

		n			
(x, y)	occupancy	2048	4096	8192	16384
(1,32)	0.25	5.94	44.13	397.46	4396.39
(1,64)	0.50	5.62	43.69	441.52	6026.91
(1,128)	1.00	6.67	51.15	525.42	7341.49
(1,256)	1.00	6.53	51.80	594.71	8612.61
(1,512)	1.00	6.97	55.55	662.66	9373.16
(8,8)	0.50	1.18	7.95	64.49	540.03
(16,16)	1.00	0.76	5.59	46.75	420.82
(16,32)	1.00	0.74	5.64	49.37	499.72
(32,2)	0.50	0.65	5.25	44.59	372.27
(32,16)	1.00	0.74	5.53	43.50	361.24
(32,1)	0.25	0.73	5.79	45.09	356.42
(64,1)	0.50	0.64	5.16	41.50	360.11
(128,1)	1.00	0.72	5.30	40.90	336.87
(256,1)	1.00	0.71	5.23	40.86	327.64
(512,1)	1.00	0.76	5.51	42.30	327.68

TABLE 11.4: C1060 runtimes (seconds) for GPU1 for different block dimensions.

transferred between the device memory and the SMs (we shall see shortly that the amount of data transferred may vary from one access to the next) for each chosen (x, y). Recall that the latency of a device-memory access is 400 to 600 cycles [1]. The warp scheduler is often able to hide much or all of this latency by scheduling the execution of warps that are ready to do arithmetic while other warps are stalled waiting for a device-memory access to complete. To succeed with this strategy, there must, of course, be ready warps waiting to execute. So, it often pays to minimize the number of device-memory accesses as well as the amount (or volume) of data transferred.

11.7.1.5   Number of device-memory accesses

For devices with compute capability 1.3, device-memory accesses are scheduled on a per-half-warp basis.[3] When accessing 4-byte data, a 128-byte segment of device memory may be accessed with a single device-memory transaction. If all data to be accessed by a half warp lie in the same 128-byte segment of device memory, a single memory transaction suffices. If the data to be accessed lie in (say) three different segments, then three transactions are required, and each incurs the 400- to 600-cycle latency. Since, for compute capability 1.3, device-memory accesses are scheduled a half-warp at a time, to service the device-memory accesses of a common instruction over all 32 threads of a warp requires at least two transactions. Devices with compute capability 2.0 schedule device-memory accesses on a per-warp basis. So, if each thread of a warp accesses a 4-byte word and all threads access data that lie in the same 128-byte segment, a single memory transaction suffices. Since devices with compute capability 2.0 cache device-memory data in the L1 and L2 cache using cache lines that are 128 bytes wide, the throughput of a memory transaction is much higher whenever there is a cache hit [1].

To analyze the number of device-memory transactions done by `GPU1` and subsequent GPU codes, we assume that the arrays a, b, and c are all segment aligned (i.e., each begins at the boundary of a 128-byte segment of device memory). Consider the statement `temp += a[i*n+k]*b[k*n+j]` for any fixed value of k. When `blockDim.x` ≥ 16 and a power of 2 (e.g., (32,1), (256,1), (16,16), (32,16)), in each iteration of the `for` loop of `GPU1`, the threads of a half-warp access the same value of a, and the accessed values of b lie in the same 128-byte segment. So, when the compute capability is 1.3, two transactions are needed to fetch the a and b values needed by a half-warp in each iteration of the `for` loop. Note that two transactions (one for a and the other for b) suffice to fetch the values needed by a full-warp when the compute capability is 2.0 and `blockDim.x` is a power of two and ≥ 32. When the compute capability is 2.0 and `blockDim.x` $= 16$, two transactions are needed to read the values of b

---

[3]The partitioning of a block of threads into half-warps and warps is done by first mapping a thread index to a number. When the block dimensions are (D_x, D_y), the thread (x, y) is mapped to the number $x + yD_x$ [1]. Suppose that threads t_1 and t_2 map to the numbers m_1 and m_2, respectively. t_1 and t_2 are in the same half-warp (warp) iff $\lfloor m_1/16 \rfloor = \lfloor m_2/16 \rfloor$ ($\lfloor m_1/32 \rfloor = \lfloor m_2/32 \rfloor$).

needed by a warp (because these lie in two different 128-byte segments), and two are needed for the two different values of a that are to be read (these also lie in two different segments).

The statement $c[i*n+j] = temp$ does a write access on c. When $blockDim.x = 16$, a warp needs two transactions regardless of the compute capability. When $blockDim.x \geq 32$ and a power of 2, two transactions are required for compute capability 1.3 and one for capability 2.0.

We note that the for loop is iterated n times and that the total number of half-warps is $n^2/16$. So, when $blockDim.x \geq 16$ and a power of 2, the total number of memory transactions is $n^3/16$ (for a) + $n^3/16$ (for b) + $n^2/16$ (for c) = $n^3/8 + n^2/16$, for compute capability 1.3. For compute capability 2.0 and $blockDim.x = 16$, the number of transactions is also $= n^3/8 + n^2/16$ because each warp accesses two values of a that lie in different segments, the b values accessed lie in two segments as do the c values. However, when $blockDim.x \geq 32$ and a power of 2, the number of transactions is $n^3/16 + n^2/32$ for compute capability 2.0. Additionally, several of the memory transactions may be serviced from cache when the compute capability is 2.0.

To simplify the remaining discussion, we limit ourselves to devices with compute capability 1.3. For such devices, GPU1 has the same number, $n^3/8 + n^2/16$, of memory transactions whenever $x \geq 16$ and a power of 2. This is true for all but six rows of Table 11.4. The first five rows of this figure have $x = 1$. When $x = 1$, the threads of a half-warp access a values that are on the same column (and so different rows) of A and, for $n > 32$, the accessed values are in different 128-byte segments. Each half-warp, therefore, makes 16 transactions to fetch the needed a values on each iteration of the for loop. On each iteration of the for loop, the threads of a half-warp fetch a common b value using a single transaction. Since the c values written by a half-warp are on the same column of C, the half-warp makes 16 write transactions. The total number of transactions is n^3 (for a) + $n^3/16$ (for b) + n^2 (for c) = $17n^3/16 + n^2$, which is substantially more than when $x \geq 16$ and a power of 2.

For the 8×8 case, a half-warp fetches, on each iteration of the for loop, two a values from different segments and eight different bs that are in the same segment. So, over all n iterations of the for loop, a half-warp makes $2n$ a transactions and n b transactions. Since there are $n^2/16$ half-warps, the number of read transactions is $n^3/8$ (for a) $+n^3/16$ (for b). The c values written out by a half-warp fall into two segments. So, the number of write transactions is $n^2/8$. The total number of device-memory transactions is $3n^3/16 + n^2/8$.

Performance is determined not just by the number of memory transactions but also by their size. Although a GPU can transfer 128 bytes of data between the host and SM in one transaction, the actual size of the transaction (i.e., the amount of data transferred) may be 32, 64, or 128 bytes. A 128-byte segment is naturally decomposed into four 32-byte subsegments or two 64-byte subsegments. If the data to be transferred lie in a single 32-byte subsegment, the transaction size is 32 bytes. Otherwise, if the data lie in a single 64-byte subsegment, the transaction size is 64 bytes. Otherwise, the transaction size is 128 bytes. For GPU1, we see that when $x \geq 16$ and a power of 2, 32-byte transactions are used for a, and 64-byte transactions are used for b and c. For the remaining cases of Table 11.4, 32-byte transactions are used for a, b, and c. Since it is possible to do four times as many 32-byte transactions (or twice as many 64-byte transactions) in the same time it takes to do one 128-byte transaction, performance may be improved by reducing the transaction size while keeping the number of transactions steady.

Average Bandwidth Utilization (ABU) and *volume* are two other important metrics with respect to device-memory transactions. Consider the case when $x \geq 16$ and a power of 2. The reads of a are accomplished using 32-byte transactions. However, each such 32-byte payload carries only 4 bytes of useful data (i.e., the solitary a value to be used by all threads of the half-warp). So, these transactions effectively utilize only 12.5% of the bandwidth required by a 32-byte transaction. This is the minimum utilization possible for a 32-byte transaction (recall that we are limiting our discussion to 4-byte data). A 64-byte transaction has at least 4 bytes of useful data in each of its two 32-byte subsegments (otherwise, the transaction size would be 32 bytes), so its minimum utilization is also 12.5%. A 128-byte transaction, on the other hand, may have only 4 bytes of useful data in each of its two 64-byte subsegments, yielding a utilization of only 6.25%.

When $x \geq 16$ and a power of 2, the $n^3/16$ transactions on a have a utilization of 12.5%, the $n^3/16$ transaction on b have a utilization of 100% as do the $n^2/16$ transactions on c. For large n, we may ignore the transaction on c and compute the average bandwidth utilization of a transaction as $(12.5 + 100)/2 = 56.25\%$. When $x = 1$, the utilization of each transaction is 12.5%, so the ABU is 12.5%. For the 8×8 case, the $n^3/8$ transactions on a have a utilization of 12.5%, while the utilization of the $n^3/16$ transactions on b is 100%, so the ABU is $(2 * 12.5 + 100)/3 = 41.7\%$.

The volume of data transfer between the SMs and device memory is the sum of the number of transactions of each size multiplied by the transaction size. So, when $x \geq 16$ and a power of 2, the volume is $n^3/16 * 32 + n^3/16 * 64 + n^2/16 * 64 = 6n^3 + 4n^2$ bytes. Data volume divided by the bandwidth between device memory and the SMs is a lower bound on the time spent transferring data between device memory and the SMs. This also is a lower bound on the time for the entire computation. It often pays to reduce the volume of data transfer.

Table 11.5 summarizes our analysis of device-memory transactions made by GPU1 for different block dimensions.

(x, y)	Transactions			Volume	ABU
	Total	32-byte	64-byte		
$(1, *)$	$17n^3/16 + n^2$	$17n^3/16 + n^2$	0	$34n^3 + 32n^2$	12.5%
$(8, 8)$	$3n^3/16 + n^2/8$	$3n^3/16 + n^2/8$	0	$6n^3 + 4n^2$	41.7%
rest	$n^3/8 + n^2/16$	$n^3/16$	$n^3/16 + n^2/16$	$6n^3 + 4n^2$	56.25%

TABLE 11.5: Device-memory transaction statistics for GPU1.

11.7.2 A Thread Computes a 1×2 Submatrix of C

11.7.2.1 Kernel code

We can reduce the number of memory transactions by having each thread read four values of a (i.e., a 1×4 submatrix) at a time using the data type float4, read two values of b (i.e., a 1×2 submatrix) at a time using the data type float2, and write two values (i.e., a 1×2 submatrix) of c at a time also using the data type float2. Now, each thread computes a 1×2 submatrix of c, so, for example, we could use a thread block with dimensions $(4, 8)$ to compute an 8×8 submatrix of C (this is equivalent to using a tile size of 8×8) as in Figure 11.6. For example, thread $(0, 0)$ computes elements $(0, 0)$ and $(0, 1)$ of the submatrix, while thread $(3, 7)$ computes elements $(7, 6)$ and $(7, 7)$ of the submatrix.

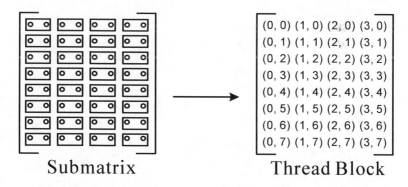

FIGURE 11.6: An 8×8 submatrix computed by a $(4, 8)$ block of threads.

The solid circles of Figure 11.7 shows the A and B values read from device memory by a thread that computes the 1×2 submatrix of C identified by solid circles. The A values are read in units of 4 as shown by the boxes in the figure; the B values are read in units of 2; and the C values are written to device memory in units of 2.

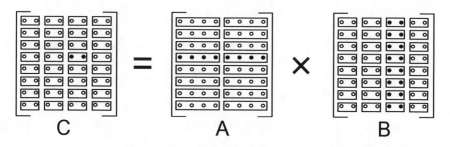

FIGURE 11.7: Computing a 1×2 submatrix of C.

```
__global__ void GPU2 (float *a, float *b, float *c, int n)
{// Thread code to compute a 1 x 2 submatrix of c
   // Cast a, b, and c into types suitable for I/O
   float4 *a4 = (float4 *) a;
   float2 *b2 = (float2 *) b;
   float2 *c2 = (float2 *) c;

   // Determine index of 1 x 2 c submatrix to compute
   int i = blockIdx.y * blockDim.y + threadIdx.y;
   int j = blockIdx.x * blockDim.x + threadIdx.x;

   int nDiv2 = n/2;
   int nDiv4 = n/4;
   int aNext = i*nDiv4;
   int bNext = j;
   float2 temp2;
   temp2.x = temp2.y = 0;
   for (int k = 0; k < nDiv4; k++)
   {
       float4 aIn = a4[aNext++];
       float2 bIn = b2[bNext];
       temp2.x += aIn.x*bIn.x; temp2.y += aIn.x*bIn.y;
       bNext += nDiv2;
       bIn = b2[bNext];
       temp2.x += aIn.y*bIn.x; temp2.y += aIn.y*bIn.y;
       bNext += nDiv2;
       bIn = b2[bNext];
       temp2.x += aIn.z*bIn.x; temp2.y += aIn.z*bIn.y;
       bNext += nDiv2;
       bIn = b2[bNext];
       temp2.x += aIn.w*bIn.x; temp2.y += aIn.w*bIn.y;
       bNext += nDiv2;
   }
   c2[i*nDiv2+j] = temp2;
}
```

LISTING 11.6: A thread computes a 1×2 submatrix of c.

Listing 11.6 gives the kernel GPU2 for a thread that computes a 1×2 submatrix of C. Suppose we index the 1×2 submatrices of C by the tuples (i, j), $0 \le i < n$ and $0 \le j < n/2$, top to bottom and left to right. Then, the 1×2 submatrix of C computed by a thread has $i = $ blockIdx.y * blockDim.y + threadIdx.y and $j = $ blockIdx.x * blockDim.x + threadIdx.x.

GPU2 uses 16 registers per thread, and each block requires 44 bytes of shared memory. Since a compute capability 1.3 GPU has 16,384 registers, it should be possible to coschedule up to 1024 threads or 32 warps on an SM and achieve an occupancy of 1.0. This happens, for example, when the block size is 16×16.

11.7.2.2 Number of device-memory accesses

Consider a single iteration of the `for` loop of GPU2. The four a values read from device memory are in the same 32-byte segment. For $(1, *)$ thread blocks, the 16 sets of a values read by a half-warp are in different 128-byte segments. So, 16 32-byte transactions are done on a on each iteration. The number of iterations is $n/4$, and the number of half-warps is $n^2/32$ (recall that each half-warp computes two c values). So, a total of $n^3/8$ 32-byte transaction are done on a. The utilization of each of these transactions is 50%. For b, GPU2 makes 32-byte transactions whose utilization is 25%. Four such transactions are made per iteration of the `for` loop for a total of n transactions per half-warp. So, the total number of b transactions is $n^3/32$. A half-warp makes 16 c transactions of size 32 bytes with 25% utilization. The total number of transactions of all types is $5n^3/32 + n^2/2$.

When 8×8 thread blocks are used, we need two transactions per iteration of the `for` loop to read the a values and four to read the b values. A half-warp also needs two transactions to write the c values. So, the number of device-memory transactions is $n^3/64$ (for a) $+ n^3/32$ (for b) $+ n^2/16$ (for c) $= 3n^3/64 + n^2/16$. The size of the a transactions is 32 bytes, and their utilization is 50%. The b and c transactions are 64 bytes each and have a utilization of 100%.

For the remaining block sizes of Table 11.4, the total number of transactions is $n^3/128$ (for a) $+ n^3/32$ (for b) $+ n^2/32$ (for c). The a transactions are 32 bytes each and have a utilization of 50%, while the b and c transactions are 128 bytes and have a utilization of 100%. Table 11.6 summarizes the device-memory transaction properties of GPU2.

(x, y)	Total	Transactions			Volume	ABU
		32-byte	64-byte	128-byte		
$(1, *)$	$5n^3/32 + n^2/2$	$5n^3/32 + n^2/2$	0	0	$5n^3 + 16n^2$	45.0%
$(8, 8)$	$3n^3/64 + n^2/16$	$n^3/64$	$n^3/32 + n^2/16$	0	$2.5n^3 + 4n^2$	83.3%
rest	$5n^3/128 + n^2/32$	$n^3/128$	0	$n^3/32 + n^2/32$	$4.25n^3 + 4n^2$	90.0%

TABLE 11.6: Device-memory transaction statistics for GPU2.

11.7.2.3 Runtime

Table 11.7 gives the time taken by GPU2 for different values of n and different block dimensions. The best performance is obtained using 8×8 blocks. This is not entirely surprising given that the volume of data movement between device memory and SMs is least when the block dimensions are $(8, 8)$ (Table 11.6). Dimensions of the form $(1, *)$ result in the maximum data volume, and this translates into the largest execution times. It is interesting that GPU2 exhibits very little variation in runtime over the different dimensions of the form $(i * 16, *)$. There is some correlation between the reduction in data volume when going from GPU1 to GPU2 and the reduction in runtime. For example, GPU1 has a volume that is 6.8 times that of GPU2 when block dimensions are of the form $(1, *)$, and for these dimensions, the runtimes for the GPU1 code are between 6.6 and 7.1 times that of GPU2 when $n = 16, 384$. For the $(8, 8)$ case, the ratio of the volumes is 2.4, and the ratio of the runtimes for $n = 16, 384$ is 2.8. For the remaining cases, the volume ratio is 1.4 and the time ratio is between 1.53 and 1.67, except for two cases where the time ratio is 1.97 and 2.3. Though not reported in Table 11.7, the dimensions $(4, 16)$ also did fairly well computing the product of two 16384×16384 matrices in 223 seconds with an occupancy of 0.50.

| (x, y) | occupancy | \multicolumn{4}{c}{n} |
		2048	4096	8192	16384
(1,32)	0.25	0.97	6.65	52.35	667.91
(1,64)	0.50	0.99	7.43	65.82	885.25
(1,128)	1.00	1.07	8.07	77.91	1058.65
(1,256)	1.00	1.11	8.79	92.86	1212.03
(1,512)	1.00	1.22	10.68	112.78	1378.81
(8,8)	0.50	0.26	2.06	17.65	194.04
(16,4)	0.50	0.41	3.29	27.23	223.48
(16,16)	1.00	0.42	3.30	27.26	213.73
(16,32)	1.00	0.42	3.34	27.38	217.04
(32,2)	0.50	0.41	3.28	26.69	217.58
(32,16)	1.00	0.42	3.33	26.88	210.48
(32,1)	0.25	0.41	3.28	26.69	229.14
(64,1)	0.50	0.41	3.28	26.72	216.23
(128,1)	1.00	0.41	3.29	26.76	216.13
(256,1)	1.00	0.42	3.30	26.77	212.96
(512,1)	1.00	0.42	3.31	26.75	211.35

TABLE 11.7: C1060 runtimes (seconds) for GPU2 for different block dimensions.

11.7.3 A Thread Computes a 1×4 Submatrix of C

11.7.3.1 Kernel code

Extending the strategy of Section 11.7.2 so that each thread computes four elements of c (say, a 1×4 submatrix) enables each thread to read four elements of b at a time using the data type `float4` and also to write four elements of c at a time. Listing 11.7 gives the resulting code, GPU3. When, for example, GPU3 is invoked with 16×16 thread blocks, each thread block computes a 16×64 submatrix of c. When the block dimensions are $(4, 16)$, a 16×16 submatrix of c is computed by each thread block.

```
__global__ void GPU3 (float *a, float *b, float *c, int n)
{// Thread code to compute a 1 x 4 submatrix of c
    // Cast a, b, and c into types suitable for I/O
    float4 *a4 = (float4 *) a;
    float4 *b4 = (float4 *) b;
    float4 *c4 = (float4 *) c;

    // Determine index of 1 x 4 c submatrix to compute
    int i = blockIdx.y * blockDim.y + threadIdx.y;
    int j = blockIdx.x * blockDim.x + threadIdx.x;

    int nDiv4 = n/4;
    int aNext = i*nDiv4;
    int bNext = j;
    float4 temp4;
    temp4.x = temp4.y = temp4.z = temp4.w = 0;
    for (int k = 0; k < nDiv4; k++)
    {
        float4 aIn = a4[aNext++];
        float4 bIn = b4[bNext];
        temp4.x += aIn.x*bIn.x; temp4.y += aIn.x*bIn.y;
        temp4.z += aIn.x*bIn.z; temp4.w += aIn.x*bIn.w;
        bNext += nDiv4;
        bIn = b4[bNext];
        temp4.x += aIn.y*bIn.x; temp4.y += aIn.y*bIn.y;
        temp4.z += aIn.y*bIn.z; temp4.w += aIn.y*bIn.w;
        bNext += nDiv4;
        bIn = b4[bNext];
        temp4.x += aIn.z*bIn.x; temp4.y += aIn.z*bIn.y;
        temp4.z += aIn.z*bIn.z; temp4.w += aIn.z*bIn.w;
        bNext += nDiv4;
        bIn = b4[bNext];
        temp4.x += aIn.w*bIn.x; temp4.y += aIn.w*bIn.y;
        temp4.z += aIn.w*bIn.z; temp4.w += aIn.w*bIn.w;
        bNext += nDiv4;
    }
    c4[i*nDiv4+j] = temp4;
}
```

LISTING 11.7: A thread computes a 1×4 submatrix of c.

GPU3 uses 21 registers per thread, while GPU2 uses 16. When the compute capability is 1.3, an occupancy of 1.0 cannot be achieved by codes that use more than 16 registers per thread as, to achieve an occupancy of 1.0, 32 warps must be coscheduled on an SM, and 32 warps with 16 registers per thread utilize all 16,384 registers available on an SM of a device with compute capability 1.3. So, for example, when GPU3 is invoked with 16×16 thread blocks, the occupancy of GPU3 is 0.5.

11.7.3.2 Runtime

Table 11.8 gives the time taken by GPU3 for different values of n. Although we experimented with many choices for block dimensions, only the times for the six best choices are reported in this figure. We remark that the remaining dimensions of the form $(*, 1)$ reported in Tables 11.4 and 11.7 performed almost as well as did $(64, 1)$, and while the remaining dimensions of the form $(1, *)$ performed better than they did for GPU2, they took between 50% to 110% more time than taken by $(1, 32)$. Of the six dimensions reported on in Table 11.8, the performance for the dimensions $(64, 1)$ and $(16, 32)$ was better for GPU2 than GPU3. The other four saw a performance improvement using the strategy of GPU3. The best performer for GPU3 is $(4, 16)$.

		n			
(x, y)	occupancy	2048	4096	8192	16384
(1,32)	0.25	0.51	3.64	29.11	336.89
(2,32)	0.50	0.29	2.14	17.27	199.92
(4,16)	0.50	0.27	2.06	16.28	130.28
(8,8)	0.50	0.38	3.01	24.10	192.71
(16,32)	0.50	0.71	5.63	43.98	351.63
(64,1)	0.50	0.69	5.50	43.97	351.67

TABLE 11.8: C1060 runtimes (seconds) for GPU3 for different block dimensions.

11.7.3.3 Number of device-memory accesses

We now derive the device-memory transaction statistics for the top three performers $((4, 16), (8, 8),$ and $(2, 32))$ of Table 11.8. In all cases, the number of half-warps is $n^2/64$. When the dimensions are $(4, 16)$, a thread makes one a access and four b accesses of device memory in each iteration of the for loop. The a accesses made by a half-warp are two float4s from different rows of A and hence from different 128-byte segments. So, a half-warp makes four 32-byte device-memory transactions per iteration of the for loop. The utilization of each access is 50%. The total number of a transactions is, therefore, $4 * n/4 * n^2/64 = n^3/64$. When the threads of a half-warp read a float4 of b, they read four float4s that lie in the same 64-byte segment. So, this read is accomplished using a single 64-byte transaction whose utilization is 100%. A thread reads four float4s of b in each iteration of the for loop. So, in each iteration of the for loop, a half-warp makes four 64-byte transactions on b, and the utilization of these transactions is 100%. Hence, the total number of b transactions is $4 * n/4 * n^2/64 = n^3/64$. For c, a half-warp makes four 64-

byte transactions with utilization 100%. The total number of c transactions is $n^2/16$. The volume and ABU may be calculated easily now.

When the dimensions are $(8, 8)$, a half-warp reads, in each iteration of the for loop, two float4s of a values that lie in different 128-byte segments. This requires two 32-byte transactions, and their utilization is 50%. So, the total number of a transactions is $2 * n/4 * n^2/64 = n^3/128$. A half-warp accesses 32 b values that are in the same 128-byte segment each time its thread reads a float4 of b. A thread does four such reads in an iteration of the for loop. So, a total of $4 * n/4 * n^2/64 = n^3/64$ 128-byte transactions are done for b. The utilization of each of these is 100%. Each half-warp uses two 128-byte transactions to write the computed c values. So, a total of $n^2/32$ 128-byte transactions are done for c, and the utilization of each of these is 100%.

The final analysis we do is the dimensions $(2, 32)$. In each iteration of the for loop, a half-warp reads eight float4s of a values that are in different 128-byte segments. So, GPU3 makes a total of $8 * n/4 * n^2/64 = n^3/32$ 32-byte transactions on a. The utilization of each of these is 50%. Further, on each iteration of the for loop, a half-warp reads $2 * 4$ float4s of bs using four 32-byte transactions whose utilization is 100%. So, a total of $4 * n/4 * n/64$ $= n^3/64$ 32-byte transactions with utilization 100% are made for b. To write the c values, a half-warp makes eight 32-byte transactions whose utilization is 100%. Table 11.9 summarizes the transaction statistics for GPU3. Of the three dimensions analyzed, $(2, 32)$ and $(4, 16)$ transfer the smallest volume of data between device memory and the SMs, and $(4, 16)$ does this transfer using a smaller number of transactions. These factors, most likely, played a significant role in causing GPU3 to exhibit its best performance when the dimensions are $(4, 16)$.

(x, y)	Total	Transactions 32-byte	64-byte	128-byte	Volume	ABU
$(4, 16)$	$n^3/32 + n^2/16$	$n^3/64$	$n^3/64 + n^2/16$	0	$1.5n^3 + 4n^2$	75.0%
$(8, 8)$	$3n^3/128 + n^2/32$	$n^3/128$	0	$n^3/64 + n^2/32$	$2.25n^3 + 4n^2$	83.3%
$(2, 32)$	$3n^3/64 + n^2/8$	$3n^3/64 + n^2/8$	0	0	$1.5n^3 + 4n^2$	66.7%

TABLE 11.9: Device-memory transaction statistics for GPU3.

11.7.4   A Thread Computes a 1×1 Submatrix of C Using Shared Memory

11.7.4.1   First kernel code and analysis

To improve on the performance of GPU2, we resort to a block matrix multiplication algorithm in which each of A, B, and C is partitioned into n^2/s^2 $s \times s$ submatrices A_{ij}, B_{ij}, and C_{ij}. We assume that s divides n. The algorithm computes C using the equation

$$C_{ij} = \sum_{0 \le k < n/s} A_{ik} * B_{kj}$$

In the strategy of this subsection, a block of threads computes one $s \times s$ submatrix of C, and each thread computes one element of this submatrix. So, we use $s \times s$ thread blocks with thread (x, y) computing element (y, x) of the submatrix (recall the difference in the convention to name threads and matrix elements). The thread block that is to compute C_{ij} executes the following pseudocode:

Step 1: Repeat Steps 2 and 3 for $0 \le k < n/s$.

Step 2: Each thread of the $s \times s$ thread block reads one value of A_{ik} and one value of B_{kj} from device memory.

Step 3: Each thread updates the element of C_{ij} it is computing by multiplying the appropriate row of A_{ik} with the appropriate column of B_{kj}. This update step accesses shared memory but not device memory.

Step 4: Each thread writes the element of C_{ij} it has computed to device memory.

Listing 11.8 gives the kernel code for the case $s = 16$. We note that GPU4 is essentially the same as Matrix Multiplication with Shared Memory in [1]. As noted in Section 11.7.1, the constraints of NVIDIA GPUs with compute capability 1.3 require us to use $s = 16$ if we are to have any prospect of achieving an occupancy of 1.0. Although, as we have seen earlier, it is possible to get better performance with a lower occupancy than a higher one, we verified experimentally that $s = 16$ gives the best performance. Further, performance is not improved by storing $as[16][t]$ and $bs[t][16]$ in shared memory for $t \ne 16$.

We now obtain the device-memory access statistics for GPU4. Since each thread computes a single value of c, the number of half-warps is $n^2/16$. In each iteration of the `for` loop, a half-warp reads 64 bytes of a values from a single 128-byte segment using a 64-byte transaction and 64 bytes of b values using another 64-byte transaction. Since the `for` loop is iterated $n/16$ times, a half-warp makes $n/8$ 64-byte transactions of a and b together. A half-warp also makes one 64-byte write transaction on c. So, the total number of device-memory transactions made by GPU4 is $n^3/128 + n^2/16$. Each of these is a 64-byte transaction with 100% utilization. The volume is $n^3/2 + 4n^2$, and the ABU is 100%. GPU4 uses 11 registers per thread and 2092 bytes of shared memory; it achieves an occupancy of 1.0.

Compared to the $(4, 16)$ case of GPU3, which results in the best performance for GPU3, GPU4 has a lower volume (67% lower), a higher ABU (33% higher), and a higher occupancy (double that of GPU3). So, we expect GPU4 to have a much better performance than GPU3.

```
__global__ void GPU4 (float *a, float *b, float *c, int n)
{// Thread code to compute an element of a 16 x 16
 // submatrix of c
   // Shared memory arrays to hold a 16 x 16 submatrix of
   // a and b
   __shared__ float as[16][16], bs[16][16];
   int nDiv16 = n/16;
   int nTimes16 = n*16;
   int aNext = (16*blockIdx.y+threadIdx.y)*n+threadIdx.x;
   int bNext = 16*blockIdx.x+threadIdx.y*n+threadIdx.x;
   float temp = 0;

   for (int u = 0; u < nDiv16; u++)
   {// Threads in a thread block collectively read a
    // 16 x 16 submatrix of a and b from device memory
    // to shared memory; each thread reads
    // 1 element of a and 1 element of b
      as[threadIdx.y][threadIdx.x] = a[aNext];
      bs[threadIdx.y][threadIdx.x] = b[bNext];
      __syncthreads(); // Wait for read to complete

      // Multiply a row of as with a column of bs
      for (int k = 0; k < 16; k++)
          temp += as[threadIdx.y][k]*bs[k][threadIdx.x];
      // Wait for all threads in thread block to complete
      __syncthreads();

      // Update to work on next submatrix of a and b
      aNext += 16;
      bNext += nTimes16;
   }
   c[(16*blockIdx.y+threadIdx.y)*n + 16*blockIdx.x
                                  + threadIdx.x] = temp;
}
```

LISTING 11.8: A $(16, 16)$ thread block computes a 16×16 submatrix of c using shared memory.

11.7.4.2 Improved kernel code

The performance of GPU4 may be improved by tuning shared-memory accesses. This tuning does not affect the volume, ABU, or occupancy of the kernel. A thread accesses a row of *as* and a column of *bs*. Although the CUDA manual does not document a cache for shared memory, it appears that it is faster for a thread to access data by rows rather than by columns. We can take advantage of this asymmetry in access time by storing in *bs* the transpose of the submatrix of *b* that is read from device memory. While this speeds access for a single thread, it results in shared-memory access conflicts for a half-warp of threads. To see why this is so, we need to be familiar with the organization of shared memory on a GPU.

On a device with compute capability 1.3, shared memory is divided into 16 banks using a round-robin allocation of 4-byte words. So, in the shared-memory array *as* of type `float`, $as[i]$ and $as[j]$ are in the same bank iff $i \bmod 16 = j \bmod 16$. (A device of compute capability 2.0 has 32 banks of shared memory.) Shared memory read/writes from different banks can be serviced at the same time. When two or more threads need to access the same bank, there is a shared-memory conflict and the read/writes get serialized into a number of conflict-free read/writes. So, performance is maximized when there are no bank conflicts.

When *a* and *b* are read from device memory and written to shared memory by a half-warp in GPU4, the half-warp accesses 16 adjacent elements of *as* and *bs*. These lie in different banks of shared memory and so, the shared-memory accesses are conflict-free. Similarly, the reads of *bs* by a half-warp in the inner `for` loop are conflict-free. Since all threads in a half-warp read the same *as* in the inner `for` loop, all threads in the half-warp get this *as* value, making a common access to shared memory. If we modify GPU4 to save the transpose of the *b* submatrix in *bs* by changing the statement

```
bs[threadIdx.y][threadIdx.x] = b[bNext];
```

to

```
bs[threadIdx.x][threadIdx.y] = b[bNext];
```

and, correspondingly, change the statement

```
temp += as[threadIdx.y][k]*bs[k][threadIdx.x];
```

to

```
temp += as[threadIdx.y][k]*bs[threadIdx.x][k];
```

then threads in a half-warp write and read *bs* values that are in the same column of *bs*. Since the values in a column are in the same bank of shared memory (assuming the rows of *bs* are assigned to a contiguous block of $16 \times 16 \times 4 = 1024$ bytes of shared memory), the 16 threads of the half-warp need to read 16 different *bs* values from the same bank (note that elements of a column of *bs* are 16 4-byte words apart and so are in the same bank). Figure 11.8 shows the shared-memory bank mapping for $bs[16][16]$. The resulting bank conflict is serialized into 16 shared-memory accesses.

To avoid bank conflicts in the reading of *bs* values, we define *bs* to be a 16×17 array. Now, elements in a column are 17 4-byte words apart and so are in different banks (see Figure 11.9).

Modifying GPU4 to store the transpose of submatrices of *b* in *bs* and defining *bs* so that elements of a column are in different banks of shared memory gives us GPU5 (Listing 11.9).

Table 11.10 gives the time taken by GPU4 and GPU5 for different values of n. We see that for $n = 16{,}384$, the shared-memory optimization done by GPU5

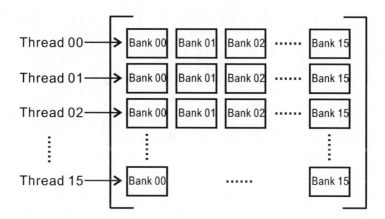

FIGURE 11.8: Bank mapping for $bs[16][16]$.

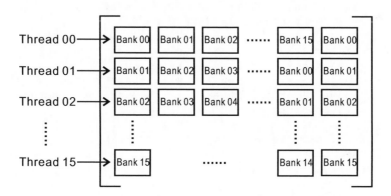

FIGURE 11.9: Bank mapping for $bs[16][17]$.

has reduced runtime by a little over 4%. For this value of n, GPU4 achieves a speedup of a little over 2.8 relative to the best case $((4, 16))$ for GPU3, while GPU5 achieves a speedup of almost 3.0. Amazingly, an additional speedup of almost 2 is possible!

	2048	4096	8192	16384
GPU4	0.09	0.69	5.54	46.19
GPU5	0.08	0.60	4.84	44.13

TABLE 11.10: C1060 runtimes for GPU4 and GPU5.

```
__global__ void GPU5 (float *a, float *b, float *c, int n)
{// Thread code to compute an element of a 16 x 16
 // submatrix of c
   // Shared memory arrays to hold a submatrix of a and b
   __shared__ float as[16][16], bs[16][17];
   int nDiv16 = n/16;
   int nTimes16 = n*16;
   int aNext = (16*blockIdx.y+threadIdx.y)*n+threadIdx.x;
   int bNext = 16*blockIdx.x+threadIdx.y*n+threadIdx.x;
   float temp = 0;

   for (int u = 0; u < nDiv16; u++)
   {// Threads in a thread block collectively read a
    // 16 x 16 submatrix of a and b from device memory
    // to shared memory; each thread reads
    // 1 element of a and 1 element of b
      as[threadIdx.y][threadIdx.x] = a[aNext];
      bs[threadIdx.x][threadIdx.y] = b[bNext];
      __syncthreads(); // Wait for read to complete

      // Multiply a row of as with a column of bs
      for (int k = 0; k < 16; k++)
         temp += as[threadIdx.y][k]*bs[threadIdx.x][k];
      // Wait for all threads in thread block to complete
      __syncthreads();

      // Update to work on next submatrix of a and b
      aNext += 16;
      bNext += nTimes16;
   }
   c[(16*blockIdx.y+threadIdx.y)*n + 16*blockIdx.x
                          + threadIdx.x] = temp;
}
```

LISTING 11.9: Same as GPU4 except that the transpose of the *b* submatrix is stored in *bs*.

11.7.5 A Thread Computes a 16×1 Submatrix of C Using Shared Memory

11.7.5.1 First kernel code and analysis

To improve the performance of the matrix-multiply kernel further, we increase the computational load per thread to better mask device-memory accesses with arithmetic operations. In particular, each thread will compute 16 values of *c*, all on the same column. Though we can have a thread compute fewer or more values of *c*, 16 gave best performance. Since a thread will com-

```
__device__ void update1(float *a, float b, float *c)
{
    for (int i = 0; i < 16; i++)
        c[i] += a[i] * b;
}
```

LISTING 11.10: Updating c values.

pute its assigned 16 values of c incrementally, we allocate a thread with 16 registers $cr[0 : 15]$ to store the incremental values computed so far. When done, the thread will write its 16 computed values to device memory.

The incremental computation of the c values is done using an 8×16 block of threads (i.e., the block dimensions are $(16, 8)$). The 8×16 thread block first reads a 16×32 submatrix of a from device memory and stores its transpose into a two-dimensional shared-memory array $as[32][17]$. To do this, each thread must read in four values of a; the four values read in are from two adjacent columns of the a submatrix. Additionally, a half-warp reads 32 adjacent values of a that lie in the same 128-byte segment. Using a 32×17 array rather than a 32×16 array avoids bank conflicts when writing to shared memory, and storing the transpose derives cache-like benefits as a thread will access as by row (see Section 11.7.4). The code of Listing 11.10 updates the value of each of the 16 cs being computed by a thread by adding in an $a[i] * b$ value. When invoked, a is a pointer to a row of as, which corresponds to a column of the 16×32 submatrix of a that was read in. Listings 11.11 and 11.12 give the complete code to multiply two $n \times n$ matrices using the described strategy.

To determine the device-memory statistics, we note that a thread computes 16 c values. So, a half-warp computes 256 c values. Therefore, the number of half-warps is $n^2/256$. In each iteration of the `for i` loop, the threads of a half-warp use two 128-byte transactions to read the required a values. The total number of transactions on a is $n^2/256 * 2 * n/32 = n^3/4096$. Each of these 128-byte transactions has a utilization of 100%. In each iteration of the `for i` loop, a half-warp makes 32 64-byte transactions on b. The total number of b transactions is therefore $n^2/256 * 32 * n/32 = n^3/256$. Each of these has 100% utilization. A half-warp makes 16 64-byte device-memory transactions to write out the 256 c values it computes. Therefore, the number of device-memory write-transactions for c is $n^2/16$, and each has a utilization of 100%. Combining the transactions for a, b, and c, we see that GPU6 makes a total of $17n^3/4096 + n^2/16$ device-memory transactions; the volume is $9n^3/32 + 4n^2$; and the ABU is 100%. Compared to GPU4 and GPU5, GPU6 makes about 47% fewer transactions and generates about 44% less volume. All three have the same ABU.

```
__global__ void GPU6 (float *a, float *b, float *c, int n)
{// Thread code to compute one column of a 16 x 128
 // submatrix of c
 // Use shared memory to hold the transpose of a 16 x 32
 // submatrix of a
   __shared__ float as[32][17];

   // Registers for column of c submatrix
   float cr[16] = {0,0,0,0,0,0,0,0,0,0,0,0,0,0,0,0};
   int nDiv32 = n/32;
   int sRow = threadIdx.y;
   int sCol = threadIdx.x;
   int sCol2 = sCol*2;
   int sCol2Plus1 = sCol2+1;
   int tid = sRow*16+sCol;
   int aNext = (16*blockIdx.y+sRow)*n+sCol*2;
   int bNext = 128*blockIdx.x+tid;
   int sRowPlus8 = sRow+8;
   int nTimes8 = 8*n;

   a += aNext;
   b += bNext;

   int i, j;
   float2 temp;

    for (i = 0; i< nDiv32; i++)
    {// Threads in a thread block collectively read a 16 x 32
     // submatrix of a from device memory to shared memory
        temp = *(float2 *)a;
        as[sCol2][sRow] = temp.x;
        as[sCol2Plus1][sRow] = temp.y;
        temp = *(float2 *)(a+nTimes8);
        as[sCol2][sRowPlus8] = temp.x;
        as[sCol2Plus1][sRowPlus8] = temp.y;
        __syncthreads(); // Wait for read to complete

        #pragma unroll
        for (j = 0; j < 32; j++)
        {
            float br = b[0];
            b += n;
            update1 (&as[j][0], br, cr);
        }

        a += 32;
        __syncthreads(); // Wait for computation to complete
    }
```

LISTING 11.11: A $(16, 8)$ thread block computes a 16×128 submatrix of C using shared memory (Part a).

```
// Output cr[]
int cNext = 16*blockIdx.y*n + 128*blockIdx.x + tid;
c += cNext;

for (int i = 0; i < 16; i++)
{
    c[0] = cr[i];
    c += n;
}
}
```

LISTING 11.12: A $(16, 8)$ thread block computes a 16×128 submatrix of C using shared memory (Part b).

11.7.5.2 Second kernel code

Successive iterations of the `for` loop of Listing 11.12 need to wait for the preceding read of $b[0]$ to complete. Further, even after the read completes, there is a 24-cycle between the time a value is written to a register and the time this value can be used [7]. To reduce the impact of these delays, we read in several values of b from device memory and use these b values in round-robin fashion. Listing 11.13 does this using 4 values of b in each round. Although we could use this strategy with a different number of bs such as 2 and 8, we found that 4 gives best performance. The code of this listing assumes that `nTimes2`, `nTimes3`, and `nTimes4` have, respectively, been defined to be $2 * n$, $3 * n$, and $4 * n$. GPU6 is modified to use four values of b per round as in Listing 11.13 and is referred to as GPU7.

11.7.5.3 Final kernel code

As in GPU6 and GPU7, our final matrix-multiply code, GPU8 (Listings 11.15 and 11.16), uses $(16, 8)$ thread blocks. However, a thread block now reads a 16×64 submatrix of a rather than a 16×32 submatrix from device memory to shared memory. Each half-warp reads the 64 a values in a row of the 16×64 submatrix, which lie in two adjacent 128-byte segments of device memory, using two 128-byte transactions. To accomplish this, each thread reads a 1×4 submatrix of a using the data type `float4`. The 16×64 a submatrix that is input from device memory may be viewed as a 16×16 matrix in which each element is a 1×4 vector. The transpose of this 16×16 matrix of vectors is stored in the array $as[16][65]$ with each 1×4 vector using four adjacent elements of a row of as. This mapping ensures that the 16 elements in each column of the 16×64 submatrix of a that is input from device memory are stored in different banks of shared memory. So, the writes to shared memory done by a half-warp of GPU8 are conflict-free. Further, by storing the transpose of a 16×16 matrix of 1×4 vectors rather than the transpose of a 16×64 matrix of scalars, we are able to do the writes to shared memory using `float4`s rather

```
float br0 = b[0];
float br1 = b[n];
float br2 = b[nTimes2];
float br3 = b[nTimes3];

#pragma unroll
for (j = 0; j< 7; j++)
{
    b += nTimes4;
    update1 (&as[j*4][0], br0, cr); br0 = b[0];
    update1 (&as[j*4+1][0], br1, cr); br1 = b[n];
    update1 (&as[j*4+2][0], br2, cr); br2 = b[nTimes2];
    update1 (&as[j*4+3][0], br3, cr); br3 = b[nTimes3];
}

b += nTimes4;
update1 (&as[28][0], br0, cr);
update1 (&as[29][0], br1, cr);
update1 (&as[30][0], br2, cr);
update1 (&as[31][0], br3, cr);
```

LISTING 11.13: Listing 11.12 modified to handle four bs in round-robin fashion.

than floats as is done in GPU6 and GPU7. This reduces the time to write to shared memory. The scheme used to map a 16×64 submatrix of a into a 16×65 array as necessitates the use of a slightly different update method, update2 (Listing 11.14).

The number of half-warps is $n^2/256$. In each iteration of the for i loop, a half-warp makes four 128-byte transactions to read in a values and 64 64-byte transactions to read in b values. GPU8 makes $n^3/4096$ 128-byte device-memory transactions on a and $n^3/256$ 64-byte transactions on b. Additionally, $n^2/16$ 64-byte transactions are made on c. Each transactions has 100% utilization. So, the device memory statistics for GPU8 are the same as for GPU6 and GPU7.

```
__device__ void update2(float *a, float b, float *c)
{
    for (int i = 0; i < 16; i++)
        c[i] += a[i * 4] * b;
}
```

LISTING 11.14: Updating c values when as read using float2.

```
__global__ void GPU8 (float *a, float *b, float *c, int n)
{// Thread code to compute one column of a 16 x 128
 // submatrix of c
 // Use shared memory to hold the transpose of a
 // 16 x 64 submatrix of 1 x 4 sub-vectors of a
    __shared__ float as[16][65];

    // Registers for column of c submatrix
    float cr[16] = {0,0,0,0,0,0,0,0,0,0,0,0,0,0,0,0};
    int nDiv64 = n/64;
    int sRow = threadIdx.y;
    int sRow4 = sRow*4;
    int sCol = threadIdx.x;
    int tid = sRow*16+sCol.x;
    int aNext = (16*blockIdx.y+sRow)*n+sCol*4;
    int bNext = 128*blockIdx.x + tid;
    int cNext = 16*blockIdx.y*n + 128*blockIdx.x + tid;
    int nTimes2 = 2*n;
    int nTimes3 = 3*n;
    int nTimes4 = 4*n;

    a += aNext;
    b += bNext;
    c += cNext;

    float4 *a4 = (float4 *)a;

    for (int i = 0; i < nDiv64; i++)
    {
        *( (float4 *)(&as[sCol][sRow4]) ) = a4[0];
        *( (float4 *)(&as[sCol][sRow4+32]) ) = a4[nTimes2];
        __syncthreads(); // Wait for read to complete

        float br0 = b[0];
        float br1 = b[n];
        float br2 = b[nTimes2];
        float br3 = b[nTimes3];
        b += nTimes4;

        #pragma unroll
        for (int k = 0; k < 15; k++)
        {
            update2 (&as[k][0], br0, cr); br0 = b[0];
            update2 (&as[k][1], br1, cr); br1 = b[n];
            update2 (&as[k][2], br2, cr); br2 = b[nTimes2];
            update2 (&as[k][3], br3, cr); br3 = b[nTimes3];
            b+= nTimes4;
        }
```

LISTING 11.15: GPU7 modified to handle four bs in round-robin fashion and read a 16×32 submatrix of a (Part a).

```
            update2 (&as[15][0], br0, cr);
            update2 (&as[15][1], br1, cr);
            update2 (&as[15][2], br2, cr);
            update2 (&as[15][3], br3, cr);


            a4 += 16;
            __syncthreads(); // Wait for computation to complete
    }

    for (int j = 0; j < 16; j++)
    {
        c[0] = cr[j];
        c += n;
    }
}
```

LISTING 11.16: GPU7 modified to handle four bs in round-robin fashion and read a 16×32 submatrix of a (Part b).

11.8   A Comparison

11.8.1   GPU Kernels

To compare the eight GPU kernels, GPU1 through GPU8, we use, for each, the block dimensions that yielded best performance. Specifically, for GPU1 we use $(256, 1)$ (i.e., blockDim.x $= 256$ and blockDim.y $= 1$) thread blocks; for GPU2 we use $(8, 8)$ blocks; for GPU3, we use $(4, 16)$ blocks; for GPU4 and GPU5, we use $(16, 16)$; and for GPU6, GPU7, and GPU8, we use $(16, 8)$ blocks. In our comparisons, we included also the code sgemm that is in the MAGMA BLAS 0.2 library [8] and is based on the algorithm of Volkov and Demmel [9] as well as the matrix multiply in the CUDA CUBLAS 3.0 library [10]. The sgemm code, which assumes a column-major mapping of matrices into one-dimensional arrays, is very similar to GPU7 in its strategy and uses $(16, 4)$ blocks. Since we do not have access to the source code for CUBLAS, we do not know precisely the multiplication strategy it uses. However, given the proximity of its performance to that of sgemm, it is very likely that sgemm and CUBLAS use the same strategy.

Table 11.11 gives the number of device-memory transactions, volume, and ABU for each of our eight GPU matrix-multiply kernels as well as the sgemm kernel. We are unable to provide data for CUBLAS as we do not have access to its source code. Figures 11.10 and 11.11 plot the number of transactions (in billions) and the volume for $16,384 \times 16,384$ matrices. We note that each refinement of the kernel resulted in either a reduction or no change in the total number of transactions and in the volume of data moved between the device memory and the SMs. So, there is a good correlation between performance and number of transactions as well as volume. This isn't too surprising; when there are more transactions, there are more latencies to hide, and the volume divided by the bandwidth between device memory and the SMs is a lower bound on the time needed to do the data transfer. We note also that GPU6, GPU7, and GPU8 make 15% fewer transactions than made by sgemm and generate 10% lower volume.

Version	Transactions	Volume	ABU
GPU1	$n^3/8 + n^2/16$	$6n^3 + 4n^2$	56%
GPU2	$3n^3/64 + n^2/16$	$2.5n^3 + 4n^2$	83%
GPU3	$n^3/32 + n^2/16$	$1.5n^3 + 4n^2$	75%
GPU4	$n^3/128 + n^2/16$	$0.5n^3 + 4n^2$	100%
GPU5	$n^3/128 + n^2/16$	$0.5n^3 + 4n^2$	100%
GPU6	$17n^3/4096 + n^2/16$	$9n^3/32 + 4n^2$	100%
GPU7	$17n^3/4096 + n^2/16$	$9n^3/32 + 4n^2$	100%
GPU8	$17n^3/4096 + n^2/16$	$9n^3/32 + 4n^2$	100%
sgemm	$5n^3/1024 + n^2/16$	$5n^3/16 + 4n^2$	100%

TABLE 11.11: Device-memory statistics for compute capability 1.3.

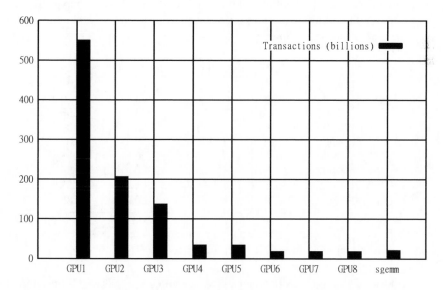

FIGURE 11.10: Transactions (in billions) for 16,384 × 16,384 matrices.

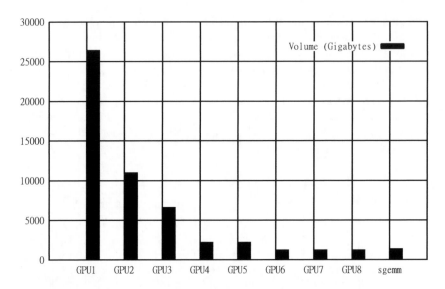

FIGURE 11.11: Volume (in GBs) for 16,384 × 16,384 matrices.

Table 11.12 gives the block dimensions, number of registers per thread, shared memory used by an SM, and the occupancy for the different kernels. We see that although high occupancy is needed to hide device-memory latency, high occupancy does not necessarily translate into better performance. In particular, our best performers, GPU6, GPU7, GPU8, and sgemm, have the lowest occupancy. The number of transactions and volume are a better indicator of performance than is occupancy.

	Block	Registers	Memory	Occupancy
GPU1	$(256, 1)$	10	44	1.00
GPU2	$(8,8)$	16	44	0.50
GPU3	$(4, 16)$	21	44	0.50
GPU4	$(16, 16)$	11	2092	1.00
GPU5	$(16, 16)$	11	2156	1.00
GPU6	$(16, 8)$	38	2220	0.38
GPU7	$(16, 8)$	38	2220	0.38
GPU8	$(16, 8)$	38	4204	0.38
sgemm	$(16, 4)$	35	1160	0.38

TABLE 11.12: Number of registers per thread, shared memory (bytes) per block, and occupancy for GPU matrix multiply methods.

Tables 11.13–11.15, respectively, give the runtime, effective GFops, and efficiency of our kernels. Figures 11.12–11.14 plot these metrics for 16,384×16,384 matrices. For the GFlop rate computation, we used $2n^3 - n$ as the total number of floating-point operations (multiplies and adds) done by a kernel and divided this by the runtime. The efficiency was measured relative to the manufacturers stated peak rate of 933 GFlops for the C1060. That is, we divided the GFlop rate given in Table 11.14 by 933 to come up with the efficiency given in Table 11.15. It is interesting to note that, for $n = 16,384$, there is a speedup of 14 when going from the most simple adaptation of SingleCoreIJK (i.e., GPU1) to the most efficient adaptation GPU8. Further, GPU6, GPU7, GPU8, sgemm, and CUBLAS all take about the same time, with GPU8 being the fastest and sgemm the slowest of the five. GPU8 is faster than CUBLAS by 2% and faster than sgemm by 3% when $n = 16,384$. This rather small reduction in runtime relative to CUBLAS and sgemm despite the 15% reduction in number of transactions and 10% reduction in volume suggests that CUBLAS and sgemm do a very good job of masking device-memory latency. These five kernels are able to achieve a GFlop rate of over 370 GFlops when $n = 16,384$. Since there is no opportunity to use SFs in a matrix multiply, the theoretical peak is actually only 622 GFlops and not 933 GFlops. Using 622 GFlops as the theoretical maximum rate, the efficiency of the five kernels GPU6, GPU7, GPU8, sgemm, and CUBLAS becomes about 60%. In Figure 11.12, we show also the theoretical minimum time needed to move the required volume of data between device memory and the SMs. This theoretical minimum was computed by dividing the data volume by the theoretical peak transfer rate of 102 GB/sec. Since we do not have the volume data for CUBLAS, the theoretical minimum data transfer time for CUBLAS is not shown.

	2048	4096	8192	16384
GPU1	0.71	5.24	40.85	327.77
GPU2	0.27	2.06	17.65	193.02
GPU3	0.27	2.05	16.28	130.25
GPU4	0.09	0.69	5.54	46.18
GPU5	0.08	0.60	4.84	44.12
GPU6	0.05	0.37	2.91	23.28
GPU7	0.05	0.37	2.91	23.26
GPU8	0.05	0.36	2.88	22.97
sgemm	0.05	0.37	2.97	23.70
CUBLAS	0.05	0.37	2.95	23.53

TABLE 11.13: Runtime (seconds) for GPU matrix multiply methods.

Figures 11.15 and 11.16, respectively, show the runtime of our GPU kernels as a function of the number of the device-memory transactions and the volume of data transferred between device memory and the SMs. CUBLAS is not included in these figures as we do not have the transactions and volume data for CUBLAS. These figures show a fairly strong correlation between runtime and transactions as well as between runtime and volume.

	2048	4096	8192	16384
GPU1	24	26	27	27
GPU2	64	67	62	46
GPU3	64	67	68	68
GPU4	197	199	198	190
GPU5	226	229	227	199
GPU6	373	377	377	378
GPU7	373	377	378	378
GPU8	373	381	382	383
sgemm	358	368	371	371
CUBLAS	354	371	373	374

TABLE 11.14: GFlops for GPU matrix multiply methods.

	2048	4096	8192	16384
GPU1	0.03	0.03	0.03	0.03
GPU2	0.07	0.07	0.07	0.05
GPU3	0.07	0.07	0.07	0.07
GPU4	0.21	0.21	0.21	0.20
GPU5	0.24	0.25	0.24	0.21
GPU6	0.40	0.40	0.40	0.41
GPU7	0.40	0.40	0.41	0.41
GPU8	0.40	0.41	0.41	0.41
sgemm	0.38	0.39	0.40	0.40
CUBLAS	0.38	0.40	0.40	0.40

TABLE 11.15: Efficiency of GPU matrix multiply methods.

11.8.2   Comparison with Single-Core and Quadcore Code

Because of the excessive time required to multiply large matrices on our Xeon quadcore host, we obtained runtimes for the single- and quadcore host codes only for $n \leq 4096$ (Tables 11.2 and 11.3). For $n = 2048$ and $n = 4096$, our fastest GPU code, GPU8 achieved speedups of 8998 and 11,183 relative to SingleCoreIKJ and 364 and 407 relative to the 4-thread OpenMP version of SingleCoreIKJ (Listing 11.3). Figure 11.17 shows the speedup achieved by our various matrix-multiply codes relative to SingleCoreIJK for $n = 4096$ transactions (in billions) and the volume for $16,384 \times 16,384$ matrices.

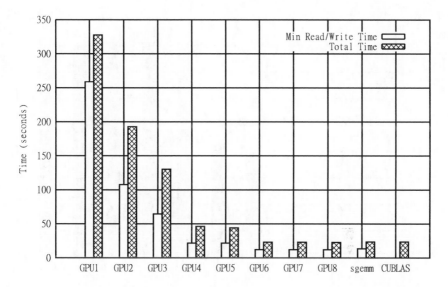

FIGURE 11.12: Runtimes (in seconds) for $16{,}384 \times 16{,}384$ matrices.

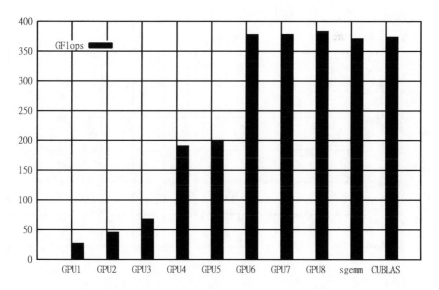

FIGURE 11.13: GFlops for $16{,}384 \times 16{,}384$ matrices.

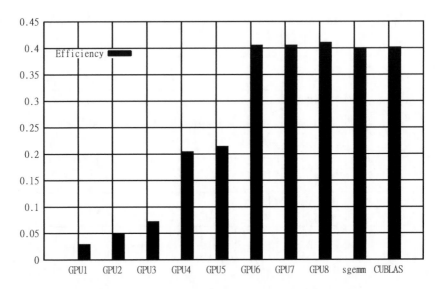

FIGURE 11.14: Efficiency for 16,384 × 16,384 matrices.

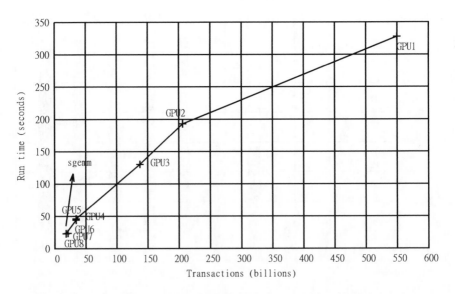

FIGURE 11.15: Time versus transactions for 16,384 × 16,384 matrices.

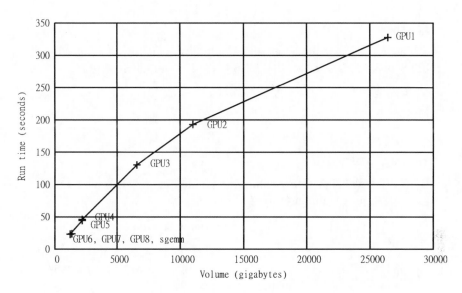

FIGURE 11.16: Time versus volume for 16,384 × 16,384 matrices.

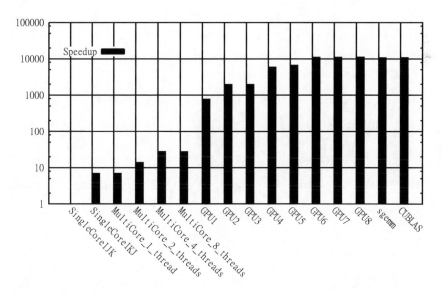

FIGURE 11.17: Speedup relative to `SingleCoreIJK` for 4096 × 4096 matrices.

Acknowledgements

This work was completed with funding provided by the following grants: NETS0963812, NETS1115194, CNS-0963812, and CNS-1115184.

Bibliography

[1] NVIDIA CUDA C Programming Guide. http://docs.nvidia.com/cuda/cuda-c-programming-guide/, 2010.

[2] NVIDIA Tesla. Website: http://en.wikipedia.org/wiki/Nvidia_Tesla, 2012.

[3] NVIDIA Tesla C2050. http://www.nvidia.com/docs/IO/43395/NV_DS_Tesla_C2050_C2070_jul10_lores.pdf, 2010.

[4] CUDA Occupancy Calculator. http://developer.download.nvidia.com/compute/cuda/CUDA_Occupancy_calculator.xls, 2013.

[5] S. Sahni. *Data Structures, Algorithms, and Applications in C++*. Silicon Press, 2nd edition, 2005.

[6] OpenMP. http://www.openmp.org/wp/, 2012.

[7] NVIDIA CUDA C Best Practices Guide, Version 3.2. http://developer.nvidia.com/object/gpucomputing.html, 2010.

[8] MAGMA BLAS. http://icl.cs.utk.edu/magma, 2011.

[9] Vasily Volkov and James W. Demmel. Benchmarking GPUs to tune dense linear algebra. In *Proceedings of the 2008 ACM/IEEE Conference on Supercomputing*, pages 31:1–31:11, Piscataway, NJ, 2008. IEEE Press.

[10] CUBLAS. http://developer.download.nvidia.com/compute/cuda/3_0/toolkit/docs/CUBLAS_Library_3.0.pdf, 2013.

Chapter 12

Backprojection Algorithms for Multicore and GPU Architectures

William Chapman

University of Florida

Sanjay Ranka

University of Florida

Sartaj Sahni

University of Florida

Mark Schmalz

University of Florida

Linda Moore

ARFL/RYAP

Uttam Majumder

ARFL/RYAP

Bracy Elton

Dynamics Research Corporation

12.1   Summary of Backprojection

Backprojection is an algorithmic technique that generates two-dimensional images from synthetic aperture radar data. The data input into the Backprojection algorithm are usually collected by airborne sensors circling around a target area, which emit a series of radar pulses, then receive and record the reflected temporal response. A single pulse provides information about the intensity of reflectors at many distances from the pulse location. Reflectors far from the pulse emitter will appear later in the received response than proximal reflectors. The time t at which a reflector's contribution will appear in the response can be predicted using the speed of light c and the distance d from the pulse emitter, as

$$t = \frac{d}{c}$$

The relationship above facilitates the division of each pulse response into discrete time intervals or range bins, which correspond to the average pulse in each discretized time interval. For example, consider the diagram of Figure 12.1, where a pulse is transmitted toward a target area containing a single point reflector.

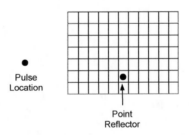

FIGURE 12.1: Point reflector in a target area.

An ideal pulse response for this arrangement consists of a single range bin with a high response intensity, where all other bins have zero intensity. Note that this type of response contains enough information to infer that one or more reflectors occur at a known distance from the pulse location. Thus, the corresponding image of a target area appears as shown in Figure 12.2.

The collection of a second pulse at a different sensor location (shown with respect to Figure 12.2 as the superposition of two pulse returns in Figure 12.3) can help determine the presence of one or more reflectors at known distances from both pulse locations. If only one point reflector is known to be present, then a likely position of the reflector can often be inferred. However, in the general case of numerous reflectors, such an inference is not necessarily correct.

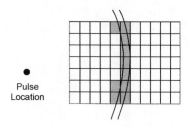

FIGURE 12.2: Single pulse view of point reflector.

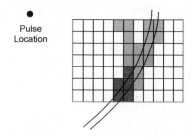

FIGURE 12.3: Two pulse view of point reflector.

The Backprojection algorithm applies this superposition process to numerous pulses to produce a clear reconstruction of the target area. For each pixel in the output image and for each pulse that contributes to that pixel, the range bin corresponding to the given pixel is computed; then, the value of that range bin is summed to the output image [5, 4, 3, 7, 6].

This simple Backprojection algorithm presents both data movement and computational challenges. With respect to data movement, the need for high-resolution output images implies a large number of pulses divided into a large number of range bins. In practical implementations, neither the pulse data nor the output image can be stored in the GPU memory that sits closest to processing cores, referred to as shared memory or L1 cache. Thus, an efficient implementation of Backprojection on a GPU must ensure that the correct subset of pulse data and image data is available in shared memory when needed.

In addition to data-movement overhead, primary computational cost is incurred by summation of each pulse's contribution to each pixel of the output image. This accumulation process begins with a range calculation that determines (1) the distance from the pixel to the pulse location, then (2) the weighted average of the two range bins that most closely correspond to this distance. A phase-correction step that handles pixels occurring at fractional multiples of the radar frequency band is applied to this weighted average before it is summed to the output image. In a Backprojection implementation we are aware of, these operations require 43 floating-point operations per pixel, per pulse. The high-level structure of Backprojection is captured by the following equation:

$$\text{image}[x][y] = \sum_{i=0}^{|p|} (p_i[\text{bin}(i,x,y)]w0(i,x,y) \qquad (12.1)$$
$$+ \ p_i[\text{bin}(i,x,y)+1]w1(i,x,y))\text{dphase}(i,x,y)$$

where image is the image data; x and y are the x and y locations of a pixel; p is the pulse-response data; p_i is the response from pulse i; $\text{bin}(i,x,y)$ is a function that returns the range bin corresponding to the distance of pixel x,y from pulse i; $w0$ and $w1$ are coefficients used to interpolate the response for pixels that fall between range bins; and dphase is a function that produces the phase offset of pixel x,y with respect to pulse i [6, 7].

12.2   Partitioning Backprojection for Implementation on a GPU

The key challenge of Backprojection is data size: both the input data (array of pulses (rows) and range bins (columns)) and the output image are often too large to fit in the GPU's shared memory or L1 cache. Consequently, a partitioning scheme is required that facilitates access locality on both data structures.

The most natural approach partitions along the rows (along pulses) of the echo data matrix. A block of pulses is sent to each multiprocessor, which produces an output image that corresponds to the contribution of its pulses. Then, the resulting output images are summed to create a reconstructed image. This approach, shown schematically in Figure 12.4, exploits Backprojection's natural parallelism with respect to pulses, as each pixel can be thought of as the sum of the contribution of each pulse.

Although having the advantage of simplicity, this approach is confounded by the size of the output image. Partitioning along the pulse dimension requires that each partition maintain a copy of the output image, or share access to a single copy. The former approach is prohibited by the small size of GPU shared memory, while the latter suffers the performance penalties of global memory.

A second approach is not inhibited by these complications. Partitioning the image into two-dimensional (2D) tiles allows small parts of both the output image and pulse data to be cached together in shared memory. This follows from the fact that the number of pulses required for rendering an image tile is linearly related to the width and height of the tile. The partitioning is shown in Figure 12.5.

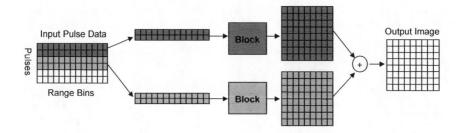

FIGURE 12.4: Partitioning Backprojection along pulse data.

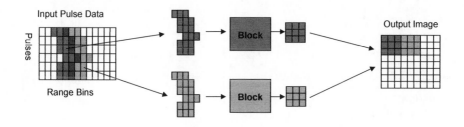

FIGURE 12.5: Tile partitioning of backprojection.

The key process of selecting a correct tile size is examined in greater detail in Section 12.4.1.

12.3 Single-Core Backprojection

The single-core C code to implement Backprojection is a straightforward implementation of the algorithm shown in the introduction and is presented below. Note that an improvement in performance can be realized by iterating over the pulses in the outer loop instead of the inner loop. Unfortunately, this does not facilitate block parallelization, so pulses are iterated inside the main loop for consistency with other code blocks.

The following data structures are input parameters to this algorithm: X and Y, the real coordinates corresponding to each pixel in the target area; r0_data, the range of the first bin in the pulse data; f0_data, frequency information about radar pulse; x_obs, y_obs, z_obs, the real coordinates corresponding to each pulse; pulse_data, the complex pulse-response data; and c4df, a constant that contains information necessary to compute the bin index and phase corrector for each pixel. The algorithm produces a single output, image, which contains the complex response projected onto each pixel.

```
for(j=0; j < image_height; j++){                              tmpRH = static_cast<int>(tmpRF / pi);
    for(k=0; k < image_width; k++){                           tmpRG = tmpRF - static_cast<float>(tmpRH)
        //get position of pixel                                   * ((float)pi);
        pixel0_X = X[k]; //load pixel x-coordinate;           tmpRF = tanf((float)tmpRG);
        pixel0_Y = Y[j]; //load pixel y-coordinate;           tmpRG = tmpRF*tmpRF + 1.0f;
        //these values can be computed outside the            f2.x = (2.0f - tmpRG) / tmpRG;
        //pulse loop                                          f2.y = (2.0f * tmpRF) / tmpRG;
        pre1 = (num_bins-1)/(2.0*c4df);                       //sum phase corrected result to the image
        pre2 = 4.0f*pi/sol/2.0f;                              pixel.x += (tmpRD*f2.x-tmpRE*f2.y);
        pre3 = c4df*num_bins/(num_bins-1.0f);                 pixel.y += (tmpRE*f2.x+tmpRD*f2.y);
        // loop through all pulses for each subtile       }
        for (i=0;i<num_pulses;i++){                   }
            //determine range to pulse location      //write back to the image
            r0 = r0_data[i]; //load r0                image_data[(int)
            Rstart=r0-pre3;                               (k+j*image_width)].x=(float)pixel.x;
            //load pulse location x,y,z               image_data[(int)
            tmpRA = pixel0_X - x_obs[i];                  (k+j*image_width)].y=(float)pixel.y;
            R = tmpRA*tmpRA;                      }
            tmpRA = pixel0_Y-y_obs[i];        }
            R += tmpRA*tmpRA;
            tmpRA=z_obs[i];
            R += tmpRA*tmpRA;
            //compute the range
            R=sqrt(R);
            //compute bin index and coefficients
            binFloat = (R-Rstart)*pre1;
            binFloor = (int)binFloat;
            w2 = binFloat - binFloor;
            w1 = 1.0f-w2;
            //validate bin index
            if(binFloor+1 < num_bins && binFloor > 0){
                //add pulse offset to memory address
                binFloor += i*num_bins;
                //load the frequency
                f0 = f0_data[i];
                //retrieve the bins
                tmpRB=pulse_data[(int)(binFloor)];
                tmpRC=pulse_data[(int)(binFloor+1)];
                //extract components from bins
                tmpRF=(float)tmpRB.x; //bin1 real
                tmpRG=(float)tmpRB.y; //bin1 imaginary
                tmpRD=(float)tmpRC.x; //bin2 real
                tmpRE=(float)tmpRC.y; //bin2 imaginary
                //compute the non phase corrected sum
                tmpRD = w1*tmpRF + w2*tmpRD;
                tmpRE = w1*tmpRG + w2*tmpRE;
                //populate the phase corrector (f2)
                tmpRF = pre2*f0*((float)R);
```

LISTING 12.1: Single-core C implementation of backprojection.

This code computes a 4096×4096 pixel radar image in 1625.7 seconds, at a throughput of 0.44 Gflop/s on a 6 Core 3.3 GHz Intel(R) CoreTM i7 CPU with a 12 MB Cache. [6, 7]

12.3.1   Single-Core Cache-Aware Backprojection

Improvements to the single-core code can be realized by considering the size of the CPU cache, then implementing a tiled partitioning scheme that ensures access locality on both pulse data and image data. Each partition in this scheme is visited sequentially, allowing data required for each partition to be stored entirely in cache until starting the next partition. This cache-aware strategy improves the performance of Backprojection by reducing the number of cache misses and, in turn, the number of system memory accesses.

The implementation of this scheme only requires a modification to the outer loop of Listing 12.1. In particular, rather than loop through the pixels in an entire output image first along one dimension (e.g., y-axis), then along the other dimension (e.g., x-axis), the approach shown in Listing 12.2 visits the pixels and pulses in blocks.

```
for(pb=0; pb < num_pulses; pb += pulse_block_size){
    for(by=0; by < blockCountY; by++){
        for(bx=0; bx < blockCountX; bx++){
            for(j=by*BLOCK_SIZE; j < (by+1)*BLOCK_SIZE; j++){
                for(k=bx*BLOCK_SIZE; k < (bx+1)*BLOCK_SIZE; k++){

                    //The main body of the loop is consistent with Listing 12.1
                    //except for the following change to the inner for loop.

                    // loop through a pulse block for each sub tile
                    for (i=pb; i < pb+pulse_block_size ; i++){
                        ...
                    }
                }
            }
        }
    }
}
```

LISTING 12.2: Single-core cache-aware implementation of Backprojection.

Pulse Block Size	Tile Size (px)	Latency (seconds)	Throughput (Gflop/s)
1	4	1922.45	0.35
1	8	1933.31	0.35
1	16	1964.24	0.34
1	32	1984.44	0.34
10	4	1536.37	0.44
10	8	1538.51	0.44
10	16	1545.15	0.43
10	32	1553.52	0.43
20	4	1506.62	0.45
20	8	1509.02	0.45
20	16	1512.09	0.44
20	32	1517.19	0.44
30	4	1798.66	0.37
30	8	1803.19	0.37
30	16	1805.68	0.37
30	32	1805.76	0.37

TABLE 12.1: Single-core, cache-aware Backprojection performance.

```
for(pb=0; pb < num_pulses; pb += pulse_block_size){
   for(by=0; by < blockCountY; by++){
      for(bx=0; bx < blockCountX; bx++){
         if( (by*blockCountX+bx)%4 != id) continue ;
         for(j=by*BLOCK_SIZE; j < (by+1)*BLOCK_SIZE; j++){
            for(k=bx*BLOCK_SIZE; k < (bx+1)*BLOCK_SIZE; k++){

               //The main body of the loop is consistent with Listing 12.2

            }
         }
      }
   }
}
```

LISTING 12.3: Modification of code in Listing 12.2 to realize multithreaded execution.

12.3.2   Multicore Cache-Aware Backprojection

A multithreaded implementation of the code in Listing 12.2 can exploit a CPU's ability to hide memory latency by overlapping memory access with computation and can also utilize multiple cores. The corresponding extension of the code in Listing 12.2 is trivial, as a block partitioning scheme has already been implemented. Pthreads can be used to launch N instances of the code, where each instance is provided with a thread ID that can be used to identify itself. Inside the main block loop, a single line of additional code can then skip to process each Nth block. This modification is shown in Listing 12.3, with experimental performance figures listed in Table 12.2. A speedup of more than the number of processing cores was obtained by launching more threads than cores, thus allowing the CPU to mask memory access by overlapping it with computation.

12.4   GPU Backprojection

12.4.1   Tiled Partitioning

An initial implementation of Backprojection on the GPU is straightforward using the tiled partitioning scheme described in Section 12.2. Each block of threads on the GPU corresponds to one image tile, where each thread corresponds to one pixel. The pulse data, image, and other data structures are transferred to the GPU using multiple invocations of the Nvidia CUDA opera-

Pulse Block Size	Tile Size (px)	Latency (seconds)	Throughput (Gflop/s)
1	4	230.89	2.91
1	8	217.63	3.09
1	16	216.08	3.11
1	32	216.85	3.10
10	4	190.10	3.53
10	8	185.15	3.63
10	16	189.56	3.54
10	32	187.52	3.58
20	4	187.24	3.59
20	8	185.92	3.61
20	16	186.11	3.61
20	32	186.55	3.60
30	4	222.42	3.02
30	8	223.82	3.00
30	16	225.83	2.98
30	32	224.47	2.99

TABLE 12.2: Multithreaded, cache-aware Backprojection performance.

tion cudaMemcpy prior to launching the kernel. The pulse data and image are stored directly in global memory, while the other data structures are stored in texture memory to improve memory read performance. These latter structures are read-only and are comparatively small.

The kernel code for a GPU implementation of Backprojection is given in Listing 12.4.

The code that invokes this kernel is quite long due to the number of data structures that must be transferred to device memory. The code illustrated in Listing 12.5, which has been truncated for compactness, demonstrates the required steps. Firstly, memory is allocated on the device to hold the data structures, which are then copied from host to device memory. Pointers to the pulse data (d_pulse_data) and image data (d_image) are passed as arguments to the kernel, while the remaining structures are passed as textures.

The kernel code in Listing 12.5 contains several optimizations that are instrumental in achieving high performance, which are listed as follows.

1. **Utilization of L1 Cache Rather than Shared Memory**

   The code in Listing 12.5 was developed for use on an Nvidia Tesla C2050, which was the first Nvidia GPU to provide developers with a traditional L1 and L2 cache hierarchy for reducing global memory access costs. Previous GPUs included only a shared memory which had to be manually controlled by the code developer. Operating this memory structure required the addition of code into the kernel that populated and maintained the state of the cache, which incurred additional costs in clock

```
__global__ void bploop(float2 *pulse_data, float2     //verify the bin is in the subset of the pulse
*image_data,float c4df, int num_pulses,int            //data this thread can access
num_bins,int image_width,int image_height){           if(binFloor+1 < num_bins && binFloor > 0){
                                                         binFloor += i*EXTERNAL_PHD_CACHE;
  float r0, f0, pixel0_Y, pixel0_X, w1, w2,
    binFloat, binFloor, pre1, pre2, pre3, R,             //load f0 for the pulse
    Rstart, tmpRA, tmpRD, tmpRE, tmpRF, tmpRG;           f0=tex1Dfetch(tex_f0_data,i);
  float2 pixel0, f2, tmpRC, tmpRB;
  int i, j, k, tmpRH;                                    //read the bins
  int2 tmpRI;                                            tmpRB = pulse_data[(int)(binFloor)];
                                                         tmpRC = pulse_data[(int)(binFloor+1.0f)];
  //the pixel that corresponds to this thread            tmpRF = tmpRB.x; //bin1.real
  j = (threadIdx.x + blockIdx.x *                        tmpRG = tmpRB.y; //bin1.imag
    INTERNAL_BLOCK_SIZE);                                tmpRD = tmpRC.x; //bin2.real
  k = (threadIdx.y + blockIdx.y *                        tmpRE = tmpRC.y; //bin2.imag
    INTERNAL_BLOCK_SIZE);
                                                         //compute f1, the non phase corrected sum
  //load pixel location                                  tmpRD = w1*tmpRF + w2*tmpRD;
  tmpRI=tex1Dfetch(tex_X,j+0);                            tmpRE = w1*tmpRG + w2*tmpRE;
  pixel0_X=(float) __hiloint2double(tmpRI.y,tmpRI.x);
  tmpRI=tex1Dfetch(tex_Y,k+0);                            //populate the phase corrector (f2)
  pixel0_Y=(float) __hiloint2double(tmpRI.y,tmpRI.x);     tmpRF = pre2 * f0 * R;
                                                         tmpRH = static_cast<int>(tmpRF / pi);
  //precompute constants to be used inside the loop       tmpRG = tmpRF-static_cast<float>(tmpRH*pi);
  pre1 = (num_bins-1)/(2.0*c4df);                         tmpRF = __tanf(tmpRG);
  pre2 = 4.0f*pi/sol/2.0f;                                tmpRG = tmpRF * tmpRF + 1.0f;
  pre3 = c4df*num_bins/(num_bins-1.0f);                   f2.x = (2.0f - tmpRG) / tmpRG;
                                                         f2.y = (2.0f * tmpRF) / tmpRG;
  // loop through all pulses for each subtile
  for(i=0; i<num_pulses; i++){                            //sum phase corrected result to the image
                                                         pixel0.x += tmpRD*f2.x - tmpRE*f2.y;
    //load r0                                             pixel0.y += tmpRE*f2.x + tmpRD*f2.y;
    r0=tex1Dfetch(tex_r0_data,i);                       }
    Rstart=r0-pre3;                                     }

    //determine range to pulse location                //write back to global memory for pixel
    tmpRA = pixel0_X - tex1Dfetch(tex_x_obs,i);        image_data[(int)(k+j*image_height)].x=pixel0.x;
    R = tmpRA * tmpRA;                                 image_data[(int)(k+j*image_height)].y=pixel0.y;
    tmpRA = pixel0_Y - tex1Dfetch(tex_y_obs,i);      }
    R += tmpRA * tmpRA;
    tmpRA = tex1Dfetch(tex_z_obs,i);
    R += tmpRA * tmpRA;
    R = sqrt(R);

    //compute bin index
    binFloat = (R - Rstart) * pre1;
    binFloor = (int) binFloat;
    w2 = binFloat - binFloor;
    w1 = 1.0 - w2;
```

LISTING 12.4: GPU implementation of Backprojection kernel.

cycles and registers. Because the Tesla C2050's L1 and L2 caches are managed transparently by hardware, they do not incur a performance penalty to initiate or maintain. The effective utilization of this cache hierarchy is achieved by the same technique traditionally applied to other processing architectures. For all thread blocks that are active on each multiprocessor, memory accesses are ordered in such a way to ensure that proximal reads and writes are localized to a region of memory that is smaller than the cache size divided by the number of active blocks [1, 2]. This achieves a high cache-hit ratio. In Backprojection, access locality is attained by implementing the tiled partitioning scheme de-

```
float2* d_pulse_data;
float* d_image;
float* d_r0;
...
cudaMalloc((void**) &d_pulse_data, 2*sizeof(float)*num_pulses*num_bins);
cudaMalloc((void**) &d_image, 2*sizeof(float)*BLOCK_SIZE*BLOCK_SIZE);
cudaMalloc((void**) &d_r0, sizeof(float)*num_pulses);
..
cudaMalloc((void**) &d_r0, sizeof(float)*num_pulses);
cudaMemcpy(d_image, &(image[startImageAddress]),
    2*sizeof(float)*BLOCK_SIZE*BLOCK_SIZE,cudaMemcpyHostToDevice);
cudaMemcpyAsync(d_r0, r0, sizeof(float)*BLOCK_SIZE,
    cudaMemcpyHostToDevice);

cudaBindTexture(0, tex_r0, d_r0, sizeof(float)*PULSE_BLOCK_SIZE);

dim3 threads(BLOCK_SIZE, BLOCK_SIZE);
dim3 grid(image_width / threads.x, image_height / threads.y);

bploop<<<grid, threads, 0>>>((float2 *)d_pulse_data, (float *)d_image,
    c4df, num_pulses, num_bins, image_width, image_height);

cudaMemcpy(&(image[0]), d_image, 2*sizeof(float)*BLOCK_SIZE*BLOCK_SIZE,
    cudaMemcpyDeviceToHost);
```

LISTING 12.5: GPU implementation of Backprojection host.

scribed in Section 12.2, where the tile size is sufficiently small to permit all pulse data and all image data for each tile to fit inside the L1 cache.

2. **Reuse of Variable Names**

   Minimizing the number of registers allocated per thread is essential to achieving high occupancy, a condition where many thread blocks are scheduled for execution on each streaming multiprocessor at the same time. In most instances, the CUDA register allocator is effective at analyzing the kernel and determining the lowest number of registers required by the program. However, in some instances, the CUDA compiler is not able to detect when it is safe to reuse a register that is storing old data. Explicit overwriting of the variable with new data ensures that its register will be used to hold the new value.

3. **Computation of Values Outside Loops**

   This optimization is not specific to the GPU but is useful because moving expensive operations outside loops often has a positive impact on performance. In the code above, this was done for the variables pre1, pre2, and pre3. On a GPU, the cost of this optimization is seen in register utilization. Values computed outside a loop must be maintained in register memory, and their registers cannot be released until the loop has completed. It is always useful to weight the spatial cost of a single register against the temporal cost of recomputing the value inside the loop.

Recall that the success of a GPU implementation of Backprojection depends on the data-partitioning strategy. From the principle of locality, within a group of neighboring pixels, the collection of range bins the pixels correspond to are neighbors or near-neighbors. This access locality permits a subregion of the image and a subregion of the pulse data to be transferred to shared memory or L1 cache to facilitate complete rendering of the pixel with the given set of pulses.

This technique is advantageous because it allows a GPU to transfer large blocks of pulse-data memory at one time rather than requiring transfer of a single bin each time it is needed. More importantly, this approach also permits frequently accessed values to be loaded from global memory once, then read from cache for all subsequent read operations (called load-once read-many).

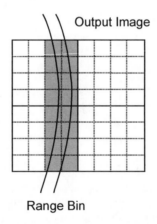

FIGURE 12.6: Pulse-data reuse increases with tile size.

Figure 12.6 helps the reader understand this notion of range-bin reuse. For example, let a target area be partitioned into nine tiles or blocks. The bin corresponding to the illustrated pulse location is used by five blocks a total of 12 times. If blocking was not used, this bin would need to be loaded from global memory each of the 12 times it was used, yielding 12 memory accesses. In contrast, the load-once read-many blocking strategy permits the same image to be reconstructed using only five global memory accesses, with the remaining seven read operations resulting in a cache hit.

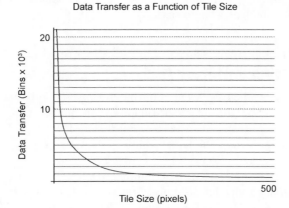

FIGURE 12.7: Data transfer cost decreases with tile size.

In the general case where blocking is used on high-resolution images having many blocks, the amount of data transfer required to render an image can be modeled as function of tile size. A smaller block size results in less block reuse, thus increasing communication cost. This effect is shown in Figure 12.7, using simulated low-resolution image data (500 × 500 pixels) with 4096 range bins. The two graphs in Figure 12.7 denote upper and lower bounds on the memory-access requirements for a given block (tile) size. The bounds differ because pulse look angle can vary with respect to block orientation.

Similar to the preceding discussion, a larger block size reduces communication cost but decreases the number of blocks and the degree of parallelism. Large block sizes can cause low occupancy if register utilization is high or if pulse data and image data exceed available storage in a block's region of L1 cache and shared memory.

Performance results are shown in Table 12.3 as a function of tile size varying from 4 × 4 pixels to 32 × 32 pixels, for a 4096 × 4096 pixel output image. These results are consistent with the observation that higher tile sizes lead to lower latencies, subject to the constraints of register availability and memory. There does not seem to be a substantial benefit going from 16 × 16 to

Tile Size (px)	Latency (seconds)	Throughput (Gflop/s)
4 × 4	14.9	44.1
8 × 8	4.79	131.8
16 × 16	3.71	181.1
32 × 32	3.68	182.4

TABLE 12.3: Performance of tiled Backprojection.

32×32, suggesting that memory access due to tile size is not significantly limiting performance at these sizes [7].

12.4.2   Overlapping Host–Device Communication with Computation

While improvements presented in the previous section are impressive when compared to the CPU implementation, one can further increase performance by reducing the latency incurred by transferring the pulse data from host to device before launching the kernel. Since pulse data is a large structure, this transfer latency can account for a significant portion of the application runtime. Fortunately, the Backprojection algorithm sequentially uses each pulse from the response without needing to revisit previous pulses or look ahead to subsequent pulses. This permits the pulse data to be treated as a data stream, so pulses can be copied into device global memory while the kernel is running. This overlap allows masking of nearly all the pulse-data transfer latency.

Modern GPU devices support the overlap of communication and computation via a CUDA structure called a stream, which is a set of GPU instructions, including kernel invocations and memory-transfer requests, and is executed sequentially. Such CUDA operation streams should not be confused with the data streams mentioned in the previous section. Within a CUDA stream, operations cannot be overlapped so must be executed in the order they were placed in the stream. However, between different streams, operations may overlap. For example, a kernel from one stream can be running while an asynchronous memory transfer in another stream is running [2].

The approach used to achieve this overlap with Backprojection (1) transmits a block of pulses to global memory; (2) launches a kernel that generates an output image from these pulses while a second block of pulses is being copied to another location in global memory; (3) leaves the image in global memory; and (4) launches a kernel on a new block of pulses.

The implementation of CUDA stream-based processing requires only one change to the kernel code shown in Listing 12.4. Previously, a single kernel call processed all pulses in the response, so there was no need to read in the pixel value currently being computed before beginning execution; it was safe to assume the pixel's value was 0. In the CUDA stream implementation, each kernel must read the current pixel value from global memory before entering the main pulse loop. This modification is shown in Listing 12.6. The corresponding modifications to the host code are given in Listing 12.7.

In this code, the host begins by allocating two blocks of memory in the device for pulse data, denoted by d_phd0 and d_phd1, and one block of memory for the output image. Additional memory is allocated for smaller supporting data structures that will later be assigned to textures. CUDA streams are created to hold each memory-transfer and kernel-invocation pair. The primary loop iterates through the pulse blocks, starting at an index value of -1 and stopping at the block count. During the first pass through the loop, the first

```
__global__ void bploop(float2 *pulse_data, float2 *image_data,float c4df,
    int num_pulses,int num_bins,int image_width,int image_height){
    float r0, f0, pixel0_Y, pixel0_X, w1, w2, binFloat, binFloor, pre1,
    pre2, pre3, R, Rstart, tmpRA, tmpRD, tmpRE, tmpRF, tmpRG;
    float2 pixel0, f2, tmpRC, tmpRB;
    int i, j, k, tmpRH;
    int2 tmpRI;

    //determine the pixel that corresponds to this thread
    j = (threadIdx.x+blockIdx.x*INTERNAL_BLOCK_SIZE);
    k = (threadIdx.y+blockIdx.y*INTERNAL_BLOCK_SIZE);

    //read the image from global memory
    pixel0.x = image_data[(int)(k+0+j*image_height)].x;
    pixel0.y = image_data[(int)(k+0+j*image_height)].y;

    for (i=0;i<num_pulses;i++){
        //compute the contribution of each pulse in the block
        //(No changes to the loop body from Figure 8.)
    }

    //write back to global memory for pixel
    image_data[(int)(k+j*image_height)].x=(float)pixel0.x;
    image_data[(int)(k+j*image_height)].y=(float)pixel0.y;
}
```

LISTING 12.6: Tiled Backprojection kernel with overlapped communication.

block of pulses is transferred to the device. During the second pass, the pulse data for the second block is sent, and a kernel is invoked on the first block. Execution continues in this manner until the last iteration, when a kernel is launched on the last block, and no data are sent to the device. This technique allows all but the transfer of the first block to be masked by computation time. This fact encourages the selection of small block sizes. However, extremely small block sizes are disadvantageous when the latency associated with launching a kernel exceeds the latency for transferring a single block. This breakeven point is hardware dependent.

Performance results are shown in Table 12.4 as a function of pulse block size varying from 10 to 200 for a 4096 × 4096 pixel output image. The image tile size is held constant at 32 × 32 pixels, which was the best-performing tile size from previous tests (Table 12.3), yielding 182.4 Gflop/s. These results demonstrate the observation that, at very low pulse-block sizes, latency is high due to communication overhead associated with launching a large number of kernels. At large block sizes, transfer cost for the first pulse-block reduces the amount of feasible overlapping. The highest performance was observed at a pulse-block size of 100, which increased throughput from 182.4 Gflop/s to 202.1 Gflop/s, an improvement of 10.8% [7].

```
// allocate device memory for 2 copies of pulse array
float2* d_pulse_data0;
cudaMalloc((void**) &d_pulse_data0,2*sizeof(float)*
    PULSE_BLOCK_SIZE*EXTERNAL_PHD_CACHE);
float2* d_pulse_data1;
cutilSafeCall(cudaMalloc((void**) &d_pulse_data1, 2*
    sizeof(float)*PULSE_BLOCK_SIZE*EXTERNAL_PHD_CACHE));

//allocate memory for other structures used to store
data about the pulses
float* d_x_obs;
cudaMalloc((void**) &d_x_obs, sizeof(float)*
    PULSE_BLOCK_SIZE);
...

//allocate image memory
float* d_image;
cudaMalloc((void**) &d_image, 2*sizeof(float)*
    image_width*image_width);

//zero the image
cudaMemset(d_image, 0, 2*sizeof(float)*
    image_width*image_width);

//make a stream for each block of pulses. The stream
//will later include a memory transfer instruction and
//a kernel invocation.
streams = (cudaStream_t*) malloc(sizeof(cudaStream_t)*
    num_pulses/PULSE_BLOCK_SIZE);

for(k=0; k < num_pulses/PULSE_BLOCK_SIZE; k++)
    cudaStreamCreate(&(streams[k]));

//set up the kernel image partitioning
dim3 threads(BLOCK_SIZE, BLOCK_SIZE);
dim3 grid(image_width / threads.x, image_width /
    threads.y);

// loop through all pulse blocks
for(k=-1; k<render_num_pulses/PULSE_BLOCK_SIZE; k++){
    //determine if this is an even or odd block index
    pm = ((k+1) % 2 == 0);
    //launch a kernel if this is not the first
    //iteration of the loop
    if(k >= 0){
        // copy small host data structures to device
        cudaMemcpyAsync(d_x_obs,&(x_obs[PULSE_BLOCK_SIZE
            *k]), sizeof(float)*PULSE_BLOCK_SIZE,
            cudaMemcpyHostToDevice,streams[k]);
        ...
```

```
    // bind the data to textures
    cudaBindTexture(0, tex_x_obs, d_x_obs,
        sizeof(float)*PULSE_BLOCK_SIZE);
    ...
    // execute the kernel
    bploop<<< grid, threads, 0, streams[k]>>>
        ((float2 *)(pm?d_pulse_data1:d_pulse_data0),
        (float *)d_image,c4df, PULSE_BLOCK_SIZE,
        num_bins,image_width, image_height);
}

//send data for the next pulse block if we are not
//on the last pulse block
if(k < num_pulses/PULSE_BLOCK_SIZE-1){
    int kplus1 = k+1;
    //send the buffer to the device
    cudaMemcpyAsync( pm ? &(d_pulse_data0[0]):
        &(d_pulse_data1[0]),
        &(pulse_data[num_bins*kplus1*
        PULSE_BLOCK_SIZE*sizeof(float2)]),
        num_bins*PULSE_BLOCK_SIZE*sizeof(float2),
        cudaMemcpyHostToDevice,streams[kplus1]);
}
// wait to start the next iteration of the loop
// until both the kernel and memory transfer are
// complete
cudaThreadSynchronize();
}

// copy image from device to host
cudaMemcpy(&(image[startImageAddress]), d_image,
    2*sizeof(float)*image_width*image_width,
    cudaMemcpyDeviceToHost);
```

LISTING 12.7: Tiled Backprojection host code with overlapped communication.

12.4.3   Improving Register Usage

The partitioning scheme described in Section 12.4.1 dictates that each kernel should process one pixel. It was subsequently determined that optimum tile size is 32×32 pixels, for a total of 1024 kernels per block. In Section 12.4.2, we used streams to overlap computation and communication to attain an improvement in performance. This decreased total latency, but did not impact kernel performance.

Pulse Block Size (px)	Latency (seconds)	Throughput (Gflop/s)
10	3.865	173.8
20	3.547	189.4
50	3.369	199.4
100	3.324	202.1
125	3.327	201.9
200	3.346	200.8
500	3.483	192.9

TABLE 12.4: Tiled Backprojection kernel with overlapped communication performance.

Additional improvement can be realized by increasing occupancy attained by our current scheme. In particular, decreasing the resource utilization of a single thread block allows more thread blocks to be active on a streaming multiprocessor, thus supporting reduction of global memory access times [2].

Compiling the CUDA code using `--ptx-options=-v` provides information about the resources being used by each thread, for example:

```
ptxas info: Compiling entry function '...' for 'sm_20'
ptxas info: Used 21 registers, 68 bytes cmem[0],
            8 bytes cmem[16]
```

Here, Line 2 indicates that each thread uses 21 registers. Recalling that there are 1024 registers per block, each block thus requires 21,504 registers on a streaming multiprocessor, which is more than half of the 32,768 registers available per streaming multiprocessor. For this reason, only one block can be active at a time, and global memory access cannot be masked.

Although it is possible to reduce the number of registers per block by reducing block size, Table 12.4 illustrates that optimal block size is 32 × 32 pixels. This implies that any benefits from higher occupancy are overcome by the higher communication costs associated with dividing the image into smaller tiles. Fortunately, it is possible to maintain a large tile size while decreasing the number of registers per block, which is accomplished by increasing the number of pixels per thread.

The success of this scheme depends on the notion that, although each thread uses 21 registers, many of those registers are not storing information about the pixel, but about the state of the thread. This includes values computed outside the body of the primary loop (to reduce computation latency), as well as index variables and other intermediate values associated with running the algorithm. One could infer that if each thread processed two pixels, the tile size would double, but the number of registers per thread would increase by a smaller factor. This would allow tile size to be maintained at a large value while also increasing occupancy.

To implement this modification, the kernel must be modified to compute two pixels concurrently. This involves declaring an extra variable `pixel_01` to store pixel state, and adding code anywhere image data is accessed or used to ensure the computation runs twice. It is tempting to wrap the entire application in a large loop where the index is varied from 0 to 1, but this is not an acceptable solution. GPU registers are non-addressable from the perspective of the developer, meaning there is no way to access a different register based on the value of the loop variable, except by using conditional logic. This would deteriorate performance, as it would violate the single-instruction, multiple-data principle. Such a loop would also require an extra register to maintain the value of the loop variable.

Listing 12.8 illustrates kernel modification for processing two pixels per thread. Implementing this strategy requires few modifications to the host code. Only the grid needs to be corrected since each block now computes twice as many pixels as it has threads. The modified code is given in Listing 12.9.

Performance results are shown in Table 12.5 for a 4096 × 4096 pixel output image. Pulse block size is fixed at 100, and block size is fixed at 16 × 16 pixels, with the number of pixels per thread being varied from 2 to 4 in steps of 2.

Pixels Per Thread	Registers Per Thread	Tile Size	Latency (seconds)	Throughput (Gflop/s)
2	25	16 × 32	2.98	221.3
4	32	32 × 32	3.01	219.5

TABLE 12.5: Tiled Backprojection with multiple pixels per thread performance.

These results show that increasing the number of pixels per thread and increasing occupancy significantly improves performance. For the case of two pixels per thread, each block used 6400 registers, permitting five blocks to be active at a time. When the number of pixels was increased to four, register usage increased to 8192, supporting only four active blocks. This resulted in a small performance decrease, as the smaller number of blocks available for scheduling per multiprocessor offset the decrease in communication cost.

```
__global__ void bploop(float2 *pulse_data, float2
  *image_data, float c4df, int num_pulses, int
  num_bins,int image_width,int image_height){
  float r0,f0, pixel0_Y, pixel0_X, pixel1_Y, w1, w2,
  tmpRF,tmpRG, f1, f2, binFloor, pre2,
  pre3, tmpRA, pre1, R, Rstart;
  float2 tmpRC, tmpRB, pixel0, pixel1;
  int i,j,k,tmpRH;
  int2 tmpRI;
  //the pixel that corresponds to this thread
  j = (threadIdx.x+blockIdx.x*INTERNAL_BLOCK_SIZE);
  k = (threadIdx.y+blockIdx.y*INTERNAL_BLOCK_SIZE);
  //multiply the pixel y index number by 2
  k *= 2;
  //Get the current pixel value from global memory.
  //Necessary because we stream the pulse blocks to
  //the kernel.
  image_00.x=image_data[(int)(k+0+j*nyout)].x;
  image_00.y=image_data[(int)(k+0+j*nyout)].y;
  image_01.x=image_data[(int)(k+1+(j+0)*nyout)].x;
  image_01.y=image_data[(int)(k+1+(j+0)*nyout)].y;
  //read the pixels location
  tmpRI=tex1Dfetch(tex_X1,j+0);
  pixel0_X=(float) __hiloint2double(tmpRI.y,
    tmpRI.x);
  tmpRI=tex1Dfetch(tex_Y1,k+0);
  pixel0_Y=(float) __hiloint2double(tmpRI.y,
    tmpRI.x);
  tmpRI=tex1Dfetch(tex_Y1,k+1);
  pixel1_Y=(float) __hiloint2double(tmpRI.y,
    tmpRI.x);
  // precompute constants for use in the main loop
  pre1 = (num_bins-1)/(2.0*c4df);
  pre2 = 4.0f*pi/sol/2.0f;
  pre3 = c4df*num_bins/(num_bins-1.0f);
  // loop through all pulses for each subtile
  for (i=0;i<num_pulses;i++){
    //load r0
    r0=tex1Dfetch(tex_r0_data,i);
    Rstart=r0-pre3;
    //determine range to pulse location
    tmpRA = pixel0_X-tex1Dfetch(tex_x_obs,i);
    R = tmpRA*tmpRA;
    tmpRA = pixel0_Y-tex1Dfetch(tex_y_obs,i);
    R += tmpRA*tmpRA;
    tmpRA=tex1Dfetch(tex_z_obs,i);
    R += tmpRA*tmpRA;
    R=sqrt(R);
    //compute bin index
    binFloor = (R-Rstart)*pre1;
    binFloor=(int)binFloor;
    w2=binFloat-binFloor;
    w1=1.0f-w2;
    //verify the bin is in the subset of the pulse
    //data this thread can access
    if(binFloor+1 < num_bins && binFloor > 0){
      binFloor += i*EXTERNAL_PHD_CACHE;
      f0=tex1Dfetch(tex_f0_data,i);
      //read the bins
      tmpRB=pulse_data[(int)(binFloor)];
      tmpRC=pulse_data[(int)(binFloor+1.0f)];
      tmpRF=(float)tmpRB.x; //bin1.real
      tmpRG=(float)tmpRB.y; //bin1.imag
      tmpRD=(float)tmpRC.x; //bin2.real
      tmpRE=(float)tmpRC.y; //bin2.imag

      //compute f1, the non phase corrected sum
      tmpRD = w1*tmpRF + w2*tmpRD;
      tmpRE = w1*tmpRG + w2*tmpRE;
```

```
      //populate the phase corrector (f2)
      tmpRF = pre2*f0*((float)R);
      tmpRH = static_cast<int>(tmpRF /
        ((float)pi));
      tmpRF = tmpRF - static_cast<float>(tmpRH) *
        ((float)pi);
      tmpRF = __tanf((float)tmpRG);
      tmpRG = tmpRF*tmpRF + 1.0f;
      f2.x = (2.0f - tmpRG) / tmpRG;
      f2.y = (2.0f * tmpRF) / tmpRG;
      //sum phase corrected result to the image
      pixel0.x += (tmpRD*f2.x-tmpRE*f2.y);
      pixel0.y += (tmpRE*f2.x+tmpRD*f2.y);
  }

  /*** Begin Second Pixel Computation***/

  //determine range to pulse location
  tmpRA = pixel1_X-tex1Dfetch(tex_x_obs,i);
  R = tmpRA*tmpRA;
  tmpRA = pixel1_Y-tex1Dfetch(tex_y_obs,i);
  R += tmpRA*tmpRA;
  tmpRA=tex1Dfetch(tex_z_obs,i);
  R += tmpRA*tmpRA;
  R=sqrt(R);
  //compute bin index
  binFloat = (R-Rstart)*pre1;
  binFloor=(int)binFloat;
  w2=binFloat-binFloor;
  w1=1.0-w2;
  //verify the bin is in the subset of the pulse
  //data this thread can access
  if(binFloor+1 < num_bins && binFloor > 0){
    binFloor += i*EXTERNAL_PHD_CACHE;
    f0=tex1Dfetch(tex_f0,i);
    //read the bins
    tmpRB=pulse_data[(int)(binFloor)];
    tmpRC=pulse_data[(int)(binFloor+1.0f)];
    tmpRF=(float)tmpRB.x; //bin1.real
    tmpRG=(float)tmpRB.y; //bin1.imag
    tmpRD=(float)tmpRC.x; //bin2.real
    tmpRE=(float)tmpRC.y; //bin2.imag
    //compute f1, the non phase corrected sum
    tmpRD = w1*tmpRF + w2*tmpRD;
    tmpRE = w1*tmpRG + w2*tmpRE;
    //populate the phase corrector (f2)
    tmpRF = pre2*f0*(R);
    tmpRH = static_cast<int>(tmpRF / pi);
    tmpRF = tmpRF - static_cast<float>(tmpRH) *
      pi;
    tmpRF = __tanf(tmpRG);
    tmpRG = tmpRfloat*tmpRF + 1.0f;
    f2.x = (2.0f - tmpRG) / tmpRG;
    f2.y = (2.0f * tmpRF) / tmpRG;
    //sum phase corrected result to the image
    pixel0.x += tmpRD*f2.x-tmpRE*f2.y;
    pixel0.y += tmpRE*f2.x+tmpRD*f2.y;
  }
  /*** End Second Pixel Computation***/
}
//write back to global memory for pixel
image_data[(int)(k+0+j*image_height)].x
  =(float)pixel0.x;
image_data[(int)(k+0+j*image_height)].y
  =(float)pixel0.y;
image_data[(int)(k+1+j*image_height)].x
  =(float)pixel1.x;
image_data[(int)(k+1+j*image_height)].y
  =(float)pixel1.y;
}
```

LISTING 12.8: Tiled Backprojection kernel with two pixels per thread, kernel code.

```
dim3 threads(BLOCK_SIZE, BLOCK_SIZE);
dim3 grid(image_width / threads.x, image_width / threads.y);
grid.y /= 2;
```

LISTING 12.9: Tiled Backprojection kernel with multiple pixels per thread, host code.

12.5   Conclusion

Given a single Nvidia Tesla C2050 GPU, it is possible to obtain significant performance improvements over traditional CPU-based cache-aware implementations of Backprojection coded in C. A tiled partitioning scheme ensures locality of access for input and output data within each partition. This facilitates an efficient, straightforward parallelization of Backprojection, where each kernel is responsible for each pixel, and each thread block is responsible for each image tile. Such an implementation is capable of attaining 182.4 Gflop/s. Additional performance is obtained by noting that the latency required to transfer pulse data from host to device can be largely overlapped with the processing of those data. This increases the throughput to 202.1 Gflop/s. Revisiting the parallelization scheme and increasing the number of pixels per thread permits the tile size to be increased without increasing the size of the thread block. This results in register usage that increases as a sublinear function of tile size, and additional active blocks to be scheduled on each streaming multiprocessor at a time. This occupancy increases the throughput to 221.3 Gflop/s.

Acknowledgements

This work was completed with funding provided by the following grants: NETS0963812, NETS1115194, CNS-0963812, and CNS-1115184.

Bibliography

[1] NVIDIA Introduction to CUDA, 2008.

[2] *NVIDIA Programming Guide—Version 2.2*, 2009.

[3] Curtis H. Casteel and LeRoy A. Gorham Jr. and Michael J. Minardi and Steven M. Scarborough and Kiranmai D. Naidu and Uttam K. Majumder. A challenge problem for 2D/3D imaging of targets from a volumetric data set in an urban environment. *Proc. of SPIE* Vol. 6568 65680D-1, 2007.

[4] Lars M. H. Ulander and Hans Hellsten and Gunnar Stenstrom. Synthetic-aperture radar processing using fast factorized Backprojection. *IEEE Transactions on Aerospace and Electronic Systems*, Vol. 39 No. 3: 760–776, 2003.

[5] Mita D. Desai and W. Kenneth Jenkins. Convolution Backprojection image reconstruction for spotlight mode synthetic aperture radar. *IEEE Transactions on Image Processing*, Vol. 1 No. 4: 505–517, 1992.

[6] William Chapman, Sanjay Ranka, Sartaj Sahni, Mark Schmalz. Parallel processing techniques for the processing of synthetic aperture radar data on FPGAs. *Proc. ISSPIT 2010*, pp. 17–22, 2010.

[7] William Chapman, Sanjay Ranka, Sartaj Sahni et al. Parallel processing techniques for the processing of synthetic aperture radar data on GPUs. *Proc. of IEEE International Symposium on Signal Processing and Information Technology 2011*, Bilbao, Spain, Dec. 14–17, 2011.

Index